NIELS BOHR
COLLECTED WORKS
VOLUME 12

NIELS BOHR ARRANGING THE ANNUAL STAFF PHOTOGRAPH AT HIS INSTITUTE FOR THEORETICAL PHYSICS, COPENHAGEN, 1960.

NIELS BOHR

COLLECTED WORKS

GENERAL EDITOR

FINN AASERUD

THE NIELS BOHR ARCHIVE, COPENHAGEN

VOLUME 12
POPULARIZATION AND PEOPLE
(1911–1962)

EDITED BY

FINN AASERUD

2007

ELSEVIER

AMSTERDAM · BOSTON · HEIDELBERG · LONDON · NEW YORK · OXFORD · PARIS
SAN DIEGO · SAN FRANCISCO · SINGAPORE · SYDNEY · TOKYO

ELSEVIER B.V.
Radarweg 29
P.O. Box 211, 1000 AE Amsterdam
The Netherlands

Library of Congress Catalog Card Number: 70-126498
Collected Works: ISBN-13: 978-0-7204-1800-2, ISBN-10: 0-7204-1800-3
Volume 12: ISBN-13: 978-0-444-52946-6, ISBN-10: 0-444-52946-2

⊗ The paper used in this publication meets the requirements of ANSI/NISO Z39.48-1992 (Permanence of Paper).

Printed and bound in the United Kingdom

Transferred to Digital Print 2009

FOREWORD

The present volume, "Popularization and People", completes the publication of the Niels Bohr Collected Works. Although some of the most important of Bohr's writings under this heading have been printed in earlier volumes (and for this reason are not reproduced here), many items of interest remain.

Bohr always considered it an important task to spread the message of modern physics to a broad audience. His three published collections of articles (the last published posthumously) presenting the epistemological implications of the new physics testify to this.[1] Intellectually demanding and densely written, these articles are hardly popularizations by today's standards. By contrast, some of Bohr's speeches and articles commissioned for special social and political events do represent genuine popularization.[2] This is also the case for the majority of the writings in Part I of the present volume, which were commissioned particularly for the purpose of popularization. Bohr readily consented to such requests in spite of his workload in physics and in the administration of his institute.[3]

Because the Collected Works aim at completeness with regard to Bohr's publications, Volume 12 must necessarily contain all of his published writings not included in earlier volumes. Consequently, Bohr's contributions to the field of superconductivity, which belong to his work in physics, find their place

---

[1] The three books, originally published in 1934, 1958 and 1963, respectively, have been reprinted as *The Philosophical Writings of Niels Bohr*, Vols. I–III, Ox Bow Press, Woodbridge, Connecticut 1987. The first two books are described in the Introduction to Part I, refs. 37 and 39. The articles are spread over several volumes of the Collected Works, notably Vols. 6, 7 and 10.

[2] See, in particular, Vol. 11, Part II.

[3] See Bohr's articles for the Encyclopædia Britannica (pp. [41] ff.), his 1949 radio speech for gymnasium students (pp. [57] ff.) and his forewords to books by other authors seeking to make science understandable to the general public (pp. [73] ff.).

here. They are not entirely incongruous, however, as they illustrate Bohr's ever-present concern with overview and synthesis. So does Bohr's one and only known book review, which was also written as an overview of a field in physics.[4]

Bohr's celebrations of colleagues, friends and family in Part II constitute the bulk of the present volume. Taken together, they provide an intriguing, though by no means complete, overview of Bohr's rich social network and, more than any of his other publications, serve to introduce Niels Bohr as a person.

The inclusion of so many publications enforced a strong limitation on the reproduction of Bohr's frequently relevant and interesting unpublished writings. Exceptions to this rule in Parts I and II were made only for superconductivity[5] and autobiography.[6] Some of Bohr's extensive correspondence with the people celebrated appeared in earlier volumes and is cited in the footnotes to the introduction to Part II. In addition, Part III comprises a few previously unpublished letters relating to the main body of the volume.[7] These letters shed light on the origin of some of Bohr's writings and further illustrate his relationship with the people dealt with in his publications. As in earlier volumes, an "Inventory of Relevant Manuscripts in the Niels Bohr Archive" is included as an appendix.[8]

Because the Collected Works are arranged according to themes which sometimes overlap, it may be difficult to establish in which volume a particular item can be found. To remedy this situation, Volume 12 concludes with a chronological bibliography of Bohr's publications, with reference to where they appear in the Collected Works.[9]

* * *

Since the first volume appeared in 1972, the Niels Bohr Collected Works have attained a history in their own right. The project was conceived by Léon Rosenfeld, physicist, historian of science and Bohr's close collaborator over a long period of years, who was also the first General Editor of the series. Some years after Rosenfeld's death in 1974 Erik Rüdinger took over as General Editor, starting with Volume 5.[10] Rüdinger served until 1989, the last four

---

[4] See below, pp. [19] ff.
[5] See below, pp. [21] ff. and [29] ff.
[6] See below, pp. [137] ff. and [197] ff.
[7] See below, pp. [489] ff.
[8] See below, pp. [521] ff.
[9] See below, pp. [535] ff.
[10] For a lucid description of general editorial policy and practice, see E. Rüdinger, *General Editor's Preface*, Vol. 5 (1984), pp. V–VII.

years also as Director of the Niels Bohr Archive (NBA), which was formally established on the centennial of Bohr's birth. Over the years, many prominent representatives of the several fields to which Bohr contributed have edited individual volumes in the series. They all deserve thanks for their considerable and entirely voluntary effort. A complete list of the Niels Bohr Collected Works, which also names the Editors is provided on p. XVII.

Too many people have been involved in too many ways in this last volume to make it possible to thank them all. My thanks go first to my predecessor both as General Editor of the Collected Works and as Director of the NBA. Erik Rüdinger generously shared his experience from his work on earlier volumes and from his original plans for the last volume of the series. His considerable help and advice regarding all aspects of the present volume has also been invaluable. The same combination of general advice with close attention to detail regarding selection of material, writing of introductions and translation of Bohr's words was provided by Hilde Levi († 2003), Knud Max Møller († 2004), Aage Bohr and Jørgen Kalckar. Conversations with Niels Bohr's two other surviving sons, Hans and Ernest, were also very useful. The subject matter, argument and language of the introductions have gained substantially from the careful reading by John L. Heilbron. I am deeply indebted to Gordon Baym, who has provided invaluable expert help with the presentation and translation of Bohr's work on superconductivity.

As in earlier volumes, the translation of Bohr's texts was a joint effort by Felicity Pors and the Editor. Hilde Levi and Helle Bonaparte made preliminary translations of some of the articles, and Erik Rüdinger and Jørgen Kalckar suggested revisions. Predrag Cvitanović and Lana Bosanac have translated from the Slovenian and checked Bohr's English-language manuscript against the version reproduced in a Yugoslavian newspaper.[11] As before, the main general challenge of translation was to retain Bohr's style, including his typically long sentences, while making the English idiomatic. We have corrected capricious misspellings but retained typical ones. Misspellings of names have been kept in the original texts but corrected in the translations.

Some of the publications were difficult to identify and locate. I would like to extend special thanks to Birte Christensen-Dalsgaard of the State Library (Statsbiblioteket) in Århus, who provided invaluable help with the collection of Danish newspapers kept there.

The explanatory footnotes needed to place Bohr's writings in context for English-speaking readers have profited from the input of Felicity Pors. Birgitte

---

[11] See below, pp. [227] ff.

Rüsz Andreasen of the Horsens Public Library, Karl Grandin of the Center for History of Science of the Royal Swedish Academy of Sciences and Andreas Tjerneld of the Swedish Biographical Encyclopaedia helped with special problems. To avoid duplication, the notes appear only in the translation when a non-English contribution is reproduced in facsimile, and only in the original when it is transcribed. In the latter case, the footnote numbers are repeated in the translation.

During my time as General Editor of the Collected Works the text has been typed and formatted electronically on the mainframe computer at the Niels Bohr Institute, which over the years has gone through substantial hardware and software changes affecting our work. I am grateful for the help and advice given throughout, on workdays and holidays alike, by Björn Nilsson, the Institute's incomparable computer expert.

The index developed gradually in parallel with other efforts. Helle Bonaparte started work on the index when we still expected that the material published in Volumes 11 and 12 would occupy a single last volume. Christina Olausson continued and finished it. I thank them both.

My closest collaborators continue to be the staff at the NBA, Felicity Pors and Anne Lis Rasmussen. Lis has attended to the meticulous task of entering the index terms electronically, typed all non-facsimile material and prepared draft versions of the inventory of manuscripts and the bibliography of Bohr's publications. Felicity has contributed to all aspects of the volume, from the translations and introductions to the identification of articles and location of photographs. I thank Felicity and Lis for their great dedication.

We are most grateful for the financial support for work on Volumes 11 and 12 from the Lounsbery Foundation, New York, and from a bequest by Hilde Levi.

Elsevier, represented by Publisher Carl Schwarz, has always been supportive as well as patient about missed deadlines. Betsy Lightfoot at Elsevier has handled the final layout of Volume 12 as brilliantly as she did that of Volumes 10 and 11. Collaboration with Zigmas Kryžius at VTEX, the Lithuanian company that also printed Volume 11, has been excellent throughout.

I would like once more to thank the NBA's board of directors for its constant trust and encouragement and my wife, Gro Synnøve Næs, and our children, Andreas and Karen, for their patience and never-failing support. It has been a pleasure and a privilege to carry this great work to its conclusion.

Finn Aaserud
The Niels Bohr Archive
June 2006

CONTENTS

PART I: OVERVIEW AND POPULARIZATION

PART II: PEOPLE

CONTENTS

PART III: SELECTED CORRESPONDENCE

Correspondents

INVENTORY OF RELEVANT MANUSCRIPTS
IN THE NIELS BOHR ARCHIVE

BIBLIOGRAPHY OF NIELS BOHR'S PUBLICATIONS
REPRODUCED IN THE COLLECTED WORKS

INDEX

THE VOLUMES OF THE
NIELS BOHR COLLECTED WORKS

In addition to the Editor of each individual volume (see below), a General Editor has overseen the preparation of the entire series of the *Niels Bohr Collected Works*. The General Editors have been: Léon Rosenfeld (1904–1974) (Volumes 1 to 3), Erik Rüdinger (Volumes 5 to 9, Volume 7 jointly with Finn Aaserud) and Finn Aaserud (Volumes 10 to 12). All volumes are published by North-Holland/Elsevier.

Vol. 1, *Early Work (1905–1911)* (ed. J. Rud Nielsen), 1972.

Vol. 2, *Work on Atomic Physics (1912–1917)* (ed. Ulrich Hoyer), 1981.

Vol. 3, *The Correspondence Principle (1918–1923)* (ed. J. Rud Nielsen), 1976.

Vol. 4, *The Periodic System (1920–1923)* (ed. J. Rud Nielsen), 1977.

Vol. 5, *The Emergence of Quantum Mechanics (mainly 1924–1926)* (ed. Klaus Stolzenburg), 1984.

Vol. 6, *Foundations of Quantum Physics I (1926–1932)* (ed. Jørgen Kalckar), 1985.

Vol. 7, *Foundations of Quantum Physics II (1933–1958)* (ed. Jørgen Kalckar), 1996.

Vol. 8, *The Penetration of Charged Particles Through Matter (1912–1954)* (ed. Jens Thorsen), 1987.

Vol. 9, *Nuclear Physics (1929–1952)* (ed. Rudolf Peierls), 1986.

Vol. 10, *Complementarity Beyond Physics (1928–1962)* (ed. David Favrholdt), 1999.

Vol. 11, *The Political Arena (1934–1961)* (ed. Finn Aaserud), 2005.

Vol. 12, *Popularization and People (1911–1962)* (ed. Finn Aaserud), 2007.

ABBREVIATED TITLES OF PERIODICALS

Ann. d. Phys.	Annalen der Physik (Leipzig)
Brit. Ass. Adv. Sci., Report	Report of the British Association for the Advancement of Science (London)
Ency. Brit.	Encyclopædia Britannica
Fys. Tidsskr.	Fysisk Tidsskrift (Copenhagen)
ICSU Review	International Council of Scientific Unions Review (Amsterdam)
J. Chem. Soc.	Journal of the Chemical Society (London)
Kgl. Dan. Vid. Selsk. Skr.	Det Kongelige Danske Videnskabernes Selskabs Skrifter. Naturvidenskabelig og mathematisk Afdeling (Copenhagen)
Mat. Tidsskr.	Matematisk Tidsskrift (Copenhagen)
Mat.–Fys. Medd. Dan. Vidensk. Selsk.	Matematisk–fysiske Meddelelser udgivet af Det Kongelige Danske Videnskabernes Selskab (Copenhagen)
Medd. Kgl. Vet. Akad., Nobel Inst.	Meddelanden från Kungliga Vetenskapsakademiens Nobelinstitut (Stockholm)
Naturwiss.	Die Naturwissenschaften (Berlin)
Ned. T. Natuurk.	Nederlandsch Tijdschrift voor Natuurkunde (Amsterdam)

Overs. Dan. Vidensk. Selsk. Forh.	Oversigt over Det Kongelige Danske Videnskabernes Selskabs Forhandlinger (Copenhagen)
Overs. Dan. Vidensk. Selsk. Virks.	Oversigt over Det Kongelige Danske Videnskabernes Selskabs Virksomhed (Copenhagen)
Phil. Mag.	Philosophical Magazine (London)
Phil. Trans. Roy. Soc.	Philosophical Transactions of the Royal Society (London)
Phil. Sci.	Philosophy of Science (East Lansing, Michigan)
Phys. Rev.	Physical Review (New York)
Proc. Am. Phil. Soc.	Proceedings of the American Philosophical Society (Philadelphia)
Proc. Cambr. Phil. Soc.	Proceedings of the Cambridge Philosophical Society
Proc. Phys. Soc.	Proceedings of the Physical Society (London)
Proc. Roy. Soc.	Proceedings of the Royal Society (London)
Rev. Mod. Phys.	Reviews of Modern Physics (New York)
Sci. Am.	Scientific American (New York)
Z. Phys.	Zeitschrift für Physik (Braunschweig)

OTHER ABBREVIATIONS

AHQP	Archives for the History of Quantum Physics
BMSS	Bohr Manuscripts, AHQP
BSC $(x.y)$	Bohr Scientific Correspondence, AHQP, section y on mf. x
BSC-Supp.	Bohr Scientific Correspondence, Supplement, NBA
BGC	Bohr General Correspondence, NBA
mf.	microfilm
NBA	Niels Bohr Archive, Copenhagen
Suppl.	Supplement
UN	United Nations
UNESCO	United Nations Educational, Scientific and Cultural Organization
Vol.	Niels Bohr Collected Works, Volume

ACKNOWLEDGEMENTS

For some of the published work in this volume, the editor and publisher were unfortunately unable to trace the copyright holders and thus to enter a formal request for permission to reproduce the material. These contributions were nevertheless considered sufficiently important to be reprinted without further delay. The effort to identify the original copyright holders will be continued.

The following articles by N. Bohr in Nature, "Modern Electrical Theory" [book review], **95** (1915) 420–421; "Sir J.J. Thomson's Seventieth Birthday", **118** (1926) 879; "Sir Ernest Rutherford, O.M., P.R.S.", (Suppl.) **118** (18 Dec 1926) 51–52; [Obituary for Rutherford], **140** (1937) 752–753; [Obituary for Rutherford], (Suppl.) **140** (1937) 1048–1049, are reprinted by permission of Macmillan Publishers Ltd.

N. Bohr, [Discussion Contribution on Superconductivity] in "Proceedings of the Lorentz Kamerlingh Onnes Memorial Conference, Leiden University 22–26 June 1953", Stichting Physica, Amsterdam 1953, pp. 761–762; N. Bohr, "Hendrik Anthony Kramers †", Ned. T. Natuurk. **18** (1952) 161–166, are reprinted by permission of Nederlandse Natuurkundige Vereniging.

N. Bohr, "Atom" in Encyclopædia Britannica, 14th edition, Vol. 2, London and New York 1929, pp. 642–648; N. Bohr "Matter, Structure of" in Encyclopædia Britannica Book of the Year 1938, pp. 403–404; N. Bohr, "Matter, Structure of" in Encyclopædia Britannica Book of the Year 1939, pp. 409–410; are reprinted by permission of Encyclopædia Britannica.

The following articles by Niels Bohr in Berlingske Tidende, Copenhagen: "Atomerne og vor erkendelse", 2 April 1949; "Professor Niels Bjerrum fylder 50 Aar", 9 March 1929, and the article by Bohr, "Professor Martin

Knudsen", 14 February 1941; as well as P. Vinding, "Samtale med Niels Bohr", 2 October 1935, both in Berlingske Aftenavis, are reprinted by permission of Berlingske Tidende.

The following articles by N. Bohr in books published by the Danish publishing house Gyldendal, Copenhagen: "Forord" in George Gamow, "Mr. Tompkins i Drømmeland", 1942, pp. 7–8; "Forord" in J. Robert Oppenheimer, "Naturvidenskab og Livsforståelse", 1960, pp. i–ii; "Den danske redaktionskomités forord" in I.B. Cohen, "Fysikkens Gennembrud", 1962, are reprinted by permission of Gyldendal.

The following articles by N. Bohr published by Cambridge University Press, Cambridge: "Preface to the 1961 reissue" in N. Bohr, "Atomic Theory and the Description of Nature", 1961, p. vi; "Newton's Principles and Modern Atomic Mechanics" in "The Royal Society Newton Tercentenary Celebrations 15–19 July 1946", 1947, pp. 56–61; [Two speeches for Rutherford, 1932] in A.S. Eve, "Rutherford: Being the Life and Letters of the Rt Hon. Lord Rutherford, O.M.", 1939, pp. 361–363, are reprinted by permission of Cambridge University Press.

N. Bohr: [Autobiography] in "Festskrift udgivet af Kjøbenhavns Universitet i Anledning af Universitetets Aarsfest, November 1911", Schultz, Copenhagen 1911, pp. 76–77, is reprinted by permission of Schultz Information.

N. Bohr, "Niels Bohr" in "Les Prix Nobel en 1921–1922", P.A. Norstedt, Stockholm 1923, pp. 126–127, is reprinted by permission of the Nobel Foundation.

The following articles by N. Bohr published by Berlingske Bogtrykkeri, Copenhagen, [Autobiography] in "Studenterne MCMIII: Personalhistoriske Oplysninger", 1928, p. 275; [Foreword] in J. Lehmann, "Da Nærumgaard blev børnehjem i 1908", 1958, p. i, are reprinted by permission of Berlingske Tidende.

The following articles in Politiken: M. Bonnesen, "Niels Bohr om spøg og alvor i videnskaben", 17 April 1949; N. Bohr, "Professor Sir Ernest Rutherford og hans Betydning for Fysikens nyere Udvikling", 18 September 1920; N. Bohr, "Prof. Martin Knudsen død i går", 28 May 1949; N. Bohr, "En Personlighed i dansk Fysik" [H.M. Hansen], 7 September 1946; N. Bohr, "Et lysende forbillede for os alle" [H.M. Hansen], 14 June 1956; N. Bohr, "Ved Hendrik Anton Kramers' død", 27 April 1952; N. Bohr, "Magister Fritz Kalckar", 7 January 1938; N. Bohr, "Han traadte hjælpende til hvor uret blev begaaet:

Mindeord af professor Niels Bohr" [A. Friis], 7 October 1949; N. Bohr, "Hans minde en kilde til mod og styrke" [N. Bjerrum], 1 October 1958; N. Bohr, "Afsked med Sveriges Gesandt i København" [C.F.H. Hamilton], 15 November 1941, are reprinted by permission of Politiken.

N. Bohr, "Selvbiografi af æresdoktoren", Acta Jutlandica **28** (1956) 135–138, is reprinted by permission of the University of Aarhus.

N. Bohr, [Tribute to Tesla] in "Centenary of the Birth of Nikola Tesla 1856– 1956", Nikola Tesla Museum, Belgrade 1959, pp. 46–47, is reprinted by permission of the Nikola Tesla Museum.

N. Bohr, [Tribute to Bering] in "Vitus Bering 1741–1941", H. Hagerup, Copenhagen 1942, pp. 49–53; N. Bohr, "Min Genbo" in "Halfdan Hendriksen: En dansk Købmand og Politiker", Aschehoug, Copenhagen 1956, pp. 171–172, are reprinted by permission of Aschehoug Dansk Forlag.

N. Bohr, "Forord" in "Johan Nicolai Madvig: Et mindeskrift", Royal Danish Academy of Sciences and Letters and the Carlsberg Foundation, Copenhagen 1955, p. vii, is reprinted by permission of the Royal Danish Academy of Sciences and Letters.

N. Bohr, [Tribute to Rutherford] in "Hommage à Lord Rutherford, sept huit et neuf novembre MCMXLVII", La fédération mondiale des travailleurs scientifiques, Paris [1948], pp. 15–16, is reprinted by permission of La fédération mondiale des travailleurs scientifiques.

N. Bohr, "Henry Gwyn Jeffreys Moseley", Phil. Mag. **31** (1916) 173–176, is reprinted by permission of the publisher, Taylor & Francis, Ltd., http://www.tandf.co.uk/journals.

The following contributions by N. Bohr in Overs. Dan. Vidensk. Selsk. Virks.: Mødet den 19. Oktober 1945 [M. Knudsen's retirement as Secretary of the Royal Danish Academy], Juni 1945 – Maj 1946, pp. 31–32; [Obituary for M. Knudsen], Juni 1949 – Maj 1950, pp. 61–65, are reprinted by permission of the Royal Danish Academy of Sciences and Letters.

The following articles by N. Bohr in Fys. Tidsskr., "Martin Knudsen 15.2.1871– 27.5.1949", **47** (1949) 145–147; [Obituary for H.M Hansen], **54** (1956) 97; "Ebbe Kjeld Rasmussen: 12. april 1901 – 9. oktober 1959", **58** (1960) 1–2; "Kirstine Meyer, f. Bjerrum: 12. Oktober 1861 – 28. September 1941", **39**

(1941) 113–115, are reprinted by permission of Selskabet for Naturlærens Udbredelse.

The following articles by N. Bohr in Naturwiss.: "Friedrich Paschen zum siebzigsten Geburtstag", **23** (1935) 73; "Sommerfeld und die Atomtheorie", **16** (1928) 1036, are reprinted by permission of the publisher, Springer Verlag.

N. Bohr, "The Internationalist [A. Einstein]", UNESCO Courier **2** (No. 2, March 1949), 1, 7 is reprinted by permission of the UNESCO Courier.

N. Bohr, "Mindeord [Albert Einstein]", Børsen, 19 April 1955, is reprinted by permission of Dagbladet Børsen A/S.

N. Bohr, "Albert Einstein 1879–1955", Sci. Am. **192** (1955) 31 is reprinted by permission of Scientific American, Inc.

N. Bohr, "Forord" in "Hanna Adler og hendes skole", Gad, Copenhagen 1959, pp. 7–9, is reprinted by permission of Gad Publishers.

N. Bohr, "Tale ved Mindehøjtideligheden for Ole Chievitz 31. December 1946", Ord och Bild **55** (1947) 49–53, is reprinted by permission of Ord och Bild.

N. Bohr, "Et frugtbart livsværk" in "Noter til en mand: Til Jens Rosenkjærs 70-aars dag" (eds. J. Bomholt and J. Jørgensen), Det Danske Forlag, Copenhagen 1953, p. 79, is reprinted by permission of Fællesforeningen for Danmarks Brugsforeninger (FDB).

N. Bohr, "Mindeord" in "Bogen om Peter Freuchen" (eds. P. Freuchen, I. Freuchen and H. Larsen), Fremad, Copenhagen 1958, p. 180, is reprinted by permission of Fremad Forlag.

PART I

OVERVIEW AND POPULARIZATION

INTRODUCTION

by

FINN AASERUD

A major driving force behind Niels Bohr's long career in theoretical physics was his striving for overview and unity in his own field as well as in its broader philosophical and practical implications. This disposition played a crucial role in Bohr's success as a teacher and inspirer of generations of physicists. It is expressed in several of his publications, most of which are reproduced in earlier volumes of the Niels Bohr Collected Works, notably in Volumes 6, 7 and 10, and in Part 2 of Volume 11. Bohr's striving for synthesis may serve as a common denominator for the remainder of Bohr's publications, which are reproduced in this last volume of his Collected Works. The striving is especially evident in the publications presented here directed to fellow physicists, the general public and gymnasium students.

1. ELECTRON THEORY AND SUPERCONDUCTIVITY

The first four writings reproduced below – the only known book review by Bohr, which dates from 1915, an unpublished set of proofs, a manuscript and a published discussion contribution at a conference – were directed to physicists. The first item reviews a book-length account of the electron theory of matter – the subject of Bohr's 1911 doctor's thesis – while the others discuss the phenomenon of superconductivity, which was discovered in 1911 and which

[3]

refers to the ability of certain metals at temperatures close to absolute zero to conduct electricity without resistance.[1]

The review concerns "The Electron Theory of Matter", a book written by Bohr's esteemed colleague and senior by six years, Owen Richardson.[2] It was based on a series of lectures for graduate students at Princeton University, where Richardson taught from 1906 to 1913, before he returned to King's College, London, to take up a professorship there. Bohr liked the book. He wrote to his brother Harald, "I have been busy reading an excellent book by Richardson which I am to review in 'Nature'. It is a text book on the entire electron theory. I am learning much from it, and I am looking forward to studying some parts of it more closely."

Bohr used the opportunity of the review to express his view on the state of theoretical physics more broadly. Only a few years earlier, Bohr wrote, classical electromagnetic theory had been thought to constitute "a final accomplishment of ordinary mechanics, and there appeared to be no limit to the application of the general principles of the theory."[3] Now the situation had changed completely:

> "If at present we may speak of a programme for the future development, it would, perhaps, be to examine the constitution of the special atomic systems actually existing, and then, by means of the directly observable properties of matter, possibly to deduce the general principles. If so, the evolution would be exactly the reverse of that anticipated."

Niels to Harald Bohr,
15 Apr 15
Danish
Full text, Vol. 1, p. [576]
Translation, Vol. 1, p. [577]

[1] This section has been written partly on the basis of a manuscript on Bohr's contributions to superconductivity kindly supplied by Gordon Baym, which is deposited at the NBA. Baym's manuscript, which outlines the scientific discussion in detail, draws in turn on the review of the development of the quantum theory of solids contained in G. Baym and L. Hoddeson, *The development of the quantum mechanical electron theory of metals: 1900–28*, Proc. Roy. Soc. **A371** (1980) 8–23, and in G. Baym, L. Hoddeson, and M. Eckert, *The development of the quantum mechanical electron theory of metals: 1928–33*, Rev. Mod. Phys. **59** (1987) 287–327. See also L. Hoddeson, H. Schubert, S.J. Heims and G. Baym, *Collective Phenomena* in *Out of the Crystal Maze: Chapters from the History of Solid-State Physics* (eds. L. Hoddeson, E. Braun, J. Teichmann and S. Weart). The Editor is grateful for Professor Baym's generous assistance.

[2] O.W. Richardson, *The Electron Theory of Matter*, Cambridge University Press, Cambridge 1914. Cf. Bohr's master's and doctor's theses on the same topic completed, respectively, in 1909 and 1911 and reproduced in Vol. 1, pp. [131]–[161] (English translation of the master's thesis) and pp. [163]–[301], [302]–[395] (Danish original and English translation of the doctor's thesis).

[3] N. Bohr, *Modern Electrical Theory*, Nature **95** (1915) 420–421. Reproduced on pp. [19] ff. The review is not written under Bohr's full name, but has the initials "N.B." I am grateful to Ole Knudsen, retired historian of physics at the University of Aarhus, for calling my attention to the review.

Half a decade later Bohr would establish his institute in Copenhagen on the basis of precisely such an experiment-driven conception of theoretical physics.

Although Richardson mentioned superconductivity,[4] Bohr did not refer to it in his review. Nevertheless, as it was directly related to the problems Bohr had discussed in his 1911 doctor's thesis, there can be no doubt that the discovery had caught Bohr's attention. Indeed, a manuscript reproduced in the Bohr Collected Works shows that he touched upon the phenomenon in a lecture at the University of Copenhagen in 1914.[5] The understanding of superconductivity became a lifelong challenge to Bohr, even though his only publication on the subject is a brief discussion contribution at a conference at Leiden University in June 1953 in memory of the Dutch physicists Hendrik Antoon Lorentz, a pioneer in the electron theory of metals, and Heike Kamerlingh Onnes, who discovered superconductivity.[6] Bohr's publication, however, has a long prehistory.

A quantum theory of the electric conductivity of metals was developed by Felix Bloch in 1928 at Werner Heisenberg's institute in Leipzig. Bloch based the theory on his recognition that individual electrons can propagate freely through a perfectly periodic lattice. It turned out, however, that the complete disappearance of resistance at low temperatures could not be explained on this basis. Thus a major riddle remained.

During his half-year long stay in Copenhagen from October 1931 Bohr stimulated Bloch to take up a problem concerning atomic collisions in the penetration of charged particles through matter. This was a major theme for Bohr throughout his career. Although being new to the field, Bloch was able to solve the problem to Bohr's great satisfaction. Looking back more than thirty years later, Bloch recalled that this influence from Bohr induced him to become "more interested in the basic phenomena underlying some problem than just in producing results."[7]

The year 1932 is known as the "miraculous year" of nuclear physics. While Bohr followed this development closely, his main thinking on the nucleus in

---

[4] Richardson, *Electron Theory*, ref. 2, pp. 423–425.

[5] N. Bohr, *Lectures on the Electron Theory of Metals*, translation of notes, Vol. 1, pp. [445]–[471], on p. [447].

[6] N. Bohr, [Discussion Contribution on Superconductivity], *Proceedings of the Lorentz Kamerlingh Onnes Memorial Conference, Leiden University 22–26 June 1953*, Stichting Physica, Amsterdam 1953, pp. 761–762. Reproduced on pp. [37] ff.

[7] Interview with Bloch 15 May 1967, AHQP. Quoted in J. Thorsen, *Introduction*, Vol. 8, pp. [203]–[265], on p. [218]. The entire Vol. 8 is devoted to "The Penetration of Charged Particles Through Matter (1912–1954)".

1932 was a continuation of a radical suggestion, which he had first proposed some years before, that energy conservation may not be upheld in nuclear processes.[8] It was only later that Bohr would turn his full attention to nuclear physics, establishing major experimental facilities in the field at his institute.

In 1932 Bohr's main scientific concerns lay elsewhere. In addition to being occupied with the collision problems, he was busy collaborating with his younger Belgian colleague Léon Rosenfeld on the measurability of the electromagnetic field, which had become a major challenge in quantum field theory.[9] He was also working on his seminal lecture "Light and Life".[10] It was in the midst of all these activities that Bohr took up the problem of superconductivity in earnest. He sought to develop a theory based on the notion of a transition to a new phase in which the conduction electrons as a whole can propagate through the metal in a coordinated motion as a macroscopic quantum effect.[11]

In June 1932 Bohr wrote to Bloch about his new thoughts on superconductivity, expressing eagerness to discuss "an idea regarding superconductivity, which I have got and cannot let go of, although I am far from understanding

Bohr to Bloch,
15 Jun 32
Danish
Full text on p. [498]
Translation on p. [500]

---

[8] R. Peierls, *Introduction*, Vol. 9, pp. [3]–[83], on pp. [4]–[14]. Bohr's radical suggestion is most clearly expressed in a manuscript from 1929, *β-ray spectra and energy conservation*, reproduced in Vol. 9, pp. [85]–[89]. By 1936 he had changed his mind entirely, rejecting, in an article reproduced in Vol. 5, pp. [213]–[216], non-conservation of energy on the basis of experimental verification of the neutrino hypothesis. In Vol. 5 it is furthermore documented that also several years earlier Bohr had called for a renunciation of energy conservation, at that time in physical processes involving electromagnetic radiation. The idea was published in 1924 in a joint article with Hendrik Kramers and John Slater, reproduced in Vol. 5, pp. [99]–[118]. This earlier attempt to renounce energy conservation was also short-lived, as Bohr gave it up early the following year on account of new experimental evidence establishing the role of the photon, notably the experiments of the German physicists Walther Bothe and Hans Geiger.

[9] The findings were first published in an abstract in the Proceedings of the Danish Academy of Sciences and Letters: N. Bohr and L. Rosenfeld, *The Limited Measurability of Electromagnetic Fields of Force* (Abstract), Overs. Dan. Vidensk. Selsk. Virks. Juni 1932 – Maj 1933, p. 3, which was printed in English in Nature **132** (1933) 75. An elaborate presentation was subsequently published in German in the Communications of the Danish Academy: N. Bohr and L. Rosenfeld, *On the Question of the Measurability of the Electromagnetic Field Quantities*, Mat.–Fys. Medd. Dan. Vidensk. Selsk. **12**, No. 8 (1933). The two publications are reproduced in Vol. 7, respectively on pp. [53]–[55] (Danish and English versions), and on [55]–[121] (German version) and [123]–[166] (English version). For the background, see J. Kalckar, *Introduction*, Vol. 7, pp. [3]–[51], on pp. [3]–[39].

[10] N. Bohr, *Light and Life*, Nature **131** (1933) 421–423, 457–459. Reproduced in Vol. 10, pp. [27]–[35]. For the background, see D. Favrholdt, *Introduction*, Vol. 10, pp. [3]–[26], on pp. [7] ff.

[11] See the articles by Baym et al., ref. 1, as well as P.F. Dahl, *Superconductivity: Its Historical Roots and Development from Mercury to the Ceramic Oxides*, American Institute of Physics, New York 1992, p. 152.

the transition between superconductivity and normal conduction of electricity." After briefly outlining his theory, Bohr continued:

> "The assumption that superconductivity concerns a coordinated motion of the entire electron lattice is of course old,[12] but just as it was first with quantum mechanics that it was possible to bring the conception of the presence of 'free' electrons in metals in closer connection with experience, so it also appears that it is first through quantum mechanics that one can understand ... how the two lattices can move through each other without resistance and appreciable deformation."

In June Bohr submitted a manuscript on his new theory to "Die Naturwissenschaften" and received the proofs a month later.[13] He continued to discuss his ideas with many of his younger colleagues, in particular in correspondence and conversations with Bloch. While acknowledging the potential value of Bohr's ideas, Bloch saw a number of difficulties confronting the theory as it stood. Bohr too was uncertain about the implementation of his new idea and the ability of the theory to account for certain experimental findings. He kept postponing the return of the proofs, vacillating into 1933 about whether or not to publish. In the meantime, another younger colleague, Ralph de Laer Kronig, had published ideas which to some extent resembled Bohr's, and there ensued extensive discussions in Copenhagen between Bohr, Bloch, Kronig and Rosenfeld. In early 1933 Bohr wrote an addendum comparing his and Kronig's contributions, with the intention of publishing it together with the delayed original article.[14]

In the end, neither the proofs nor the addendum reached publication, partly because of Bohr's uncertainty about their contents, partly because of the many other ideas he was pursuing. For example, on the same day in late December

---

[12] Bohr seems here to be referring to the work of Frederick Lindemann, later Lord Cherwell. See p. [26], ref. 5.

[13] N. Bohr, *Zur Frage der Supraleitung*, proofs for article in Die Naturwissenschaften, BMSS. See *Inventory of relevant manuscripts in the Niels Bohr Archive*, below, p. [525], folder 8. The original proofs in German and a translation into English are reproduced on pp. [21] ff. Although this article and the subsequently written supplementary note (ref. 14) were never published, they are included here because they convey Bohr's thoughts on a subject that occupied him deeply.

[14] N. Bohr, *On Superconductivity*, manuscript in Danish, BMSS. See *Inventory*, ref. 13, p. [525], folder 8. The Danish original and a translation into English are reproduced on pp. [29] ff. The date of the document is uncertain. It is based on the facts that Bohr expressed great interest in the superconductivity issue into 1933 and that "1933" is written by hand on top of an early English translation of the manuscript.

1932 Bohr wrote to Kronig[15] and Bloch,[16] explaining his lack of communication on the issue of superconductivity by, respectively, his strong involvement with the measurement problem and his effort to prepare the "Light and Life" lecture for publication.

Although the results of Bohr's thoughts on superconductivity in the early 1930s were never published, his efforts constitute an instructive example of his way of discussing and collaborating with younger colleagues. Thus Bohr's influence in physics went far beyond his publications. The ideas on superconductivity discussed in Copenhagen may well have played a role when the field was further developed by the younger generation of physicists.

In an unpublished note written many years later in connection with the preparation of Volume 9 of the Collected Works, Rudolf Peierls, who edited the volume, described Bohr's 1932 contribution to superconductivity as follows:[17]

> "... it was right to withdraw the paper. Yet we see that Bohr's intuition had grasped a number of important points. He saw that the sharpness of the transition [to superconductivity] must mean a phase change, and that this must take place whether or not there is a current flowing. He saw that it would therefore be a serious difficulty if the transition temperature depended on the frequency of the current with which the conductivity was tested. He saw that the skin effect could offer a way out of this dilemma."

Bohr's published contribution of 1953 was the first of four comments in response to a lecture by the German physicist Heinrich Fröhlich on "Superconductivity and Lattice Vibrations";[18] the other commentators were Heisenberg, Bloch and Hendrik Brugt Gerhard Casimir, another regular visitor to Bohr's institute. Bohr began his statement by describing himself as "one who remembers how the problems of metallic conduction were discussed from the time of Lorentz and Kamerlingh Onnes".[19] His one-page contribution to the discussion refers to Fröhlich's lecture only in passing and takes up without mathematics the general questions in his unpublished 1932 paper: the difference between the states of normal conductivity and superconductivity and the transition between them. Bohr pointed especially to the effect of impurities, which in metals that do not become superconducting give rise to a residual resistance at the

[15] Bohr to Kronig, 27 December 1932, BSC (22.4).
[16] Bohr to Bloch, 27 December 1932, BSC (17.3).
[17] R. Peierls, undated notes, NBA.
[18] *Proceedings*, ref. 6, pp. 755–761.
[19] Bohr, [Discussion Contribution], ref. 6.

lowest temperatures, but which in a superconductor merely produce a reduction of the transition temperature below which the resistance vanishes completely. Bohr expressed his conviction that an understanding of superconductivity would require new concepts and repeated his suggestion that the electrons might be able to propagate through the lattice in a coordinated motion. While recognizing that his attempt to describe the new phase was still of a preliminary character, he thus remained convinced that the possibility of a coordinated motion as a novel quantum mechanical effect was at the core of the phenomenon of superconductivity.

Bohr concluded:

> "The purpose of my remarks is merely to raise the question whether the explanation of superconductivity, notwithstanding the wide scope of application of the statistical methods, should not rather demand an exploration of the limitations of these methods."

This is an echo of Bohr's appeal to fellow physicists in the review from 1915 of Richardson's book, that conflict between theory and experiment ought to serve as an incentive for what he frequently called the "renunciation" of established concepts and for a radical rethinking of the theoretical basis. This was precisely the way of doing physics that Bohr had been practising in his own research.

At the time of Bohr's discussion contribution at Leiden, no specific mechanism had yet been found for understanding the stability of a supercurrent as envisaged by Bohr. Only four years later, however, John Bardeen, Leon Cooper and J. Robert Schrieffer introduced the new idea of electron pairing, which became the basis for a successful theory of superconductivity. Bohr was not fully convinced by the new theory and continued his own approach, as borne out by numerous notes in relation to the subject prepared together with Rosenfeld and other younger colleagues.[20] In his biographical sketch of Bohr starting out the Collected Works, Rosenfeld, who was the instigator and first General Editor of the series, described these efforts by Bohr as follows:[21]

> "Only much later [than 1943], during the last two summers of his life, did he for a while manage to concentrate again on a deep-lying phenomenon very near to those with which he had started his scientific career: the superconductivity of metals, in which the quantum of action manifests itself, so to speak, by macroscopic effects; he tried, without success, to put the somewhat abstract theory of these effects on a more physical basis."

---

[20] BMSS. See *Inventory*, ref. 13, p. [534], folder 62 and p. [532], folder 54.
[21] L. Rosenfeld, *Biographical Sketch*, Vol. 1, pp. XVII–XLVIII, on pp. XLIV–XLV.

2. ENCYCLOPEDIA CONTRIBUTIONS

In the course of his career Bohr made several efforts to present modern physics to a general audience. As documented in Part 2 of Volume 11, he frequently made such presentations in lectures or articles prepared for a political or social occasion. Bohr also presented popularizations of his science in contexts specially designed for the purpose.

After his theory of the atom had become recognized as a major breakthrough, not least with the award of the Nobel Prize for Physics in 1922, Bohr became widely known to the larger public interested in science. Capitalizing on his reputation, the editors of Encyclopædia Britannica asked him to prepare an entry on the "Atom". Bohr's article first appeared in the thirteenth edition of the Britannica, published in 1926, which like the twelfth edition, published from 1921 to 1922, consisted of a three-volume supplement to the 29-volume eleventh edition published from 1910 to 1911.[22]

In his article, Bohr describes the advance in the understanding of atomic theory represented by quantum theory after the electron and the nucleus had been established as the atomic constituents. He gives an account of the unravelling of the spectral laws for the radiation from atoms and of the chemical properties of the elements as embodied in the periodic system, and of the bearing of these developments on the representation of the atom in terms of the stationary states of the individual electrons.

Bohr wrote at a critical juncture in the development of quantum physics. When submitting his contribution on 24 March 1926, he wrote to T.C. Hodson of the Britannica that he had "found the work more difficult than ... anticipated due to the rapid progress within this field of science." Thus Bohr was able to refer to Werner Heisenberg's recently published matrix theory of quantum mechanics, but not to Erwin Schrödinger's alternative wave theory published a year later and subsequently proved by Schrödinger and others to be equivalent to Heisenberg's theory.[23]

Bohr to Hodson,
24 Mar 26
English
Full text on p. [508]

While revising his article for the next edition of the Britannica, as he had been encouraged to do by the publishers, he wrote to his friend and colleague Edward Neville da Costa Andrade, who was editor at the Britannica:

Bohr to Andrade,
18 Feb 28
English
Full text on p. [494]

"In the last days I have tried to do my best as regards the revision of the article for the Encyclopaedia Britannica. Still I have found it difficult due to

---

[22] N. Bohr, *Atom*, Encyclopædia Britannica, 13th edition, Suppl., Vol. 1, London and New York 1926, pp. 262–267. Reproduced in Vol. 4, pp. [657]–[663].

[23] Bohr himself deals with this development in *The Genesis of Quantum Mechanics* in *Essays 1958–1962 on Atomic Physics and Human Knowledge*, Interscience Publishers, New York 1963, pp. 74–78. Reproduced in Vol. 10, pp. [421]–[428].

the rapid progress of the subject to re-write the article according to present views. As the old article still indicates the limits of representing the subject when use is only made of the elementary classical concepts familiar to the general reader, I have therefore found it most convenient to return to my old proposal of writing a separate addendum."

The addendum comprised a full additional page on "Recent Progress" in which Bohr first referred briefly to the rapid development of matrix mechanics and then at greater length to Louis de Broglie and Schrödinger's wave mechanics with its striking success in elucidating atomic structure. In spite of these achievements, however, the wave approach could not, according to Bohr, provide a complete basis for the new physics, which required the recognition of the wave–particle dualism.

To bring his contribution up to date, Bohr mentioned Wolfgang Pauli's exclusion principle and concluded with the following words: "Quite recently even a successful attack on the fundamental problem of the origin of the so-called electron spin has been made by [P.A.M.] Dirac, whose work has opened new prospects." The reproduction of the original article preceding the new addendum has a new introductory paragraph by the editors and some minor editorial changes. It was published in the fourteenth edition of Encyclopædia Britannica, the first completely new edition after the eleventh, which with its twenty-four volumes comprised shorter articles and fewer volumes than its predecessors. Bohr's contribution is reproduced in its entirety below.[24]

Between the publication of its thirteenth and fourteenth editions, Encyclopædia Britannica had moved its operations from Britain to the United States. This involved several innovations, not least the publication of the Encyclopædia Britannica Book of the Year, which appeared for the first time in 1938. Bohr contributed both to this and the next volume in the series.

Bohr's entries were entitled "Matter, Structure of".[25] Again, the articles reflect the rapid development of physical theory and experiment at the time. After a recapitulation of the contents of his former article on the "Atom", to which he referred extensively, and a description of the several forms of "matter in bulk", Bohr turned in the first entry to the discovery in recent years of new particles (using the now obsolescent terms negaton and positon for the

---

[24] N. Bohr, *Atom*, Encyclopædia Britannica, 14th edition, Vol. 2, London and New York 1929, pp. 642–648. Reproduced on pp. [41] ff.

[25] N. Bohr, *Matter, Structure of*, Encyclopædia Britannica Book of the Year 1938, pp. 403–404. Reproduced on pp. [49] ff. N. Bohr, *Matter, Structure of*, Encyclopædia Britannica Book of the Year 1939, pp. 409–410. Reproduced on pp. [53] ff.

negative and positive electron) and the on-going revolution in nuclear physics. He concluded by referring to the special conditions for matter in the stars.

The second article reported on new developments over the last year. In one and the same paragraph, Bohr referred to the experimental confirmation of Dorothy Wrinch's "cyclol" theory of proteins and the discovery of the super-fluidity of helium at low temperatures. Further on he described his own "liquid drop" model of the atomic nucleus. Without using the term "fission" he then reported the evidence interpreted by Otto Robert Frisch and Lise Meitner on this basis "of other types of disintegration, in which a heavy nucleus is divided into two lighter nuclei". Bohr also made note of the experimental discovery of the meson particle, which physicists then identified as the particle proposed by the Japanese physicist Hideki Yukawa in 1935 to account for the short-range forces between nucleons and to provide an intermediate particle in beta decay. Typically, he concluded by looking to the future rather than celebrating the momentous development he reported: "We are here indeed concerned with the beginning of a new promising stage of the development of the atomic theory of matter."

3. SPEAKING TO GYMNASIUM STUDENTS

After having declined the year before, in early April 1949 Bohr spoke on Danish and Norwegian national radio to students in the gymnasiums preparing them for entrance to the universities. Bohr's was the last talk in a series of sixteen programmes broadcast by Danish radio in the spring of 1949 directed especially to gymnasium students. The other topics ranged from "Ethics of our time and Christianity" through "The new season at the Royal Theatre" and "New orientation in geography in the age of flight" to "English folk songs". Bohr found time both to record the talk and to prepare a written version which was published in a major Danish newspaper the day after it was broadcast.[26]

The talk is one of Bohr's finest writings devoted to the popularization of the achievements of atomic physics. As in his first two articles for Encyclopædia Britannica, his point of departure was the concept of the atom, which he now traced back to Antiquity. With due attention to the qualifications of his young audience, he described how the concept had developed from philosophical speculation to an entity subject to physics research and how, in part as a consequence of this, quantum physics had come to replace classical physics. In contrast to his presentations in the Britannica, which were limited to physics

---

[26] N. Bohr, *Atoms and Human Knowledge*, Berlingske Tidende, 2 April 1949. Reproduced and translated on pp. [57] ff.

as such, Bohr ended by introducing his interpretation of quantum physics involving "the complementarity way of arranging" our experience. He thus likened the complementary pairs of "wave" and "particle" in quantum physics to conceptual dualities in ordinary life such as thought–feeling and justice–love. In any particular instance, only one of the pair could be applied, yet both were necessary to gain a complete description of experience. Bohr concluded:

> "... the circumstance that we are forced to recognize such situations within a field such as physics, which because of its applications occupies such a large place in the school curriculum, may also be expected to influence the upbringing of young people."

To Bohr, physics in the school curriculum served not merely as a basis for understanding and applying the processes governing the nature around us. At least equally important, the lessons to be drawn from recent progress in his field were of critical philosophical and moral importance to our entire existence. This was Bohr's main message to the gymnasium students. On 10 February 1962, after television had been introduced in Denmark, Bohr spoke to school children in the new medium. His talk, however, has not been published and the notes for the talk are very preliminary.[27]

4. FOREWORDS TO BOOKS

In 1938 two of Bohr's collaborators at his institute, the theoretician Christian Møller and the experimentalist Ebbe Rasmussen,[28] published a book presenting the development of atomic physics to a scientifically interested Danish audience; a revised edition appeared in 1939.[29] The second edition was translated for the British market in 1940. The Danish version carried a jocular title that may be translated as "Atoms and Other Little Things"; the more sober English translation ran "The World and the Atom". It retained Bohr's foreword, a facsimile of which is reproduced below.[30]

Another popular book for which Bohr wrote a foreword was Gamow's perennial favourite, "Mr. Tompkins in Wonderland", first published by Cambridge University Press in 1940.[31] Gamow imagined how a person would perceive

---

[27] See *Inventory*, ref. 13, p. [534], folder 63.

[28] Bohr's obituary for Rasmussen is reproduced on pp. [379] ff.

[29] C. Møller and E. Rasmussen, *Atomer og andre Smaating*, Hirschsprung, Copenhagen 1938, 1939.

[30] N. Bohr, *Foreword* in C. Møller and E. Rasmussen, *The World and the Atom*, Allen & Unwin, London 1940, p. 9. Reproduced on pp. [73] ff.

[31] G. Gamow, *Mr. Tompkins in Wonderland*, Cambridge University Press, Cambridge 1940.

the world if the velocity of light and the quantum of action were respectively small and large enough for relativistic and quantum effects to be experienced directly. Such an approach went against Bohr's emphasis on classical physics as the means by which scientists necessarily obtained and exchanged scientific information. Nevertheless, Bohr wrote a foreword in Danish at the request of his colleague Sven Werner, who had translated the book. After a period with Bohr in Copenhagen, Werner had become the first professor of physics at the University of Aarhus. Werner motivated his request by pointing out that "the book, of course, is dedicated to Professor Bohr."[32] The Danish version of Gamow's book was published in 1942.[33]

Werner to Bohr,
9 Sep 42
Danish
Full text on p. [518]
Translation on p. [519]

The next item is not a foreword to a book properly speaking, but a greeting published in the journal "Nucleonics" on the occasion of its tenth anniversary.[34] It was written on the journal's request together with the greetings of two other Nobel Prize winners, John D. Cockcroft and Otto Hahn. For the sake of completeness all three greetings, which appeared on the same page, are reproduced below.

In 1960 Bohr wrote a foreword to a series of essays written by another close colleague, J. Robert Oppenheimer.[35] Again, the foreword was not written for the original edition but for the Danish one, translated by Bohr's assistant Aage Petersen.[36] It was the sixth book in a paperback series, previous authors in which included Bertolt Brecht, Benedetto Croce, José Ortega y Gasset and Søren Kierkegaard.

The essays in Bohr's collection "Atomic Theory and the Description of Nature", first published in 1934, are reproduced in earlier volumes of the Niels Bohr Collected Works,[37] as is the preface to the first edition of the

---

[32] Gamow dedicated the book to Lewis Carroll and Niels Bohr.

[33] N. Bohr, *Foreword* in G. Gamow, *Mr. Tompkins i Drømmeland*, Gyldendal, Copenhagen 1942, pp. 7–8. Reproduced and translated on pp. [77] ff.

[34] N. Bohr, [For the tenth anniversary of the journal "Nucleonics"], Nucleonics **15** (September 1957) 89. Reproduced on pp. [83] ff.

[35] J.R. Oppenheimer, *Science and the Common Understanding*, Simon and Schuster, New York 1953.

[36] N. Bohr, *Foreword* in J.R. Oppenheimer, *Naturvidenskab og Livsforståelse*, Gyldendal, Copenhagen 1960, pp. i–ii. Reproduced and translated on pp. [85] ff.

[37] *Introductory Survey*, Vol. 6, pp. [279]–[302]; *Atomic Theory and Mechanics*, Vol. 5, pp. [273]–[280]; *The Quantum Postulate and the Recent Development of Atomic Theory*, Vol. 6, pp. [109]–[136]; *The Quantum of Action and the Description of Nature*, Vol. 6, pp. [208]–[217]; *The Atomic Theory and the Fundamental Principles Underlying the Description of Nature*, Vol. 6, pp. [236]–[253]. Only the latter two are reproduced as facsimiles from the book; the others are facsimiles from earlier published versions.

book.[38] However, Bohr's "Preface to the 1961 Reissue" has not previously been reproduced in the Collected Works.

The long and winding road to the reissue started in 1954, when Bohr requested ten copies of the book from Cambridge University Press, only to learn that it had been out of print since 1945. Cambridge first responded that a reissue would be untimely, but expressed willingness to reconsider the matter when offered the opportunity also to publish a separate book with a new set of essays. However, the Press turned down the new book in October 1957, whereupon Bohr approached John Wiley and Sons in New York, which published "Atomic Physics and Human Knowledge" the following year.[39]

In March 1960 R.J.L. Kingsford at Cambridge University Press asked Bohr "whether you see any objection to the [1934] book being re-issued and whether you would be willing to contribute a new preface, or a brief introduction of a few pages to put the book into the context of the present day." Bohr answered: "I agree that, besides the original preface, the book be provided with a short preface to the second edition, and within a few weeks I shall send you a manuscript for it." Although Cambridge would have preferred a longer introduction than the one Bohr sent, they accepted it, and Bohr's book finally appeared in a second edition in 1961.[40]

Kingsford to Bohr,
24 Mar 60
English
Full text on p. [509]

Bohr to Kingsford,
29 Mar 60
English
Full text on p. [510]

In 1956 Jerrold Zacharias, James Killian and Francis Friedman of the Massachusetts Institute of Technology formed the Physical Science Study Committee to develop a new curriculum for high schools. It included a new "Science Studies Series of Books" in which the historian of physics I. Bernard Cohen's "The Birth of a New Physics" (1960) constituted one of the first vol-

---

[38] Vol. 6, p. [278].

[39] The information in this and the following paragraph is taken from the rich correspondence of Bohr and his collaborators with Cambridge University Press, BGC. The articles in *Atomic Physics and Human Knowledge* are reproduced in the Bohr Collected Works as follows (all but the first item as facsimiles of earlier versions): *Preface* and *Introduction*, Vol. 10, pp. [107]–[112]; *Light and Life*, Vol. 10, pp. [27]–[35]; *Biology and Atomic Physics*, Vol. 10, pp. [49]–[62]; *Natural Philosophy and Human Cultures*, Vol. 10, pp. [237]–[249]; *Discussion with Einstein on Epistemological Problems in Atomic Physics*, Vol. 7, pp. [339]–[381]; *Unity of Knowledge*, Vol. 10, pp, [79]–[98]; *Atoms and Human Knowledge*, Vol. 7, pp. [411]–[423]; *Physical Science and the Problem of Life*, Vol. 10, pp. [113]–[123].

[40] N. Bohr, *Preface to the 1961 reissue* in *Atomic Theory and the Description of Nature*, Cambridge University Press, Cambridge 1961, p. vi. Reproduced on pp. [89] ff. After having gone out of print once more, this collection of essays by Bohr, as well as the two subsequent collections published by John Wiley and Sons in 1958 and (posthumously) in 1963, have been reprinted in facsimile as *The Philosophical Writings of Niels Bohr*, Volumes 1 to 3, Ox Bow Press, Woodbridge, Connecticut 1987.

umes.[41] The series was immediately adapted to the Danish market, with Bohr giving a speech at its launching. Cohen's book, translated by Bohr's assistant Jens Bang, was the first of half a dozen volumes published in 1962, all of which contain the same foreword written by Bohr together with the three other members of the editing committee, all of them physicists: Henning Højgaard Jensen (professor at the University of Copenhagen), Aage Petersen and Søren Sikjær (professor at the Royal Danish School of Educational Studies).[42] It was one of Bohr's last pieces published in his lifetime. He died in November the same year.

---

[41] I.B. Cohen, *The Birth of the New Physics: From Copernicus to Newton*, Anchor Books, New York 1960.
[42] N. Bohr, H. Højgaard Jensen, A. Petersen and S. Sikjær, *Den danske redaktionskomités forord* in I.B. Cohen, *Fysikkens Gennembrud*, Gyldendals Kvantebøger, Copenhagen 1962, pp. 5–6. Reproduced and translated on pp. [91] ff.

The 1932 conference in Copenhagen, which took place during Bohr's main struggle with superconductivity and at which the physicists with whom Bohr discussed the matter were present. From left to right: W. Heisenberg, P. Hein, N. Bohr, L.N. Brillouin (sitting), L. Rosenfeld, M. Delbrück, W. Heitler (below Delbrück), L. Meitner and P. Ehrenfest (sitting together), F. Bloch, I. Waller, J. Solomon, E. Fues, B. Strömgren, R. de Laer Kronig, unidentified (below Kronig), G. Steensholt, H.A. Kramers (sitting), C.F. von Weizsäcker, J.P. Ambrosen, G. Beck (below Ambrosen), A. Gjelsvik (sitting), E. Buch-Andersen, H.H. Nielsen (sitting), F. Kalckar, J. Rud Nielsen, E.A. Hylleraas (sitting), R.H. Fowler (half hidden), I. Lam, P.A.M. Dirac and unidentified (both sitting), E. Rindal, C.G. Darwin, G. Lund (sitting), C. Manneback.

[17]

1. ELECTRON THEORY AND SUPERCONDUCTIVITY

I. MODERN ELECTRICAL THEORY

Nature **95** (1915) 420–421

Review of O.W. Richardson, "The Electron Theory of Matter",
Cambridge University Press, Cambridge 1914

See Introduction to Part I, p. [4].

[19]

MODERN ELECTRICAL THEORY.

The Electron Theory of Matter. By Prof. O. W. Richardson. Pp. vi+612. (Cambridge: At the University Press, 1914.) Price 18*s.* net.

THIS book is based on a series of lectures delivered by Prof. Richardson at the University of Princeton, and gives a general survey of the electron theory. The book starts with an account of the elementary principles of the theory of electricity and magnetism, and a discussion of phenomena which can be explained on the general Maxwell theory. From this we are gradually led to the discussion of such phenomena as dispersion and selective absorption, which have first found satisfactory explanations on the electron theory. Next follows a closer account of the theory of the mechanics of electrons, containing detailed considerations of the problems of electromagnetic mass, the radiation from an accelerated electron, and the properties of moving systems. This part ends with a brief account of the principles of the theory of relativity. After this we return again to the consideration of the general properties of matter, and the results deduced in the preceding chapters are employed in a discussion of the bearing of the electron theory on the problems of temperature radiation, magnetism, and the theory of metallic conduction. Finally, after an account of the theories of spectroscopic phenomena and the phenomena of radio-activity and X-rays, we are led into a discussion of the theories of the constitution of the atom.

It will be seen that the book covers a very extensive field. To give an adequate representation of the entire electron theory is naturally a task of the greatest difficulty, but the author appears to have done this in an admirable manner. Of necessity the treatment is at many points very restricted, but almost all points of general interest are considered. If any problem is treated more fully than others it is the theory of metallic conduction, as might naturally be expected from the author's own work. The exposition is throughout very clear and concise, and Prof. Richardson possesses a great gift of making even complicated arguments very easy for the reader to follow. A close connection with the latest experimental progress is everywhere maintained, and problems which involve hitherto unsolved difficulties are treated in a manner very far from being dogmatic.

In reading Prof. Richardson's book one gets ample opportunity to think about the present state of theoretical physics. The collection of the numerous brilliant achievements of the electromagnetic theory and the electron theory fills one with the greatest admiration. Still, the difficulties, first discovered with relation to the problem of temperature radiation and later in other problems, seem to be so great and of such a fundamental character that the theory will need very great alterations. Even if a way is indicated by Planck's theory no satisfactory solution of the difficulties has yet been found. In text-books only a few years old one finds great enthusiasm over what was called the future programme of the electromagnetic theory. It was believed that this theory constituted a final accomplishment of ordinary mechanics, and there appeared to be no limit to the application of the general principles of the theory. This attitude has changed most decisively. The impression obtained by reading the present book, however, is anything but merely disillusioning. Scarcely at any time has our knowledge increased so very rapidly as of late years, and, above all, we now possess much more powerful methods of experimental attack than were dreamed of a short time ago. Especially investigation of the radiation from radio-active bodies has proved most efficient in disclosing the internal structure of the atoms. If at present we may speak of a programme for the future development, it would, perhaps, be to examine the constitution of the special atomic systems actually existing, and then, by means of the directly observable properties of matter, possibly to deduce the general principles. If so, the evolution would be exactly the reverse of that anticipated.

In the present unsettled state, Prof. Richardson's book, which gives a balanced and masterly survey of a wide range of knowledge, will no doubt be especially welcome. It can be most heartily recommended, not only to students who seek an introduction to the electron theory, but to all interested in the modern development of physics.

N. B.

II. ON THE QUESTION OF SUPERCONDUCTIVITY

ZUR FRAGE DER SUPRALEITUNG
Proofs, 1932

TEXT AND TRANSLATION

See Introduction to Part I, p. [7].

[21]

Naturwissenschaften
J. Springer.

II. 7. 32   (Art. 492. Bohr)

Spamersche Buchdruckerei in Leipzig.

492
Zur Frage der Supraleitung.

Bekanntlich bietet die Deutung der metallischen Elektrizitätsleitung vom Standpunkt der klassischen Elektronentheorie prinzipielle Schwierigkeiten. Zwar lieferten unter der Annahme der freien Beweglichkeit der Elektronen im Metall schon die klassischen Methoden eine Erklärung des charakteristischen Verhältnisses zwischen Elektrizitäts- und Wärmeleitungsvermögen der Metalle und dessen Abhängigkeit von der Temperatur. Weder der normale Wert der spezifischen Wärme der Metalle noch die Kleinheit des Thomson-Effektes waren aber von dieser Auffassung aus verständlich. Wie zuerst SOMMERFELD zeigte, verschwinden jedoch diese Schwierigkeiten, wenn anstatt der klassischen Statistik eine dem PAULIschen Ausschließungsprinzip entsprechende Quantenstatistik zur Behandlung des Elektronengases in Metallen herangezogen wird. Im Anschluß zu diesem entscheidenden Fortschritt erwies es sich dann auch möglich, mit Hilfe der quantenmechanischen Methoden die weiteren augenscheinlichen Schwierigkeiten, die der Vorstellung der Bewegungsfreiheit der Metallelektronen vom klassischen Standpunkt anhaften, zu überwinden und insbesondere, wie von BLOCH nachgewiesen wurde, die rasche Zunahme der Leitfähigkeit mit abnehmender Temperatur zu erklären. Auf Grund der Vorstellung, daß die Elektronen sich im Metall unabhängig voneinander bewegen, konnte aber keine Deutung gegeben werden für das von KAMERLINGH ONNES entdeckte, plötzliche Verschwinden des Widerstandes einiger Metalle zu einer gewissen Temperatur. In dieser Verbindung dürfte es von Interesse sein, darauf hinzuweisen, daß auf Grund der Quantenmechanik der Strom sich unter Umständen auch auffassen läßt als eine gemeinsame Translationsbewegung des ganzen zusammengekoppelten Elektronensystems durch das vom Gitter der Metallionen gebildete Gerüst. Dies führt in der Tat zu einer Vorstellung der Supraleitung, die eine gewisse Ähnlichkeit hat mit einer Auffassung der metallischen Leitung, die vor mehreren Jahren besonders von LINDEMANN vertreten wurde um die oben erwähnten Schwierigkeiten der Elektronengasvorstellung umzugehen, die aber vom damaligen Standpunkt nicht durchgeführt werden konnte.

Betrachten wir ein Metallstück vom Volumen V, welches N-Elektronen enthält, wobei wir zunächst die Gitterionen als unbeweglich annehmen. Der normale stromlose Zustand des Elektronensystems wird dargestellt durch eine Wellenfunktion

$$\psi(O) = \varphi(x_1, \ldots, z_N;\ \sigma_1, \ldots, \sigma_N)\, e^{\frac{i}{n} E_0 t},$$

$$i\frac{2\pi}{h} E_0 t$$

wo $x_1, \ldots, z_N$ die Raumkoordinaten, $\sigma_1, \ldots, \sigma_N$ die Spinkoordinaten der N-Elektronen, n die durch 2π dividierte PLANCKsche Konstante und E_0 die Gesamtenergie des Elektronensystems bedeuten; ferner besitzt φ in allen Elektronenkoordinaten eine dem Raumgitter entsprechende Periodizität und ist unter Berücksichtigung der Spinkoordinaten in üblicher Weise antisymmetrisch in bezug auf jedes Elektronenpaar. Man überzeugt sich nun leicht, daß jeder solchen Lösung sich eine unendliche Schar von Lösungen anschließt, die durch folgenden approximativen Ausdruck dargestellt werden können:

$$\psi(\delta) = (\varphi + i\,\delta\,\psi_x)\, e^{\frac{i}{a}\delta (x_1 + \ldots + x_N)}\, e^{\frac{i}{n}\left(E_0 + \delta^2 \frac{n^2 N}{2 m a^2} A_x\right)t};$$

hier ist δ ein kleiner Parameter, a eine Länge von der Größenordnung der Gitterkonstanten, m die Elektronen-masse, ψ_x eine von der Ausgangslösung $\psi(o)$ abhängige, periodische und antisymmetrische Funktion und A_x eine ebenfalls von dieser Lösung abhängige Konstante. Während $\psi(o)$ stromlos war, entspricht $\psi(\delta)$ ein Strom in der x-Richtung von der mittleren Dichte

$$S_x = \delta \frac{\varepsilon \hbar N}{m a V} A_x,$$

wobei ε die Ladung eines Elektrons bezeichnet.

In bezug auf die Koordinaten eines einzelnen Elektrons ist die Lösung $\psi(\delta)$ von demselben Typus wie die von BLOCH zur Darstellung der Bewegungszustände der „freien" Elektronen herangezogen, sog. modulierten Wellen. Abgesehen von dem für unsere Betrachtung wesentlichen Umstand, daß die Lösung $\psi(\delta)$ den Zustand des Gesamtsystems der Metallelektronen darstellt, besteht ein weiterer bedeutsamer Unterschied darin, daß die Elektronenwellen, die in der betreffenden Darstellung zum resultierenden Strom hauptsächlich beitragen, wegen der benützten Dirac-Fermi-Statistik Werten von δ von der Größenordnung Eins entsprechen und daß daher die Ausdrücke für Energie und Strom in ihrer Abhängigkeit von δ mehr verwickelt sind als es den obigen Formeln entspricht, in welchen δ als eine sehr kleine Größe angenommen ist. Die Konstante A_x, die bei gegebenem δ für die Stromdichte S_x maßgebend ist, wird bestimmt durch die Kräfte, denen die Elektronen im Ionengitter ausgesetzt sind. In dem Grenzfall, wo diese Kräfte vernachlässigt werden, ist $A_x = 1$; im anderen Grenzfall, wo die Kräfte so groß sind, daß die Elektronen als an den einzelnen Gitterionen gebunden angesehen werden können, wird asymptotisch $A_x = 0$. Unter den Umständen, die in den Metallen realisiert sind, wo die Abstände zwischen den Gitterionen mit den Atomdurchmessern vergleichbar sind, dürfte jedoch A_x jedenfalls wenn man sich auf die Betrachtung der Valenzelektronen beschränkt, nie um mehr als eine oder zwei Zehnerpotenzen unter die Einheit sinken. Wegen der großen Dichte der Metallelektronen bekommt man daher einen beträchtlichen Strom bereits für sehr kleine Werte des Parameters δ. So wird, für $A_x = 10^{-2}$ und $N/V = a^{-3} = 10^{+23}$, die Stromdichte S_x bereits von der Größenordnung einer elektromagnetischen Einheit für Werte von δ so klein wie 10^{-8}. Wir erhalten also das vom klassischen Standpunkt paradoxale Resultat, daß ein Strom von der Stärke, wie sie in den Versuchen über die Supraleitung in Frage kommt, tatsächlich als eine „Verschiebung" des Gesamtsystems der Elektronen durch das Metallgitter ohne Störung ihrer gegenseitigen Kopplung aufgefaßt werden kann.

Die vorangehenden Betrachtungen gelten zunächst nur beim absoluten Nullpunkt der Temperatur, wo auch die Annahme der unabhängigen Beweglichkeit der einzelnen Elektronen eine unendliche Leitfähigkeit ergibt. Nach dieser letzten Auffassung rührt ja der Widerstand her von den im Ionengitter durch die Temperaturschwingungen hervorgerufenen Unregelmäßigkeiten und läßt sich als eine Reflexion der Elektronenwellen an den longitudinalen elastischen Wellen im Gitter auffassen. Wegen der kurzen Wellenlängen, von der Größenordnung a, der stromführenden Elektronen, spielt diese Reflexion schon bei tiefen Temperaturen eine wesentliche Rolle und gibt zu einem mit einer kleinen Potenz der Temperatur proportionalen Widerstand Anlaß. Bei der der Lösung $\Psi(\delta)$ entsprechenden Darstellung des Stromes kommt aber eine derartige Impulsübertragung zwischen dem Elektronensystem und dem Ionengitter nicht zustande. Vielmehr können die Gitterschwingungen in die Beschreibung des normalen stromlosen Elektronenzustandes einbezogen werden, wenigstens wenn man von Gitterwellen absieht, deren

3

Wellenlängen vergleichbar sind mit a/δ. Für diese Auffassung spricht auch der Befund von DE HAAS und von MEISSNER, daß nicht nur reine Metalle, sondern auch Legierungen und sogar Halbleiter supraleitend werden können. Der unter gewöhnlichen Umständen auftretende Restwiderstand der Legierungen läßt sich, wie von NORDHEIM nachgewiesen, durch die Streuung der Elektronen am unregelmäßigen Gitter erklären. Auch diese Ursache des Widerstandes dürfte aber verschwinden, sobald die Darstellung des Stromes vom Typus $\Psi(\delta)$ zulässig ist.

Bei weiter steigender Temperatur müssen wir annehmen, daß die Lösung $\Psi(\delta)$ instabil wird, in dem Sinne, daß außer dem tiefsten Energiezustand des Elektronensystems auch die Mannigfaltigkeit der höheren Zustände zur Geltung kommt. Die Statistik dieser Zustände wird schließlich mit einer MAXWELLschen Verteilung der als frei zu betrachtenden Elektronen zusammenfallen, aber lange vorher wird sie mit großer Annäherung durch die FERMIsche Verteilung gegeben. Den Übergang zwischen diesen verschiedenen Stufen im einzelnen zu verfolgen, dürfte eine sehr schwierige Aufgabe sein. Der empirische Befund des sprunghaften Verschwindens der Supraleitfähigkeit bei einer gewissen Temperatur legt es aber nahe, anzunehmen, daß es beim Sprungpunkt sich um eine einem Schmelzvorgang ähnliche Erscheinung handelt, wie von Anfang an KAMERLINGH ONNES betont hat. In der Tat dürften auch einfache theoretische Überlegungen die Annahme stützen, daß im betreffenden Temperaturbereich sowohl der „fest" gekoppelte, durch $\Psi(o)$ symbolisierte, wie der durch die Fermi-Verteilung symbolisierte „flüssige" Zustand des Elektronensystems jeder für sich eine stabile Lösung darstellt, während bei gleichzeitigem Vorhandensein in anschließenden Raumgebieten des Metalls sich je nach der Temperatur der erste bzw. der zweite Zustand auf Kosten des anderen ausbreitet. Diese Betrachtungen sind aber nur qualitativer Art und, bevor sich eine mehr quantitative Beschreibung entwickeln läßt, ist es schwierig, die Stichhaltigkeit der hier vorgeschlagenen Auffassung der Supraleitungserscheinungen und deren Anwendbarkeit auf die bemerkenswerte, von KAMERLINGH ONNES und von McLENNAN entdeckte Beeinflussung des Sprengpunktes durch Magnetfelder und elektrische Wechselfelder zu beurteilen.

Bei dieser Gelegenheit möchte ich noch F. BLOCH und L. ROSENFELD für erläuternde Diskussionen meinen herzlichen Dank aussprechen.

Kopenhagen, Institut for teoretisk Fysik, Juni 1932.

N. BOHR.

TRANSLATION

[Editor's comment: Essential handwritten corrections in the proofs are reproduced in square brackets, whereas text that Bohr asked to be removed is overwritten with a horizontal line. Formulae and symbols are taken from a manuscript for the proofs in the BMSS, since the proofs are not fully consistent in this regard. Two handwritten comments in the margin, which are difficult to read and the latter of which is in Bohr's handwriting, have been included. The footnotes have been added by the editor with the kind assistance of Gordon Baym.]

On the Question of Superconductivity

As is well known, the interpretation of metallic conduction of electricity in terms of the classical electron theory meets with difficulties of principle. It is true that, assuming free mobility of the electrons in the metal, the classical methods could provide an explanation of the characteristic ratio between the electric and thermal conductivity of metals and its temperature dependence. Neither the normal value of the specific heat of metals nor the smallness of the Thomson effect[1] could, however, be understood from this point of view. As was first shown by SOMMERFELD,[2] these difficulties disappear if instead of classical statistics a quantum statistics corresponding to PAULI's exclusion principle is invoked for treating the electron gas in the metal. Following this decisive step, it also then proved possible, with the aid of quantum mechanical methods, to resolve the other apparent difficulties to which the idea of freely mobile metal electrons give rise from the classical point of view, and, in particular, as was shown by BLOCH,[3] to explain the rapid increase of conductivity with decreasing temperature. The conception of electrons moving in the metal independently of

---

[1] The production or absorption of heat when an electric current passes through a circuit of a single material with a temperature difference along its length. It was discovered by the British physicist William Thomson (Lord Kelvin, 1824–1907) in 1854.

[2] A. Sommerfeld, *Zur Elektronentheorie der Metalle*, Naturwiss. **15** (1927) 824–832.

[3] F. Bloch, *Über die Quantenmechanik der Elektronen in Kristallgittern*, Z. Phys. **52** (1928) 554–600.

[25]

each other did not, however, permit any explanation of the sudden disappearance, discovered by KAMERLINGH ONNES,[4] of the resistance of some metals at a certain temperature. In this connection it could be of interest to point out that in quantum mechanics the current may also in some circumstances be regarded as a common translatory motion of the whole coupled system of electrons through the framework formed by the lattice of metal ions. This actually leads to a conception of superconductivity which has a certain resemblance with a view of metallic conduction proposed several years ago, particularly by LINDEMANN,[5] in order to avoid the difficulties of the electron gas conception mentioned above, but which could not be carried through on the basis of the viewpoint held at the time.

Consider a piece of metal of volume V containing N electrons, assuming at first the lattice ions to be fixed. The normal current-free state of the electron system will be represented by the wave function

$$\Psi(0) = \varphi(x_1, \ldots, z_N; \sigma_1, \ldots, \sigma_N)e^{i\frac{\pi}{\hbar}E_0 t},$$

where $x_1, \ldots, z_N$ denotes the space coordinates, $\sigma_1, \ldots, \sigma_N$ the spin coordinates of the N electrons, $\hbar$ PLANCK's constant divided by 2π, and E_0 the total energy of the electron system; moreover φ has in each electron coordinate a periodicity corresponding to the space lattice, and, taking into account the spin coordinates, is antisymmetric with respect to any two electrons in the usual way. One can now easily see that each such solution corresponds to an infinite family of solutions, which can be represented by the following approximate expression:

$$\Psi(\delta) = (\varphi + i\delta\psi_x)e^{i\frac{\delta}{a}(x_1 + \cdots + x_N)}e^{\frac{i}{\hbar}(E_0 + \delta^2 \frac{\hbar^2 N}{2ma^2}A_x)t};$$

here δ is a small parameter, a a length of the order of the lattice constant, m the electron mass, ψ_x a periodic and antisymmetric function depending on the initial solution $\Psi(0)$, and A_x a constant which also depends on this solution. Whereas $\Psi(0)$ carried no current, $\Psi(\delta)$ corresponds to a current in the x-direction with the average density

$$S_x = \delta\frac{\varepsilon\hbar N}{maV}A_x,$$

where ε denotes the electron charge.

[4] H. Kamerlingh Onnes, *Further Experiments with Liquid Helium. D. On the Change of the Electrical Resistance of Pure Metals at very low temperatures. ... V. The Disappearance of the Resistance of Mercury*, Koninklijke Nederlandse Akademie van Wetenschappen Amsterdam, Proceedings of the Section of Sciences **14** (1911) 113–115.

[5] F. Lindemann, *Note on the Theory of the Metallic State*, Phil. Mag. **29** (1915) 127–141.

[With respect to the coordinates of any single electron, the solution $\Psi(\delta)$ is of the same kind as the so-called modulated waves used by BLOCH to describe the states of motion of the "free" electrons. Apart from the fact, essential for our considerations, that the solution $\Psi(\delta)$ represents the state of the combined system of all electrons, another significant difference is that the electron waves, which in the representation in question contribute the most to the resulting current, correspond, because of the use of Fermi–Dirac statistics, to values of δ of the order of unity, and that therefore the expressions for energy and current in their dependence on δ are much more complicated than suggested by the formulae given above, in which δ is assumed to be a very small quantity. The constant A_x, which for a given δ is decisive for the current density S_x, is determined by the forces to which the electrons in the lattice are subject. In the limit in which these forces are negligible, $A_x = 1$; in the opposite limit in which the forces are so strong that the electrons may be regarded as being bound to individual lattice ions, A_x approaches zero.][6] In the circumstances realized in metals, in which the distances between the lattice ions are comparable to the atomic diameters, A_x should not decrease from unity by more than one or two powers of ten, at least if we restrict the consideration to the valence electrons. Because of the high density of electrons in a metal, we then find a considerable current already for very small values of δ. Thus, for $A_x = 10^{-2}$ and $N/V = a^{-3} = 10^{23}$, the current density S_x reaches the order of an electromagnetic unit ~~already~~ [even] for δ ~~as small as~~ [of the order of] 10^{-8}. We therefore arrive at the result, paradoxical from the classical point of view, that a current of the strength relevant to the experiments on superconductivity can be conceived as a [slow] "displacement" of the electron system through the lattice without an [appreciable] disturbance of their mutual coupling.

> It is <u>not</u> the case that $\Psi(x_1) = e^{ika}\Psi(x_1 + a)$!

The preceding considerations are valid in the first place only at the absolute zero of temperature, where also the assumption of an independent mobility of each electron leads to an infinite conductivity. According to this latter view the resistance is of course due to the irregularities in the ion lattice caused by the thermal vibrations and can be understood as the reflection of the electron waves by the longitudinal elastic waves in the lattice. Because of the current-carrying electrons' short wavelengths of the order a, the reflection plays a substantial role even at low temperatures and gives rise to a resistance proportional to a low power of the temperature. For the current represented by the solution $\Psi(\delta)$ such a transfer of momentum between the ~~electron system~~ [individual electrons] and the lattice does not, however, arise. Instead, the lattice vibrations

> The Fermi distribution dealt with by Bloch.

---

[6] This section in square brackets is crossed out by hand in the proofs.

[27]

can be included in the description of the normal current-free electron state (at least when one disregards lattice waves of a wavelength comparable to a/δ). ~~In favour of this view is also~~ [This view is also in accordance with] the finding of DE HAAS and MEISSNER[7] that not only pure metals, but also alloys and even semiconductors, can become superconducting. The residual resistance of alloys occurring under ordinary circumstances can be explained, as NORDHEIM has shown,[8] by the scattering of electrons from lattice irregularities. Also this source of the resistance should disappear as soon as the representation of the current by $\Psi(\delta)$ is applicable.

On a further rise of temperature we must assume that the solution $\Psi(\delta)$ becomes unstable, in the sense that besides the lowest energy state of the electron system also the manifold of higher states comes into effect. The statistics of these states will eventually coincide with a MAXWELL distribution of electrons, considered as free, but long before that, it will be given in good approximation by the FERMI distribution. To follow the passage between these various steps might present a very difficult task. The empirical finding of the discontinuous disappearance of superconductivity makes it natural to assume that the transition point represents a phenomenon similar to the process of melting, as was emphasized from the start by KAMERLINGH ONNES.[9] Indeed, also simple theoretical considerations support the assumption that in the relevant temperature region both the "solidly" coupled state symbolized by $\Psi(0)$ and the "liquid" state symbolized by the Fermi distribution each form separately a stable solution, whereas in the simultaneous presence of both in neighbouring space regions of the metal, the former or the latter of these states will spread at the expense of the other according to the temperature. However, these considerations are only qualitative in nature, and before a more quantitative treatment can be developed, it is difficult to judge the reliability of the view of superconductivity phenomena proposed here and their applicability to explain the remarkable influence on the transition temperature of magnetic and alternating electrical fields discovered by KAMERLINGH ONNES and McLENNAN.

I would like to take the opportunity to express my sincere thanks to F. BLOCH and L. ROSENFELD for instructive discussions.

Copenhagen, Institute for Theoretical Physics, ~~June~~ [August] 1932.

N. BOHR.

[7] See P.F. Dahl, *Superconductivity: Its Historical Roots and Development from Mercury to the Ceramic Oxides*, American Institute of Physics, New York, 1992, p. 137, and the work cited there.
[8] L. Nordheim, *Zur Elektronentheorie der Metalle. I*, Ann. d. Phys. **9** (1931) 607–640; *Zur Elektronentheorie der Metalle. II*, *ibid.*, 641–678.
[9] Ref. 4.

III. ON SUPERCONDUCTIVITY

OM SUPRALEDNINGSEVNEN
Manuscript, 1933

TEXT AND TRANSLATION

See Introduction to Part I, p. [7].

[29]

Om Supraledningsevnen.

Paa Grund af visse Vanskeligheder, der har vist sig saavel for det almindelige Bevis for den omhandlede bølgemekaniske Løsnings Eksistens som ved Sammenligningen med den omhandlede Opfattelse af Supraledningsevnen og de eksperimentelle Resultater,[1] har jeg tilbageholdt Korrekturen af denne Note i adskillige Maaneder. I Mellemtiden er der imidlertid udkommet et Arbejde af Kronig,[2] hvori der fremsættes en Opfattelse af Supraledningsfænomenerne, der paa væsentlige Punkter udviser en Lighed med de i Noten fremsatte Synspunkter. Jeg har derfor tænkt, at det var hensigtsmæssigt alligevel at offentliggøre Noten i uforandret Form, og i dette Anhang at gøre Rede saavel for Ligheden og Forskellen mellem Kronigs og min Opfattelse som for de Vanskeligheder, der for Gennemførelsen af hver af disse Opfattelser endnu staar tilbage at overvinde. Ved Diskussionen af disse Spørgsmaal har jeg haft den store Fordel at kunne drøfte Sagen med Kronig personlig og tillige haft værdifuld Hjælp af Raadslagninger med Bloch og Rosenfeld.

Fælles for de to Opfattelser er, at det ved den springvise Overgang imellem den sædvanlige Ledningsevne og Supraledningen drejer sig om en Foreteelse, der kan sammenlignes med en Tilstandsændring af Elektronsystemet i Metallet uafhængig af Tilstedeværelsen af en elektrisk Strøm. Medens Kronig tænker sig denne Tilstandsændring som en Smeltning af et mere eller mindre fast Elektrongitter, hvis Eksistens og Forskydning i første Tilnærmelse betragtes som uafhængig af Metaliongitteret, drejer det sig ved den ovenfor omhandlede Opfattelse om en Tilstandsændring, der kun har en symbolsk Analogi med en ved anskuelige Rumbilleder beskrivelig Smeltningsproces. Efter denne Opfattelse vil nemlig Elektronsystemets Tilstand saavel over som under Springtemperaturen i første Tilnærmelse kunne beskrives som et fast hvilende Gitter, og Supraledningen hidrører udelukkende fra Forsvindingen af den af Temperaturen betingede forholdsvis ringe Modifikation af denne Tilstand, hvorpaa Modstanden mod den elektriske Strøm beror. Netop i denne Omstændighed ligger

---

[1] Bohr probably here refers to J.C. McLennan, A.C. Burton, A. Pitt and J. Wilhelm, *The Phenomena of Superconductivity with Alternating Currents of High Frequency*, Proc. Roy. Soc. **A136** (1932) 52–76.

[2] R. Kronig, *Zur Theorie der Superleitfähigkeit*, Z. Phys. **78** (1932) 744–750. This paper, received on 31 August 1932, proposed that the interactions between electrons lead to their forming a rigid lattice intermeshed with the ionic lattice. The electron system can be superconducting since "in analogy with Bloch's theory for a single electron, translation [by an electron lattice constant] of the whole electron lattice can experience no resistance."

efter den paagældende Opfattelse Begrundelsen for den i Forhold til sædvanlige Smeltetemperatur for faste Legemer overordentlig lave Springtemperatur.
Paa den ene Side tillader den omhandlede Opfattelse endvidere at slippe fra
de tilsyneladende uoverstigelige Vanskeligheder, som Forskydningen af et fast
Elektrongitter gennem Iongitteret ogsaa efter Kvantemekanikken turde være til
Stede, og som staar i saa grel en Modsætning til den indbyrdes Uforskydelighed
af de modsat ladede Iongittere i en isolerende Saltkrystal. Paa den anden Side
rejser netop en Sammenligning som den sidste det Spørgsmaal, hvorfor efter
vor Opfattelse af Supraledningsevnen foruden Metallerne ikke ogsaa alle andre
Legemer opbyggede af Elektroner og Atomkerner udviser Supraledningsevne
ved tilstrækkelig lav Temperatur. Den væsentlige Begrænsning af de i Noten
anstillede teoretiske Betragtninger ligger netop i, at de ikke umiddelbart giver
noget tilfredsstillende Svar paa dette Spørgsmaal. En strømførende Løsning af
Typen 1[3] kan aabenbart kun eksistere for Legemer som Metallerne, hvor Elektronsystemet i de enkelte Atomer danner uafsluttede Systemer i kvantemekanisk
Forstand og ikke for de typiske Isolatorer, hvor alle Elektronerne i Atomerne
er bundne i afsluttede Grupper. Lignende Forhold som de af Peierls[4] ved
Diskussionen mellem Forskellen i den sædvanlige Ledningsevne for Metaller
og Isolatorer fremdragne gør sig sikkert her gældende, men en tilfredsstillende Behandling af dette Punkt er ikke hidtil lykkedes. Hvad Sammenligningen med de eksperimentelle Resultater angaar, ligger Vanskeligheden foruden
paa det nævnte Punkt i paa tilfredsstillende Maade at gøre Rede for Iagt-
tagelsesresultaterne vedrørende Springtemperaturens Ændring ved Benyttelse
af højfrekvente Vekselstrømme i Stedet for Jævnstrøm. En simpel Forbindelse
mellem Strømfrekvens og Springtemperatur vil nemlig betyde, at Supraledningsevnen ikke som antaget skulde være en af Strømmens Tilstedeværelse
uafhængig Egenskab hos Metallet ved lav Temperatur, men at den maatte
paa det nøjeste være knyttet til Strømningsmekanismen selv. Den springvise
Forsvinding af Supraledningsevnen ved en bestemt Temperatur synes imidlertid
at berede en Opfattelse som den sidstnævnte uoverstigelige Vanskeligheder, thi
selve den mekaniske Opfattelse af Varmefænomenerne tillader nemlig næppe
at forstaa nogen springvis Ændring af en fysisk Egenskab af Temperaturen,
undtagen i det Tilfælde at det drejer sig om en Ligevægt mellem to hver
for sig stabile Tilstandsformer. Under disse Omstændigheder turde det være
af Interesse at paapege, at McLennans Resultater[5] maaske kan forklares ved

---

[3] Bohr is likely referring to states of the form $\Psi(\delta)$ on p. [26].
[4] R. Peierls, *Zur Theorie der elektrischen und thermischen Leitfähigkeit von Metallen*, Ann. d.
Phys. **4** (1930) 121–148.
[5] Ref. 1.

den store Strømtæthed, hvormed man paa Grund af den saakaldte Skindeffekt[6] har at gøre med Forsøg over højfrekvente Vekselstrømme. Som det er omtalt i Noten, vil jo Muligheden for en Løsning af Typen 1 være betinget af den ringe Størrelse af Parameteren δ. Medens denne Betingelse sikkert er opfyldt i de sædvanlige Forsøg over Supraledningsevne, er det vanskeligt ud fra det foreliggende Materiale at være sikker paa, at den i tilstrækkelig Grad er opfyldt ved Forsøgene med Vekselstrømme. I denne Henseende ligger Forholdene ganske anderledes ved Forsøgene over Indflydelse af konstante Magnetfelter, og rent kvalitativt tilbyder de ogsaa den nærliggende Analogi mellem sædvanlige Smeltetemperaturers Afhængighed af Trykket en Forklaring af den iagttagne Sænkning af Springtemperaturen med Magnetfeltets Forøgelse. Denne Forklaring synes det ogsaa muligt at udbygge kvantitativt ved Sammenligning med den velkendte Formel for Trykeffekten.[7] (Sammenligningen giver saavel den rette Størrelsesorden af Effekten som dens funktionelle Afhængighed af Magnetkraften.)

---

[6] The tendency of alternating high-frequency currents to crowd towards the surface of a conducting material.

[7] The Clausius–Clapeyron equation.

TRANSLATION

On Superconductivity.

Because of certain difficulties which has appeared both regarding the general proof of the existence of the wave-mechanical solution in question and regarding the comparison between the conception of superconductivity in question and the experimental results,[1] I have held back the proofs of this note for several months. Meanwhile, however, a paper by Kronig[2] has appeared which sets out a conception of the phenomena of superconduction which in important respects resembles the points of view put forward in the note. I have therefore thought it appropriate nevertheless to publish the note in the original form, and in this appendix to deal with both the similarity and the difference between Kronig's and my conception, as well as the difficulties which have yet to be overcome for the realization of either of these conceptions. In the discussion of these questions I have had the great advantage of being able to talk the matter over with Kronig personally and also had valuable help from discussions with Bloch and Rosenfeld.

It is common to both conceptions that in the discontinuous transition between normal conductivity and superconduction one is dealing with a phenomenon which can be compared to a change of state of the electron system in the metal independent of the presence of an electric current. While Kronig considers this change of state as the melting of a more or less rigid electron lattice, whose existence and displacement is in first approximation considered as independent of the metal ion lattice, in the conception proposed above there is a question of a change of state which has only a symbolic analogy with a melting process describable in terms of visualizable spatial pictures. According to this conception the state of the electron system above as well as below the transition temperature can in first approximation be described as a fixed lattice at rest, and the superconduction stems entirely from the disappearance of the relatively small, temperature-dependent modification of this state, which is the cause of the resistance to the electric current. According to the conception in question, precisely this circumstance is the reason for the extremely low transition temperature compared to the ordinary melting temperature of solid bodies. On the one hand, the conception proposed here makes it furthermore possible to escape from the apparently insurmountable difficulties which the displacement of a rigid electron lattice through the ion lattice would meet even in quantum mechanics, and which is in such stark contrast to the mutual impossibility of displacing the oppositely charged ion lattices in an insulating salt crystal. On the other hand, precisely a comparison such as the latter gives rise to the question why, according to our conception of superconductivity, all other bodies than metals made of nuclei and electrons do not also exhibit superconductivity at sufficiently low temperature. The essential limitation of the theoretical considerations made in the note is precisely that they do not immediately give a satisfactory answer to this question. A current-carrying solution of type 1[3] can evidently only exist for bodies such as metals where the electron system in the individual atoms forms open systems in the sense of quantum mechanics, and not for the typical insulators, where all the atomic electrons are bound in closed groups. Circumstances similar to those invoked by Peierls[4] in the discussion of the difference in the ordinary conductivity in metals and insulators must surely arise here, but a satisfactory treatment of this point has not yet been found. As to comparison with experimental results, the difficulty lies, besides in the point mentioned above, in how to account in a satisfactory way for the observed results regarding the change in the transition temperature when using high-frequency alternate currents instead of a direct current. A simple relation between the frequency of the current and the transition temperature would namely mean that superconductivity cannot, as assumed, be a property of the metal at low temperature independent of the presence of the current, but that it must be most intimately coupled with the mechanism of current flow itself. However, the discontinuous disappearance of superconductivity at a definite temperature appears to create insurmountable difficulties for a conception such as the latter, since the mechanical conception of heat phenomena itself hardly allows one to understand any discontinuous change of a physical property with temperature, except in the case of an equilibrium between two states each of which is stable in itself. In these circumstances it could be of interest to point out that McLennan's results[5] may perhaps be explained by the high current density that one has to deal with in experiments with high-frequency currents due to the so-called skin effect.[6] As mentioned in the note, the possibility of a solution of type 1 will of course be conditional upon the small value of the parameter δ. While this condition is certainly satisfied in the ordinary experiments on superconductivity, it is difficult on the basis of the existing evidence to be certain that it is fulfilled to a sufficient degree in experiments with alternate currents. In this respect, the circumstances are quite different in the experiments on the effect of constant magnetic fields, and also from a purely qualitative point of view they offer, by the close analogy with the pressure dependence of the ordinary melting temperature, an explanation of the observed lowering of the transition temperature with the increase of the magnetic field. It seems also possible to extend this explanation quantitatively by comparison with the well-known formula for the pressure effect.[7] (The comparison gives both the right order of magnitude of the effect and the functional dependence on the magnetic force.)

IV. [DISCUSSION CONTRIBUTION ON SUPERCONDUCTIVITY]

"Proceedings of the Lorentz Kamerlingh Onnes Memorial Conference,
Leiden University 22–26 June 1953",
Stichting Physica, Amsterdam 1953, pp. 761–762

See Introduction to Part I, p. [5].

Participants in the Lorentz Kamerlingh Onnes Memorial Conference in Leiden, 1953. Standing, left to right: A.N. Gerritsen, A.F. van Itterbeek, H. Wergeland, J. Vlieger, K.W. Taconis, G. Borelius, A. Pais, M.J. Druyvesteyn, P.A.M. Dirac, R. de Laer Kronig, K. Mendelssohn, C.F. Squire, W.J. de Haas, L.J.F. Broer, J. de Boer, A.D. Fokker, W. Heisenberg, A.B. Pippard, N. Bohr, W.E. Lamb, D. Shoenberg, H.B.G. Casimir, J. Korringa, J. Dingle, R.E. Peierls, S.A. Wouthuysen, F.J. Belinfante, G. Källén, J. Smit, H. Fröhlich, J. de Nobel, F. Bloch, F. London, H.J. Groenewold, G.W. Rathenau, P.H.E. Meyer, J.G. Daunt, D.K.C. MacDonald, S.R. de Groot, R.M.F. Houtappel, B.R.A. Nijboer, W. Pauli, L. Rosenfeld. Sitting from left: J. Clay, M.A. Proca, S. Tomonaga, H.M. Gijsman, C.J. Gorter, B. Ferretti, J. van den Handel, G.J. van den Berg, M. Fierz.

Bohr: As one who remembers how the problems of metallic conduction were discussed from the time of L o r e n t z and K a m e r l i n g h O n n e s, I should like, although in later years I have not occupied myself actively with problems of superconductivity, to make some general remarks or rather to raise some questions concerning the present situation.

I hardly need to recall the masterly way in which L o r e n t z treated the problem of metallic conduction on the basis of classical statistical mechanics, nor how K a m e r l i n g h O n n e s interpreted his discovery of superconductivity as a change of phase in the metal leaving room for a mechanism of conduction essentially different from that which comes in consideration at higher temperatures. Of course, the foundation for the treatment of such problems has been radically changed by the advent of quantum mechanics and the formulation of the Pauli principle which is the basis for Fermi's statistical treatment of an electron gas. In particular, I most thoroughly appreciate the great importance of Fröhlich's contribution to our understanding of the interaction between the electrons through their coupling with the ion lattice.

The question I want to raise, however, is how such a treatment could explain the phenomena of superconductivity itself. As an example of the marked differences between ordinary metallic conduction and superconduction I want especially to call attention to the way in which the presence of impurities in a nearly homogeneous metal influences the resistance to the current. Indeed, one of the most successful features of the statistical theory of metallic conduction is just the explanation of the residual resistance due to impurities as arising from the scattering of the free electron waves by the foreign atoms in the lattice. In superconductivity on the contrary, the effect of impurities is merely to change the transition temperature, but below this their effect vanishes completely, showing that nothing like the scattering of free electrons takes place under such circumstances. It appears therefore to me that the mechanism of superconductivity cannot be described on the basis of the statistical methods.

In fact, as is generally recognized, we have to do in the superconductive phase with a state of the electrons which, although differing very little in energy from the normal one, exhibits a high degree of order, which abolishes all statistical features of free or coupled electron motions and rather corresponds to a wave function in configuration space, adjusted to the interaction between the electrons and the ion lattic, including the impurities, as well as to the mutual interactions between the individual electrons. In such a state a current is conceived as an adiabatic modification of the comprehensive wave function and should herefore not be accompanied by any such transfer of momentum from the electrons to the lattice as is an inherent feature of the mechanism of normal conduction. The transaction to the ordinary conducting phase should correspond to a collapse of this order on account of the increase of the irregular thermal motion of the lattice ions.

In spite of the apparent simplicity of this view, which no doubt is shared by many, a detailed treatment of the transition problem is of course beset with great difficulties, which will probably demand an asymptotic approach, involving an estimate of the order of magnitude of the number of electrons whose inclusion is relevant for the account of the characteristic features of the phenomena. The purpose of my remarks is merely to raise the question whether the explanation of superconductivity, notwithstanding the wide scope of application of the statistical methods, should not rather demand an exploration of the limitations of these methods.

2. ENCYCLOPEDIA CONTRIBUTIONS

V. ATOM

Encyclopædia Britannica, 14th edition, Vol. 2,
London and New York 1929, pp. 642–648

See Introduction to Part I, p. [11].

influenced by external agencies. This, however, will depend on the attractive force due to the nucleus which keeps the cluster together. On account of the small size of the nucleus compared with the distance apart of the electrons in the cluster, this force will to a high approximation be determined solely by the total electric charge of the nucleus. The mass of the nucleus and the way in which the charges and masses are distributed among the particles making up the nucleus itself will only have an exceedingly small influence on the behaviour of the electronic cluster.

2. To the second class belong such properties as the radioactivity of the substance. These are determined by the actual internal structure of the nucleus. In the radioactive processes we witness, in fact, explosions of the nucleus in which positive or negative particles, the so-called α and β particles, are expelled with very great velocities.

The complete independence of the two classes of properties is most strikingly shown by the existence of substances which are indistinguishable from one another by any of the ordinary physical and chemical tests, but of which the atomic weights are not the same, and whose radioactive properties are completely different. Any group of two or more such substances are called isotopes (q.v.), since they occupy the same position in the classification of the elements according to ordinary physical and chemical properties. The first evidence of their existence was found in the work of Soddy and other investigators on the chemical properties of the radioactive elements. It has been shown that isotopes are found not only among the radioactive elements, but that many of the ordinary stable elements consist of isotopes, for a large number of the latter that were previously supposed to consist of atoms all alike have been shown by Aston's investigations to be a mixture of isotopes with different atomic weights. Moreover the atomic weights of these isotopes are whole numbers, and it is because the so-called chemically pure substances are really mixtures of isotopes, that the atomic weights are not integers.

The inner structure of the nucleus is still but little understood, although a method of attack is afforded by Rutherford's experiments on the disintegration of atomic nuclei by bombardment with α particles. Indeed, these experiments may be said to have started a new epoch in natural philosophy in that for the first time the artificial transformation of one element into another has been accomplished (see TRANSMUTATION OF ELEMENTS). In what follows, however, we shall confine ourselves to a consideration of the ordinary physical and chemical properties of the elements and the attempts which have been made to explain them on the basis of the concepts just outlined.

THE RELATIONSHIPS BETWEEN THE ELEMENTS

It was recognized by Mendelejeff that when the elements are arranged in an order which is practically that of their atomic weights, their chemical and physical properties show a pronounced periodicity. A diagrammatic representation of this so-called periodic table is given in Table I., which represents in a slightly modified form an arrangement first proposed by Julius Thomsen. In the table the elements are denoted by their usual chemical symbols, and the different vertical columns indicate the so-called periods. The elements in successive columns which possess homologous chemical and physical properties are connected by lines. The meaning of the square brackets around certain series of elements in the later periods, the properties of which exhibit typical deviations from the simple periodicity in the first periods, will be mentioned below.

Radiation.—The discovery of the relationship between the elements was primarily based on a study of their chemical properties. Later it was recognised that this relationship appears also very clearly in the constitution of the radiation which the elements emit or absorb in suitable circumstances. In 1883 Balmer showed that the spectrum of hydrogen, the first element in the table, could be expressed by an extremely simple mathematical law. This so-called Balmer formula states that the frequencies ν of the lines in the spectrum are given to a close approximation by

$$\nu = R\left(\frac{1}{(n'')^2} - \frac{1}{(n')^2}\right) \qquad (1)$$

ATOM, when ordinarily used in chemistry and physics, refers to the smallest particle of an element which can exist either alone or in combination with similar particles of the same or of a different element. The atom also refers to a quantity proportional to the atomic weight of an element. According to the theory of atomism, which dates from pre-Socratic times, the atom is one of the minute indivisible particles of which the whole universe is composed. (X.)

Through the important experimental discoveries of the second half of the 19th century it became gradually clear that the atoms of the elements, far from being indivisible entities, had to be thought of as aggregates built up of separate particles. Thus from experiments on electrical discharges in rarefied gases, and especially from a closer study of the so-called cathode rays, the existence of small negatively charged particles—the mass of which was found to be about 2,000 times as small as the mass of the lightest atom, the hydrogen atom—was recognized. These small particles, which may be regarded as atoms of negative electricity, are now, following Johnstone Stoney, generally called electrons. Through the investigations of J. J. Thomson and others convincing evidence was obtained that these electrons are a constituent of every atom. On this basis a number of the general properties of matter, especially as regards the interaction between matter and radiation, receive a probable explanation.

In fact, the assumption that electrons are vibrating around positions of stable equilibrium in the atom offered a simple picture of the origin of spectral lines, which allowed the phenomena of selective absorption and dispersion to be accounted for in a natural way. Even the characteristic effect of magnetic fields on spectral lines discovered by Zeeman could, as was shown by Lorentz, be simply understood on this assumption. The origin of the forces which kept the electrons in their positions remained for a time unknown, as well as the way in which the positive electrification was distributed within the atom. From experiments on the passage through matter of the high speed particles expelled from radioactive substances, however, Rutherford was in 1911 led to the so-called nuclear model of the atom. According to this the positive electricity is concentrated within a nucleus of dimensions very small compared with the total space occupied by an atom. This nucleus is also responsible for practically the whole of the atomic mass.

Properties of the Elements.—The nuclear theory of the atom has afforded a new insight into the origin of the properties of the elements. These properties can be divided into two sharply distinguished classes.

1. To the first class belong most of the ordinary physical and chemical properties. These depend on the constitution of the electron cluster round the nucleus and on the way in which it is

where R is a constant, and where n' and n'' are whole numbers. If n'' is put equal to 2 and n' is given successively the values 3, 4, . . . the formula gives the frequencies of the series of lines in the visible part of the hydrogen spectrum. If n'' is put equal to 1 and n' equal to 2, 3, 4, . . . a series of ultra-violet lines is obtained which was discovered by Lyman in 1914. To $n''=3$, 4, . . . correspond series of infra-red hydrogen lines which also have been observed.

TABLE I. *Periodic Table of the Elements*

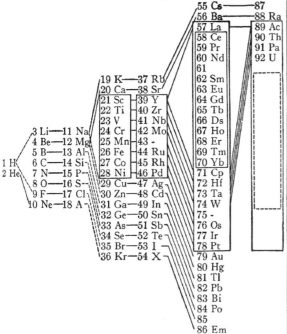

Rydberg in his famous investigation of line spectra more than 30 years ago was able to analyse in a similar way many spectra of other elements. Just as in the case of hydrogen he found that the frequencies of a line-spectrum (such as that of sodium) could be represented by a formula of the type

$$\nu = T'' - T' \qquad (2)$$

where T''. T' can be approximately represented by

$$T = \frac{R}{(n - a_\kappa)^2} \qquad (3)$$

a_κ is a constant for any one series, but takes different values $a_1, a_2 \ldots$ for the different series, while n takes a set of successive integral values. R is constant throughout for all spectra, and is the same constant as that appearing in (1); it is generally called the "Rydberg number." In many spectra the terms of most series are multiple, *i.e.*, the terms which we consider as forming a series do actually form two, three or more series corresponding to two, three or more slightly different values of a_κ. Rydberg also discovered that the spectra of elements occupying homologous positions in the periodic table were very similar to each other, a similarity which is especially pronounced as regards the multiplicity of the terms.

Moseley's Discovery.—The study of X-ray spectra made possible by the work of Laue and Bragg brought out relations of a still simpler kind between different elements. Thus Moseley (*q.v.*) in 1913 made the fundamental discovery that the X-ray spectra of all elements show a striking similarity in their structure, and that the frequencies of corresponding lines depend in a very simple way on the ordinal number of the element in the periodic table. Moreover the structure of these spectra was very like that of the hydrogen spectrum. The frequency of one of the strongest X-ray lines for the various elements could, for instance, be given approximately by

$$\nu = N^2 R \left(\frac{1}{1^2} - \frac{1}{2^2} \right) \qquad (4)$$

and that of another line by

$$\nu = N^2 R \left(\frac{1}{2^2} - \frac{1}{3^2} \right) \qquad (5)$$

where R is again the Rydberg constant and N the ordinal number of the element in the periodic table. The extreme simplicity of these formulae enabled Moseley to settle any previous uncertainty as to the order of the elements in the periodic table, and also to state definitely the empty places in the table to be filled up by elements not yet discovered.

In the nuclear model of the atom, the ordinal number of an element in the periodic table receives an extraordinarily simple interpretation. In fact, if the numerical value of the charge on an electron is taken as unity, this ordinal number, which is often called the "atomic number" (*q.v.*), can simply be identified with the magnitude of the nuclear charge. This law which was foreshadowed by J. J. Thomson's investigations of the number of electrons in the atom, as well as by Rutherford's original estimate of the charge on the atomic nucleus, was first suggested by van den Broek. It has since been established by refined measurements of the nuclear charge, and it has proved itself an unerring guide in the study of the relationship between the physical and chemical properties of the elements. This law also offers an immediate explanation of the simple rules governing the changes in the chemical properties of radioactive elements following the expulsion of a or β particles.

THE QUANTUM THEORY

The discovery of the electron and of the nucleus was based on experiments, the interpretation of which rested on applications of the classical laws of electrodynamics. As soon, however, as an attempt is made to apply these laws to the interaction of the particles within the atom, in order to account for the physical and chemical properties of the elements, we are confronted with serious difficulties. Consider the case of an atom containing one electron: it is evident that an electrodynamical system consisting of a positive nucleus and a single electron will not exhibit the peculiar stability of an actual atom. Even if the electron might be assumed to describe an elliptical orbit with the nucleus in one of the foci, there would be nothing to fix the dimensions of the orbit, so that the magnitude of the atom would be an undetermined quantity. Moreover, according to the classical theory the revolving electron would continually radiate energy in the form of electromagnetic waves of changing frequency and the electron would finally fall into the nucleus. In short, all the promising results of the classical electronic theory of matter would seem at first sight to have become illusory. It has nevertheless been possible to develop a coherent atomic theory based on this picture of the atom by the introduction of the concepts which formed the basis of the famous theory of temperature radiation developed by Planck in 1900.

This theory marked a complete departure from the ideas which had hitherto been applied to the explanation of natural phenomena, in that it ascribed to the atomic processes a certain element of discontinuity of a kind quite foreign to the laws of classical physics. One of its outstanding features is the appearance in the formulation of physical laws of a new universal constant, the so-called Planck's constant, which has the dimensions of energy multiplied by time, and which is often called the "elementary quantum of action." We shall not enter upon the form which the quantum theory exhibited in Planck's original investigations, or on the important theories developed by Einstein in 1905, in which the fertility of Planck's ideas in explaining various physical phenomena was shown in an ingenious way. We shall proceed at once to explain the form in which it has been possible to apply the quantum theory to the problem of atomic constitution. This rests upon the following two postulates:—

1. An atomic system is stable only in a certain set of states, the "stationary states," which in general corresponds to a discrete sequence of values of the energy of the atom. Every change

in this energy is associated with a complete "transition" of the atom from one stationary state to another.

2. The power of the atom to absorb and emit radiation is governed by the law that the radiation associated with a transition must be monochromatic and of frequency ν such that

$$h\nu = E_1 - E_2 \qquad (6)$$

where h is Planck's constant and E_1 and E_2 are the energies in the two stationary states concerned.

The first of these postulates aims at a definition of the inherent stability of atomic structures, manifested so clearly in a great number of chemical and physical phenomena. The second postulate, which is closely related to Einstein's law of the photoelectric effect, offers a basis for the interpretation of line spectra; it explains directly the fundamental spectral law expressed by relation (2). We see in fact that the spectral terms appearing in this relation can be identified with the energy values of the stationary states divided by h. This view of the origin of spectra has been found to agree with the experimental results obtained in the excitation of radiation. This is shown especially in the discovery of Franck and Hertz relating to impacts between free electrons and atoms. They found that an energy transfer from the electron to the atom can take place only in amounts which correspond with the energy differences of the stationary states as computed from the spectral terms.

The Hydrogen Spectrum.—From the Balmer formula (1) and the quantum theory postulates, it follows that the hydrogen atom has a single sequence of stationary states, the numerical value of the energy in the nth state being Rh/n^2. Applying this result to the nuclear model of the hydrogen atom, we may assume that this expression represents the work necessary to remove the electron from the nth state to an infinite distance from the nucleus. If the interaction of the atomic particles is to be explained upon the laws of classical mechanics, the electron in any one of the stationary states must move in an elliptical orbit about the nucleus as focus, with a major axis whose length is proportional to n^2. The state for which n is equal to 1 may be considered as the normal state of the atom, the energy then being a minimum. For this state the major axis is found to be approximately 10^{-8} centimetres. It is satisfactory that this is of the same order of magnitude as the atomic dimensions derived from experiments of various kinds. It is clear, however, from the nature of the postulates, that such a mechanical picture of the stationary states can have only a symbolic character. This is perhaps most clearly manifested by the fact that the frequencies of the orbital revolution in these pictures have no direct connection with the frequencies of the radiation emitted by the atom. Nevertheless, the attempts at visualizing the stationary states by mechanical pictures have brought to light a far-reaching analogy between the quantum theory and the classical theory. This analogy was traced by examining the radiation processes in the limit where successive stationary states differ comparatively little from each other. Here it was found that the frequencies associated with the transition from any state to the next succeeding one tend to coincide with the frequencies of revolution in these states, if the Rydberg constant appearing in the Balmer formula (1) is given by the following expression:

$$R = \frac{2\pi^2 e^4 m}{h^3} \qquad (7)$$

where e and m are the charge and mass of the electron and h is Planck's constant. This relation is actually found to be fulfilled within the limits of the experimental errors involved in the measurements of e, m and h, and seems to establish a definite relation between the spectrum and the atomic model of hydrogen.

Correspondence Principle.—The considerations just mentioned constitute an example of the application of the so-called "correspondence principle" which has played an important part in the development of the theory. This principle gives expression to the endeavour, in the laws of the atom, to trace the analogy with classical electrodynamics as far as the peculiar character of the quantum theory postulates permits. On this line much work has been done in the last few years, and quite recently in the hands

of Heisenberg has resulted in the formulation of a rational quantum kinematics and mechanics. In this theory the concepts of the classical theories are from the outset transcribed in a way appropriate to the fundamental postulates and every direct reference to mechanical pictures is discarded. Heisenberg's theory constitutes a bold departure from the classical way of describing natural phenomena but may count as a merit that it deals only with quantities open to direct observation. This theory has already given rise to various interesting and important results, and it has in particular allowed the Balmer formula to be derived without any arbitrary assumptions as to the nature of the stationary states. However, the methods of quantum mechanics have not yet been applied to the problem of the constitution of atoms containing several electrons, and in what follows we are reduced to a discussion of results which have been derived by using mechanical pictures of the stationary states. Although in this way a rigorous quantitative treatment is not obtainable it has nevertheless been possible, with the guidance of the correspondence principle, to obtain a general insight into the problem of atomic constitution.

SPECTRA OF ELEMENTS OF HIGHER ATOMIC NUMBER

The hydrogen spectrum may be considered as evidence of a step-like process in which an electron is captured and bound increasingly strongly in the field surrounding the nucleus, the stages of this process being the stationary states of the atom. Simple arguments lead to the conclusion that the stages corresponding to the binding of an electron by a nucleus of any given charge will be represented by a similar sequence of stationary states and that the energy W_n necessary to remove the electron from the nth state will be given by the expression:

$$W_n = N^2 \frac{Rh}{n^2} \qquad (8)$$

where N is the atomic number of the elements under consideration. These states may be visualized as mechanical orbits of the electron in which the major axis is N times as small as the major axis in the corresponding orbit in the hydrogen atom. The spectrum associated with the binding process under consideration is represented by the formula:

$$\nu = N^2 R \left(\frac{1}{(n'')^2} - \frac{1}{(n')^2} \right) \qquad (9)$$

For $N = 2$, this formula actually represents the spectrum which is emitted by a singly ionized helium atom, i.e., a helium atom, which has lost one of its electrons. Spectra of this type have not yet been observed for values of N larger than 2, but it will be seen that formula (9) includes the approximate formulae (4) and (5) representing the frequencies of the strongest lines in the X-ray spectra of the elements. This may be understood if we assume that an X-ray spectrum is associated which changes in the state of binding of one of the electrons in the inner region of the atom, where, at least when the atomic number is large, the force on the electron due to the nucleus will far outweigh the forces due to the other electrons, and where consequently the presence of these electrons will have a comparatively small influence on the strength of the binding.

Influence of Electrons.—In general the mutual influence of the electrons is very considerable. Consider the stages by which an electron is captured by an atom of which the nucleus already has s electrons circulating round it. In the initial stages of this process while the orbits may be supposed to have dimensions which are large compared with the orbital dimensions of the electrons previously bound, the repulsive forces from these latter electrons may be assumed to neutralise s units of the nuclear charge, and the resultant force will be approximately the same as when an electron is circulating round a nucleus of atomic number N-s. In the later stages, when the dimensions of the orbit of the new electron are smaller, the other electrons can no longer be considered to act as a single central charge, and their repulsion cannot be easily determined. Thus the conditions become more complicated, and the stationary states can no longer be treated by picturing the motion of the new electron as following a Keplerian ellipse.

It has been found, however, that many features of the resulting spectra would be explained by assuming the added electron to move in a plane central orbit consisting of a sequence of quasi-elliptic loops. In contrast to a Keplerian orbit the single loops are not closed, but the successive maximum radii will be placed at constant angular intervals on a circle with the nucleus at the centre. For such central orbits it is possible, as was first shown by Sommerfeld, to select from the continuous multitude of possible orbits a set of orbits which may be taken as representing station-ary states in the sense of the quantum theory. These states are labelled with two integral numbers; the one, denoted by n, cor-responds to the integer appearing in the Balmer formula and is called the principal quantum number. The other, denoted by k, may be called the subordinate quantum number. For any given value of n, the number k can take the values $1, 2, 3 \ldots n$, corresponding to a set of orbits with increasing minimum dis-tance from the nucleus. For a given value of k increasing values of n correspond to orbits which exhibit an increasing maximum distance from the nucleus, but which are similar in size and shape in the region where the electron comes nearest to the nucleus. For the work necessary to remove an electron in an n_k orbit completely from the nucleus, the theory leads to the following approximate expression

$$W_{n,k} = (N-s)^2 \frac{Rh}{(n-a_k)^2} \qquad (10)$$

where a depends only on the subordinate quantum number k, and approaches zero for increasing k.

If s is equal to $N-1$, we see that the $W_{n,k}$ when divided by h co-incides exactly with Rydberg's expressions (3) for the spectral terms of the ordinary series spectra of the elements. These spectra may therefore be considered as evidence of processes, represent-ing the last stage in the formation of a neutral atom, in which a nucleus of charge Ne, which holds already $N-1$ electrons bound in its field, is capturing an Nth electron. In recent years it has been found that many elements under suitable conditions besides their ordinary spectra also emit spectra for which the terms can be represented by

$$T = p^2 \frac{R}{(n-a_k)^2} \qquad (11)$$

where p may take the integral values $2, 3, 4. \ldots$ Comparing (11) with formula (10) we see that these spectra must be as-cribed to atoms, which after having lost p electrons are rebinding an electron in the field of the remaining atomic ion.

This interpretation of series spectra allows also the rules gov-erning the possible combinations of spectral terms to be explained. In fact, it has been found that only those lines appear in the spectrum for which the k-values of the spectral terms involved differ by one unit. From an investigation of the constitution of the radiation which on classical electrodynamics would be emitted from an electron performing a central motion, this rule can be shown to be a simple consequence of the correspondence principle.

Multiplex Structure.—The multiplex structure exhibited by the terms of most series spectra makes it necessary to assume that the motion of the electron involved in the emission of these spectra is somewhat more complicated than the simple central motion described above. An analysis based on the correspond-ence principle indicates that this motion may be described as a central motion on which is superposed a uniform precession of the orbital plane round an invariable axis in space. For a time, however, it seemed very difficult to obtain any closer connection between the observed structures and the above hypothesis of the constitution of the atom. In particular the remarkable analogy between the finer structures of the optical spectra and the X-ray spectra, which had been brought out by the experiments, was very puzzling. The study of the strange anomalies exhibited by the effect of a magnetic field on the components of the optical multiplets has, however, quite recently led to the view that the electron itself carries, besides its electric charge, also a magnetic moment which may be associated with a swift rotation round an axis through its centre. This new assumption allows not only the anomalous Zeeman effect to be accounted for, but affords at the same time a natural explanation for the empirical rules governing the dependency of the widths of the multiplet structures on the atomic number.

ATOMIC CONSTITUTION AND THE PERIODIC TABLE

Soon after the discovery of the electron it was recognized that the relationships between the physical and chemical properties of the elements expressed in the periodic table point towards a group-structure of the electronic distribution in the atom. Fun-damental work on these lines was done by J. J. Thomson in 1904. After the discovery of the nucleus and the simple interpretation of the atomic number given above, his work has been followed up with great success especially by Kossel and Lewis.

It is suggested that the electrons within the atom possess a tendency to form stable groups, each containing a definite number of electrons which, in the neutral state of the atom, surround the centre of the atom like successive shells or layers. An explanation of the simple valency properties holding for the second and third period of the periodic table was, for instance, obtained by assum-ing that there was a tendency to form completed shells each con-taining eight electrons. The single valency of sodium and the double valency of magnesium are ascribed to the facility with which the neutral atoms of these elements can lose one or two electrons respectively, as the atomic ions remaining would then contain completed shells only. On the other hand the double negative valency of sulphur and the single negative valency of chlorine are ascribed to the tendency of their outermost shells to take up two or one additional electrons respectively in order to form a complete shell of eight electrons, like that contained in the neutral atom of the inactive gas argon.

Spatial Arrangement of Electrons.—Attempts have been made to associate the existence of such groups with statical con-figurations of electrons possessing a high degree of symmetry. The presence of groups of eight electrons for instance has been explained as an arrangement of electrons at the corners of a cube. However suggestive these ideas have been in affording pictures of the constitution of chemical compounds, they do not allow a direct connection with other properties of the atom to be established; the main difficulty being that stable spatial arrange-ments of the electrons are incompatible with the nuclear theory of the atom. In the meantime, however, it has been possible to connect the group structure of the electronic cluster in the atom with the quantum-theory interpretation of spectra. Thus the constitution of the neutral atom in its normal state can be in-vestigated by imagining a process by which N electrons one after one are captured and bound in the field of force surrounding a nucleus of charge Ne.

To each step there corresponds a multitude of stages, i.e., stationary states, in which the electron is more and more firmly bound to the atom. The final state, in which binding is strongest, corresponds to the normal state of the atomic ion. A definite connection between the spectra and the group structure was now established by assuming that, in the normal atom only a limited number of electrons can be bound in states visualized as orbits characterized by definite values of the quantum numbers n and k. The electrons bound in orbits corresponding to a given value of n are said to form an n-quantum group, which in its finally com-pleted stage will contain n subgroups, corresponding to the possi-ble values $1, 2 \ldots n$ which k may take. For a sufficiently large nuclear charge, the strength with which the electrons in the different subgroups belonging to one and the same group are bound will be nearly equal.

In the gradual building up of the groups in atoms with increas-ing nuclear charge, it is, however, to be noted that when an n_k orbit appears for the first time in the neutral atom, the strength of the binding will depend very considerably on the value of k. This is due to the circumstance that this quantum number fixes the closest distance to which the electron may approach the nucleus. The screening of the nuclear charge by the other elec-trons in the atom may therefore be very different for orbits cor-responding to different values of k, and the effect on the strength of the binding can be so large that an orbit characterized by certain values of n and k may correspond to a stronger binding than an orbit for which n is smaller but k larger. This offers a natural explanation of one feature of the periodic table, namely that the periods grow gradually larger, while there appear sequences of elements which differ comparatively little in their chemical and physical properties. Such a sequence marks a stage in the development of an n-quantum group, which consists in the addition of a subgroup corresponding to a value of k which was previously not yet represented in that group, and which takes place after the building up of a group corresponding to a higher value of n has already begun. In fact, during the addition of the subgroup a temporary standstill will occur in the development of the latter group, the constitution of which will primarily determine the chemical affinity of the atom, since it contains the most loosely bound electrons.

In the accompanying table (Table II.) is given a summary of the structure of the normal state of the neutral atoms of the elements. The figures before the different elements are the atomic numbers, which give the total number of electrons in the neutral atom. The figures in the different columns give number of electrons in orbits corresponding to values of the principal and subordinate quantum numbers standing at the top. A comparison with the periodic table (Table I.) will show that those elements which in

chemical respect are homologous, will have the same number of electrons in the electronic groups most loosely bound, containing the so-called valence-electrons. The atoms of elements which in Table I. are enclosed in brackets possess electronic configurations in which a subgroup is being added to a group, whose principal number is less than the group containing the typical valence-electrons. An especially conspicuous example of such a completion of an inner group is offered by the elements forming the family of the rare earths. Here we witness the addition of the fourth subgroup to the 4-quantum group, which begins first in Ce (58) while the addition of the third subgroup was already finished in Ag (47).

Table II. is in general agreement not only with the optical spectral evidence but also with that in the region of X-rays. As mentioned earlier, we see in X-ray spectra a change in the binding of an electron in the interior of the atom. This takes place when, for instance, by the impact of a swiftly moving particle on the atom, an electron is removed from one of the electronic groups, and its place is taken by an electron belonging to a group for which the binding energy is smaller. As an example it may be stated that the strong X-ray whose frequency is approximately represented by formula (4) is emitted when an electron has been removed from the 1-quantum group, and one of the 2_2 electrons performs a transition so as to occupy the empty place. The line represented approximately by formula (5) originates from a

TABLE II. *Summary of the Structure of the Neutral Atoms*

	1_1	2_1	2_2	3_1	3_2	3_3	4_1	4_2	4_3	4_4	5_1	5_2	5_3	5_4	5_5	6_1	6_2	6_3	6_4	6_5	6_6	7_1	7_2
1 H	1																						
2 He	2																						
3 Li	2	1																					
4 Be	2	2																					
5 B	2	2	1																				
10 Ne	2	2	6																				
11 Na	2	2	6	1																			
12 Mg	2	2	6	2																			
13 Al	2	2	6	2	1																		
18 A	2	2	6	2	6																		
19 K	2	2	6	2	6		1																
20 Ca	2	2	6	2	6		2																
21 Sc	2	2	6	2	6	1	2																
22 Ti	2	2	6	2	6	2	2																
29 Cu	2	2	6	2	6	10	1																
30 Zn	2	2	6	2	6	10	2																
31 Ga	2	2	6	2	6	10	2	1															
36 Kr	2	2	6	2	6	10	2	6															
37 Rb	2	2	6	2	6	10	2	6			1												
38 Sr	2	2	6	2	6	10	2	6			2												
39 Y	2	2	6	2	6	10	2	6	1		2												
40 Zr	2	2	6	2	6	10	2	6	2		2												
47 Ag	2	2	6	2	6	10	2	6	10		1												
48 Cd	2	2	6	2	6	10	2	6	10		2												
49 In	2	2	6	2	6	10	2	6	10		2	1											
54 Xe	2	2	6	2	6	10	2	6	10		2	6											
55 Cs	2	2	6	2	6	10	2	6	10		2	6				1							
56 Ba	2	2	6	2	6	10	2	6	10		2	6				2							
57 La	2	2	6	2	6	10	2	6	10		2	6	1			2							
58 Ce	2	2	6	2	6	10	2	6	10	1	2	6	1			2							
59 Pr	2	2	6	2	6	10	2	6	10	2	2	6	1			2							
71 Gp	2	2	6	2	6	10	2	6	10	14	2	6	1			2							
72 Hf	2	2	6	2	6	10	2	6	10	14	2	6	2			2							
79 Au	2	2	6	2	6	10	2	6	10	14	2	6	10			1							
80 Hg	2	2	6	2	6	10	2	6	10	14	2	6	10			2							
81 Tl	2	2	6	2	6	10	2	6	10	14	2	6	10			2	1						
86 Em	2	2	6	2	6	10	2	6	10	14	2	6	10			2	6						
87 —	2	2	6	2	6	10	2	6	10	14	2	6	10			2	6					1	
88 Ra	2	2	6	2	6	10	2	6	10	14	2	6	10			2	6					2	
89 Ac	2	2	6	2	6	10	2	6	10	14	2	6	10			2	6					2	1

transition by which a 3_2 electron takes the place left open upon the removal of a 2_2 electron.

The question how many electrons there are in the various groups and subgroups has been subject to much discussion in the last few years. Table II. is the temporary result of this discussion and seems to give an adequate description of the spectral as well as the chemical evidence. It is clear that a full theoretical treatment of the problem cannot be obtained from considerations based only on the simple picture of central orbits. Such a treatment will essentially involve an examination of those features of the binding of the electrons, which appear in the multiplet structure of spectral lines. Indeed it is very probable that the idea that the electron itself has magnetic properties may give the clue to the interpretation of the empirical rules governing the number of electrons in the group structure of the atom.

Recent Progress.—Such is the outline of the theory of the atom and its structure as it stood in 1925. Since then the subject of atomic structure has undergone a remarkable development due to the establishment of rational quantum theoretical methods which enable a quantitative treatment to be given to a large number of atomic problems that, earlier, were accessible only to considerations of a more qualitative character. These methods take their origin from two sources. On the one hand the symbolic procedure of "quantum mechanics" initiated by Heisenberg, and briefly referred to above, has, thanks to the collaboration of a number of eminent physicists, developed into a structure which, as regards generality and consistency, may be compared with the theory of classical mechanics. On the other hand a new method of "wave mechanics" of great power and fertility has been developed by Schrödinger having its starting point in the pioneer work of de Broglie. (*See* QUANTUM THEORY.) This method utilizes the analogy between mechanics and optics emphasized already long ago by Hamilton. According to de Broglie, the motion of a material particle may be compared with the propagation of a train of waves, the frequency of which is related to the kinetic energy of the particle, as calculated on the relativity theory, by the general quantum relation $E=h\nu$. Indeed, this view may be considered as an inversion of the considerations by which Einstein was led to the hypothesis that the carrier of light energy had to be considered not as waves but as corpuscles—the so-called light quanta—which concentrated within a small volume contained the energy $h\nu$. Notwithstanding the indispensability of the wave theory of light for the account of ordinary optical experience, Einstein's hypothesis has proved most fruitful in explaining a number of phenomena, notably the important discovery of Compton of the change in the frequency which X-rays suffer when scattered by electrons. Similarly the view of de Broglie, strange as it is from the classical point of view, has received a striking support from the recent discovery of Davisson and Germer about the selective reflection of electrons from metal crystals. Indeed, in these experiments the electrons were found to behave as waves possessing the wave length anticipated from quantum theory.

The first indication of the importance of the wave idea in the problem of atomic constitution was the suggestion of de Broglie that the stationary states of an atom might be interpreted as an interference effect of the waves associated with a bound electron. A real advance in this direction, however, was first achieved by Schrödinger, who succeeded in replacing the classical equations of motion for the particles in the atom by a certain differential equation of a type similar to that known from the theory of elastic vibrations of solid bodies. As well known from acoustics any such vibration can be resolved into a number of purely harmonic components, representing the fundamental tones of a musical instrument. It was now found that the "characteristic solutions" of the Schrödinger wave equation, corresponding to such purely harmonic vibrations, offer a detailed interpretation of the properties of stationary states. First of all the energy values appearing in the quantum theory of spectra are obtained by multiplying the frequencies of the characteristic vibrations by Planck's constant. Next Schrödinger succeeded in associating with the solution of his wave equation a continuous distribution

of electric charge and current, which, when applied to a characteristic vibration, represents the electrostatic and magnetic properties of an atom in the corresponding stationary state. Similarly the superposition of two characteristic solutions corresponds to a continuous vibrating distribution of electric charge, which on classical electrodynamics would give rise to an emission of radiation, fulfilling the requirement of the quantum postulate and the correspondence principle as regards frequency as well as intensity and polarization.

These remarkable results have given rise to a renewed discussion regarding the physical nature of the constituents of the atom. Indeed, the view has been advocated that the wave idea offers a real picture of the atom, allowing a direct application of the methods of classical physics. On this view the wave mechanics represent a natural generalization of classical mechanics of material particles, to which it is related in the same way as the modern theory of optics based on the fundamental equations of electrodynamics is related to the more primitive theory of geometrical optics, which makes use of the idea of light rays. It would appear, however, that the situation is more complicated. Due to the very contrast between the ideas of quantum theory and the fundamental principles of classical physics, we cannot expect to be able to visualize atomic phenomena by means of our classical ideas. In the dilemma regarding the nature of light and the ultimate constituents of matter we witness a general feature of a dualism inherent in the quantum theory description. Indeed, the wave and particle ideas are both indispensable if we attempt to get a full description of experience. This situation is brought out very clearly by the recent development of the symbolic method of quantum mechanics, through which an intimate connection between the correspondence argument and Schrödinger's work is established. Just when due regard is given to the feature of dualism in question, the quantum theory can, unfamiliar as it is, still be regarded as a natural development of the ordinary description of physical phenomena.

In the problem of atomic constitution we meet with a very striking example of the dualism mentioned. Notwithstanding the wonderful power of the Schrödinger wave functions of illustrating properties of stationary states, the wave theory fails to account for the peculiar stability of these states, on which the interpretation of atomic phenomena rests so essentially. Indeed, we have here to do with the very feature of discontinuity or rather "individuality," by which the quantum theory departs from the ideas of classical physics, and of which we perhaps have the most striking example in the existence of the individual particles themselves. For the rest, the dualism of the quantum theory brings with it the conclusion that the use of the idea of stationary states excludes the possibility of following at the same time the behaviour of the single particles in the atom. Just this situation finds its adequate representation in the characteristic vibrations of the Schrödinger wave problem. This problem, in fact, is not a 3-dimensional one, as that of ordinary spatial description, but one which operates with a number of dimensions equal to the number of degrees of freedom of the whole atom. This fact has recently found an important application in the interpretation of a certain peculiar duplexity in the structure of spectra especially marked in the helium spectrum. This duplexity, which for a long time eluded explanation, has recently been explained by Heisenberg, who pointed out that we have here to do with an effect of the mutual interaction of the electrons in the atom, which exhibits a close correspondence with a classical resonance problem, but cannot be accounted for on the simple procedure of characterizing the behaviour of the individual electrons by quantum numbers. The justification of this procedure in a large number of applications rests on the circumstance that in general the resonance effect is very small, the mutual influence of the various electrons on each other being, as already described, to a close approximation to that of a conservative central field of force.

It is impossible here to give anything but a vague idea of the abundance of details regarding the physical and chemical properties of the elements which have been explained by means of the new methods of quantum theory. It may still be mentioned that the important contributions of Main Smith and Stoner to the interpretation of the periodic table—embodied already in the scheme of electron orbits given in the article—have been brought into most convincing connection with the so-called exclusion principle of Pauli and with the idea of the magnetic electron referred to already. Moreover a study of the fine structure of band spectra has led to the conclusion that the proton, or the nucleus of the hydrogen atom, also possesses an angular momentum and a magnetic moment. Quite recently even a successful attack on the fundamental problem of the origin of the so-called electron spin has been made by Dirac, whose work has opened new prospects. (*See* also ATOMIC WEIGHTS; CHEMISTRY; ELECTRICITY, CONDUCTION OF: *in Gases;* ISOTOPES; QUANTUM THEORY.)

BIBLIOGRAPHY.—E. N. da C. Andrade, *The Structure of the Atom* (1923); G. Birtwistle, *The Quantum Theory of the Atom* (1926), *The New Quantum Mechanics* (1928); N. Bohr, *The Theory of Spectra and Atomic Constitution* (1922); A. Sommerfeld, *Atomic Structure and Spectral Lines* (1923); J. D. Main Smith, *Chemistry and Atomic Structure* (1924); N. V. Sidgwick, *The Electronic Theory of Valency* (1927). (N. B.)

VI. MATTER, STRUCTURE OF

Encyclopædia Britannica Book of the Year 1938, pp. 403–404

See Introduction to Part I, p. [11].

[49]

MATTER, STRUCTURE OF. Fundamental for modern views on the structure of matter are the discoveries of the structural units of the atom, the electron, and the atomic nucleus. As is explained in greater detail in the article ' Atom ' in the *Ency. Brit.*, vol. 2, pp. 642–48 (referred to in the following as A), the nuclear model of the atom due to Lord Rutherford allows above all a simple discrimination between such properties of matter, including all ordinary physical and chemical effects, which depend primarily upon the extra-nuclear electron configuration of the atom and those which essentially involve changes of the atomic nuclei themselves, as manifested in the phenomena of natural radioactivity or of artificial nuclear transmutations initiated by impact of material particles or by radiation, the continued study of which has given such remarkable results in the past few years.

As regards the account of the former properties of matter on the basis of the nuclear atom, the incorporation of the quantum of action discovered by Planck into the mechanical treatment of the electron configuration has offered a decisive clue to the understanding of the intrinsic stability of those configurations. In particular, it has led, through the study of optical and high-frequency spectra of the elements, to a classification of the binding of the electrons in the normal state of all atoms by means of so-called quantum numbers (A, Table II), which has given a complete explanation of the relationships between the elements, as expressed in the well-known periodic table (A, Table I). The gradual development of proper methods of quantum mechanics, including the general formulation of the exclusion principle for electron systems (A, p. 648), has further allowed a complete understanding of the different types of bindings of atoms into molecules, which in the case of the so-called polar bonds are ascribed to the electrostatic forces between ions in their normal state, and in the case of homopolar bonds, are ascribed to the sharing of a pair of electrons of opposite spins in the same quantum state by two atoms.

Great progress has also been achieved in recent years in our understanding of the structure of matter in bulk, especially as regards the crystalline state. Not only has the analysis of the stationary states of crystals led to a comprehension of the optical properties of solids and of the variation of their specific heats at low temperatures, but it has even been possible to derive, on the basis of quantum mechanics, approximate expressions for the cohesive forces between atoms and molecules responsible for the elastic properties of matter in its different states of aggregation. Moreover, a general theory of the typical properties of the metallic state has been obtained by a quantum mechanical treatment of the ensemble of more or less loosely bound electrons in metals. Such a treatment, in which special regard is taken to the exclusion principle, has also given a most instructive explanation of the magnetic properties of metals, including ferromagnetism. Still, certain remarkable properties of electronic or atomic aggregates at very low temperatures, such as the supraconductivity of metals and the suprafluidity of helium which have not yet found any satisfactory explanation, seem to indicate that some essential feature of the lowest quantum states of these aggregates has as yet escaped us.

Apart from the incessant development of the explanation of the properties of matter on the basis of the original discoveries of the structural units of the atom, quite new fields of atomic theory have in recent years been opened by the discovery of further elementary particles which, like the positon and the neutron, only exceptionally appear in ordinary atomic phenomena, but which play a fundamental part in the problem of the constitution of the atomic nuclei themselves. The first isolation in cosmic rays of positive charged corpuscle of the same mass as the ordinary electron, now often called negaton, was a most striking confirmation of the rational development of the relativistic quantum theory of the electron, which led to the prediction of the possibility under certain conditions of the materialization of radiation quanta into a positon and a negaton, and of the inverse process of the annihilation of such a pair of oppositely charged electrons with appearance of radiative energy. It is just the last phenomenon which prevents the stable existence of the positon as a constituent of matter, where only ordinary electrons can be held round the positive atomic nuclei. Positons are liberated, however, as products of artificial radioactive disintegrations as frequently as negatons, the sign of the charge depending primarily on the ratio between charge and mass of the radioactive nucleus compared with that of stable nuclei. The liberation process itself must indeed in both cases be considered as the very creation of an electron as a mechanical entity, because such light particles cannot be considered as constituents of quantum mechanical systems of dimensions as small as those of atomic nuclei.

The possibility of treating nuclei as quantum mechanical systems entirely composed of heavy particles was opened by the isolation of the neutron through the study of nuclear transmutations by impact of fast material particles. In the development of a comprehensive theory of nuclear constitution based on a model of the nucleus composed of neutrons and protons in numbers indicated directly by its mass and charge we meet, however, with a problem which in some way is the inverse of that of atomic constitution, the characteristic simplicity of which lies in the possibility of identifying the forces between the constituent particles with the ordinary electric attractions and repulsions. In fact, the stability of nuclei claims the existence of forces of a novel type between protons and neutrons, appearing at small distances, and the character of which can only be gradually explored by the study of the nuclear phenomena themselves. Still, without a precise knowledge of the intranuclear forces, a most instructive explanation of the fundamental probability law of radioactive decay under emission of heavy charged particles has been obtained on quantum mechanics which permit the passage of such particles through a region outside the nucleus, where the potential energy corresponding to electrostatic repulsion is higher than the kinetic energy of the released particle, and the penetration of which would thus be impossible on classical mechanical ideas.

A more detailed theory of nuclear phenomena is, however, above all confronted with the difficulty that, owing to the close packing of the nuclear particles, no such approximative procedure as that applied with so great success to the classification of the binding of the individual electrons in the extranuclear configurations of atoms by means of quantum numbers is legitimate. On the contrary, we have in the normal and excited states of nuclei to do with types of motion of an essentially collective character, which must be quantized in a similar way as the states of molecular aggregates, as also appears clearly from the great differences between the distribution of the energy levels of nuclei compared with those of atoms. In particular has the study of nuclear transmutations revealed that the level distribution of a nucleus for high excitation is practically continuous, and that every impact between a fast particle and a nucleus leads, in the first place, to the formation of a semi-stable compound system, the eventual subsequent disintegration of which is to be considered as a separate event, independent of the first stage of the collision process. In this intermediate state the energy is stored in the compound system in a way similar to that of the thermal energy of ordinary bodies, and by introducing, in accordance with this comparison, a suitable definition of the temperature of the highly excited compound nucleus, it has been possible to treat the disintegration phenomena in close analogy to the evaporation processes of liquids and solids.

The effective temperatures involved in nuclear transmutations (thousands of million degrees centigrade) are not only extremely high compared with those available in ordinary experiments, but are even very high in comparison with those estimated generally to occur in the interior of celestial bodies like the sun, where nevertheless nuclear transformations must be expected to play an essential part as a source of the large energy continuously radiated away from the surface. Quite apart from this fascinating problem, the extreme conditions of temperature and pressure in stars confront us with new aspects of the structure of matter. Above all, stellar matter is certainly in a state in which, not only all usual molecular bonds are disrupted, but where even most electrons are entirely loosened from their ordinary bindings to atomic nuclei. The high energy of the free electrons which, together with the more or less stripped nuclei, constitute the bulk of the star, must even in some cases be expected to lend to the stellar matter new properties differing essentially from those of ordinary gaseous bodies. In cases of extreme pressures, it should even be possible that the high-speed electrons would combine with the nuclei, giving rise to an entirely new state of matter composed exclusively of free neutrons. A closer examination seems, however, to show that this peculiar state is hardly reached in the stars of the most common type, including our sun.

BIBLIOGRAPHY.—Lord Rutherford, *The Newer Alchemy* (Cambridge, 1937); N. Bohr, *The Theory of Spectra and Atomic Constitution* (Cambridge, 1924); C. G. Darwin, *The New Conceptions of Matter* (London, 1931); N. Feather, *An Introduction to Nuclear Physics* (Cambridge, 1936); A. S. Eddington, *Stars and Atoms* (Oxford, 1927). (N. B.)

VII. MATTER, STRUCTURE OF

Encyclopædia Britannica Book of the Year 1939, pp. 409–410

See Introduction to Part I, p. [11].

MATTER, STRUCTURE OF. Our knowledge of the atomic constitution of matter has recently been extended both as regards the structural properties of material bodies composed of large numbers of atoms and as regards the very foundations of the theory of the elementary particles of which atoms and especially atomic nuclei are built up.

In the first respect, the fertile study of the crystalline structure of inorganic substances has been continued, and in particular experimental and theoretical investigations of the superlattices exhibited by certain alloys have yielded very interesting and promising results.[1] Moreover, research on the peculiar crystalline properties of highly complex organic substances would seem to give support to the so-called " cyclol " theory of proteins, proposed a few years ago by D. Wrinch ; according to this theory the molecular edifices of these substances are essentially characterized by the folding of a two-dimensional fabric of carbon and nitrogen atoms into cage structures, to which are appended the various radicals determining the specific properties of the proteins concerned.[2] Quite new problems regarding the structure of the liquid state of matter have further been raised by the remarkable discovery of the so-called supra-fluidity of helium at very low temperatures. Although this phenomenon, just as the supra-conductivity of metals, so much studied in recent years, is not yet fully understood, it becomes ever clearer that here we have to do with a striking example of the predominating influence of the so-called zero-point motion of the atoms, which directly follows from the fundamental principles of quantum theory.[3]

In the field of nuclear physics, a considerable amount of experimental evidence regarding nuclear reactions initiated by collisions or radiation has been accumulating and has been found to fit into the general ideas on the structure of atomic nuclei outlined in the *Ency. Brit. Book of the Year*, 1938. In particular, the new evidence has brought further support to the view that on account of the extreme facility of energy exchange between the closely packed constituents of the nucleus, we have to do, in nuclear reactions, with a step-like process involving an intermediate state of the nucleus in which the energy is distributed among the various degrees of freedom in a way resembling the thermal agitation of a solid or a liquid body. The final course of such a reaction therefore depends on the relative probabilities of the different processes by which this energy can be released as radiation or converted into some particular type of motion leading to the disintegration of the intermediate nucleus. The ordinary course of such disintegrations, consisting in the escape of a single nuclear particle, thus claims the concentration of a considerable part of the energy available on some particle at the surface of the nucleus and resembles in this respect the evaporation of a molecule from a liquid drop.[4] Just recently, evidence has been brought forward which indicates the occurrence of other types of disintegration, in which a heavy nucleus is divided into two lighter nuclei, much in the same way as a liquid drop may be split into two droplets as a result of an external disturbance.[5] Such processes evidently claim the conversion of the energy into some particular mode of vibration of the nucleus involving a considerable deformation of the surface, and their further investigation promises most valuable information regarding the mechanism of nuclear excitation.

Apart from such studies of nuclear reactions, a great advance has recently been made as regards our knowledge and understanding of the forces acting between the particles in nuclei. On the one hand, experiments on the scattering of protons accelerated in fields of more than a million volts have led to the important conclusion that the short-range forces between nuclear particles are in the first approximation independent of the charge of these particles. On the other hand, the occurrence of such forces has been brought into a most interesting connexion with the existence of a new kind of particle, the so-called " meson," of a mass intermediate between that of the electron and that of the proton. The possible existence of such particles was already suggested several years ago by H. Yukawa in an attempt to ascribe the short-range nuclear forces to the exchange of lighter particles between the heavy nuclear constituents, and this suggestion has in the last years received most striking confirmation from experiments on cosmic rays. The penetrating power of these rays as well as the observation of particular cloud-chamber tracks produced by them offer in fact convincing evidence on the occurrence of particles of the mass predicted from the range of the nuclear forces.[6] Moreover it is found from analysis of cosmic ray observations that the mesons are not stable, but have a limited life-time of the same order of magnitude as that originally assumed by Yukawa in order to account for the beta-ray decay of nuclei, which on his conception of nuclear forces has to be ascribed to disintegrations of individual mesons released during the exchange processes.[7] We are here indeed concerned with the beginning of a new promising stage of the development of the atomic theory of matter.

BIBLIOGRAPHY. [1] A survey of these researches is to be found in *Reviews of Modern Physics* (F. Nix and W. Shockley), 10, 1, 1938 ; [2] See especially D. Wrinch and J. Langmuir, *Journal American Chemical Society*, 60, 2247, 1938 ; [3] J. Satterly, *Reviews of Modern Physics*, 8, 347, 1936 ; [4] Cf. N. Bohr, *Science*, 86, 161, 1937 ; [5] R. Frisch and L. Meitner, *Nature*, 143, 239, 1939 ; [6] D. Froman and J. Stearns, *Reviews of Modern Physics*, 10, 133, 1938 ; [7] H. Euler and W. Heisenberg, *Ergebnisse der exakten Naturwissenschaften*, 17, 1, 1938. (N. B.)

Niels Bohr speaking on the radio, 1949.

3. SPEAKING TO GYMNASIUM STUDENTS

VIII. ATOMS AND HUMAN KNOWLEDGE

ATOMERNE OG VOR ERKENDELSE
Berlingske Tidende, 2 April 1949

TEXT AND TRANSLATION

See Introduction to Part I, p. [12].

ATOMS AND HUMAN KNOWLEDGE (1949)

Versions published in Danish

A   Berlingske Tidende, 2 April 1949

B   Vor Viden **33** (1950) 123

C   "Atomer og Kerner" (eds. J. Kalckar and E. Rüdinger), Rhodos 1985, pp. 147–154

    A and *C* agree with each other. *B* is an abbreviated and slightly edited version.

[58]

Særtryk af Berlingske Tidende 2. april 1949

Niels Bohr:

Atomerne
og vor erkendelse

Efter anmodning har professor Bohr overladt os manuskriptet til det foredrag, han i gaar har holdt i skoleradioen for danske og norske gymnasieelever

DET er en stor glæde for mig at faa en saadan lejlighed til at komme i forbindelse med den ungdom i Danmark og Norge, der gennem uddannelse i gymnasierne forbereder sig paa en gerning i samfundets tjeneste. Titlen paa mit foredrag „atomerne og vor erkendelse" lyder maaske meget lærd, men jeg skal forsøge at fortælle lidt om det indblik i atomernes verden, som er vundet ved en af de mest æventyrlige rejser af menneskeaanden i ukendte egne, hvor vi hele tiden har mødt store overraskelser og maattet overvinde mange vanskeligheder. De erfaringer, vi paa denne rejse har samlet, betyder et stort fremskridt i vor beherskelse af naturen, men samtidig har vi vundet en almen menneskelig belæring, som har bud til alle, baade ældre og yngre.

Tanken om, at alle legemer, som vi ser eller føler paa, er opbygget af atomer, som ved deres lidenhed unddrager sig direkte iagttagelse, gaar jo langt tilbage, og allerede græske filosoffer som Demokrit forstod, at det for at gøre rede for elementernes bestandighed var nødvendigt at antage en grænse for stoffernes delelighed. Naar f. eks. almindeligt vand ved fordampning kan overføres i en helt anden tilstandsform, og dampen derefter igen kan fortættes til vand med ganske samme egenskaber som før, ville dette jo ikke være til at begribe, hvis vi havde at gøre med fuldstændig forvandling af stofferne, men forholdet kan forklares helt simpelt, naar vi antager, at det, der sker ved fordampningen, er blot, at de tæt sammenpakkede smaadele i vandet fjernes fra hinanden, saa de for en tid kan bevæge sig mere frit imellem hverandre, og at der ved fortætningen ikke sker andet, end at molekylerne igen pakkes sammen, og deres bevægelighed begrænses.

Saadanne forestillinger viste sig efterhaanden yderst frugtbare og oplysende paa mange forskellige omraader, og især vandt vi i begyndelsen af forrige aarhundrede gennem Daltons værk en simpel forklaring af de konstante mængdeforhold, hvori grundstofferne indgaar i kemiske forbindelser. I et grundstof antager vi jo, at alle atomerne er ens, medens vi i en kemisk forbindelse har smaadele eller molekyler, der er sammensat af atomer af forskellige grundstoffer. F. eks. er vand en forbindelse af grundstofferne brint og ilt, som vi siger paa dansk, eller vandstof og surstof paa norsk, eller med betegnelser, som bruges saa mange andre steder, og som derfor ogsaa er i færd med at blive indført hos os: hydrogen og oxygen. Et vandmolekyle bestaar af to atomer hydrogen og eet atom oxygen, men vi har ogsaa en anden forbindelse af disse grundstoffer, brintoverilte, eller hydrogenperoxyd, hvis molekyler bestaar af to hydrogenatomer og to oxygen-atomer, og ved dannelse af dette stof vil derfor for samme mængde hydrogen indgaa netop den dobbelte mængde oxygen som ved dannelsen af vandet.

Omend paa mange saadanne maader atomteorien viste sig mere og mere uundværlig og overbevisende, var det dog lige op til vort aarhundrede en almindelig anskuelse, at det aldrig ville være muligt at føre direkte bevis for atomernes eksistens, idet man gik ud fra, at vore sansers grovhed ville udelukke enhver paavisning af tilstedeværelsen og virkningen af de enkelte atomer. Under fysikkens udvikling har vi imidlertid faaet mere og mere forfinede instrumenter til hjælp for vore sanser, og f. eks. har vi allerede forlængst ved mikroskopets hjælp været i stand til at undersøge fine enkeltheder i legemernes bygning, som vi ikke kan se med det blotte øje. Jeg nævner mikroskopet, fordi det, hvor overraskende dette end lyder, har gjort det muligt for os i dag at tælle og veje atomerne. Der er dog naturligvis ikke tale om, at vi i et mikroskop kan se de enkelte atomer og maale deres afstande. Den mindste afstand, som man paa saadan maade kan maale, er nemlig givet ved bølgelængden af det lys, man benytter, og selv om lysbølgerne er umaadelig korte, mindre end en tusindedel af en millimeter, er de alligevel overordentlig lange sammenlignet med selve atomernes størrelse og afstande.

Imidlertid kan man i et almindeligt mikroskop iagttage smaalegemer, der, selv om de bestaar af et meget stort antal atomer, alligevel har en saa ringe vægt, at deres bevægelser, naar de er opslemmet i vand, vil paavirkes af stødene fra vandmolekylerne, der stadig rammer dem fra alle sider. Selv om smaalegemernes vægtfylde er større end vandets, vil de derfor ikke lægge sig paa bunden af vandbeholderen, men vil svømme omkring paa den mest uregelmæssige maade, dog saaledes at der til enhver tid vil være flere af dem nærmere bunden end i større højde over denne. Forholdene svarer her nøje til de usynlige bevægelser af molekylerne i luften, som jo paa grund af tyngdekraften ogsaa bliver tyndere og tyndere, jo mere vi fjerner os fra jorden. Ved at sammenligne den maade, hvorpaa tætheden aftager opefter i de to tilfælde, er det faktisk muligt at bestemme forholdet imellem vægten af et luftmolekyle og vægten af smaalegemer, der kan vejes direkte, og saa snart vi ved, hvor meget et enkelt luftmolekyle vejer, kan vi med det samme beregne, hvor mange molekyler der findes i en bestemt mængde luft.

En helt anden sindrig og i princippet simpel maade at tælle atomerne paa har man fundet ved at undersøge størrelsen af de elektriske ladninger, som smaalegemer kan optage. Saaledes kan man holde en ganske lille elektriseret oliedraabe svævende i luften ved at udsætte den for en elektrisk kraft, der trækker den opefter med akkurat samme styrke som den, hvormed tyngdekraften trækker den nedad. Naar man kender draabens vægt, faar man derfor et maal for dens elektriske ladning, og denne viser sig nu altid at være lig med et helt antal af en bestemt mindste enhed. En saadan begrænset delighed af elektriciteten selv var man allerede meget tidligere kommet paa spor efter gennem opdagelsen af de simple love, der gælder for de mængder af grundstoffer, der udskilles ved elektrolyse af kemiske forbindelser. For samme elektricitetsmængde udskilles nemlig mange grundstoffer i mængder, der svarer til deres atomers vægte, hvis forhold man kender fra de kemiske forbindelsers sammensætning, og dette kan jo kun tydes saaledes, at den elektriske ladning, som ethvert atom i elektrolysen fører med sig, er den samme for alle saadanne stoffer. Da vi nu f. eks. efter oliedraabemetoden nøje kender enhedsladningens absolute værdi, kan vi nemt beregne antallet af atomer, der udskilles ved elektrolyse for en given elektricitetsmængde, og vejningen af den udskilte mængde giver os derfor umiddelbart vægten af det enkelte atom.

Disse store fremskridt fandt sted i det første aarti af dette aarhundrede, men allerede forinden var den udvikling begyndt, som skulde give os indgaaende oplysninger om den maade, hvorpaa legemerne er bygget op af atomer, og endda give os kendskab til atomernes indre bygning. Det første skridt var her opdagelsen af røntgenstraalerne, som ved deres mærkelige gennemtrængningsevne, som bl. a. har givet lægevidenskaben nye hjælpemidler, idet det er muligt udefra at undersøge menneskers knogler og indre organer. Røntgenstraalerne har en lignende natur som lyset, men en mange gange kortere bølgelængde, og ved at undersøge straalernes tilbagekastning fra de lag, som dannes af atomerne i en krystal, har det været muligt at bestemme afstanden mellem atomerne indbyrdes og i enkeltheder at danne sig et billede af den maade, hvorpaa atomerne er ordnet i de forskellige krystaller. Ja, det har endda været muligt ved røntgenstraalernes hjælp at vinde oplysninger om, hvordan de enkelte atomer er ordnet inden for molekylerne af kemiske forbindelser.

Ved en række andre fysiske op-

dagelser fik man imidlertid endnu kraftigere hjælpemidler i hænde, der skulle give os kendskab til selve bestanddelene af atomerne. Et skel blev her sat, da man i elektronen lærte en fælles byggesten af alle stoffers atomer at kende. Til denne opdagelse blev man ført ved studiet af de smukke fænomener, der fremkommer ved elektriske udladninger i beholdere med stærkt fortyndet luft. I saadanne beholdere frembringes de saakaldte katodestraaler, og der, hvor disse træffer beholderens vægge, er det, at røntgenstraalerne opstaar. Da snævre bundter af katodestraaler, der kan iagttages ved deres lysende spor i luften, afbøjes i et elektrisk eller i et magnetisk felt, maatte man slutte, at disse straaler bestod af elektrisk ladede partikler, der bevæger sig med stor hastighed.

Til megen overraskelse viste det sig, at forholdet mellem katodestraalepartiklernes ladning og masse var langt større end det forhold, man finder for atomer, der udskilles ved elektrolyse, og da ladningen jo altid optræder i de samme enheder, maatte man derfor slutte, at partiklernes masse var langt mindre end atomernes masser, ja, over tusind gange mindre end massen af det letteste atom, hydrogenatomet. Det er netop katodestraalepartiklerne, som vi nu kalder for elektroner, og da disse kan frembringes ved udladninger i mange forskellige luftarter, forstod man, at elektronerne maatte findes i alle grundstoffers atomer. Længe var man imidlertid i uvidenhed om, hvad atomet ellers indeholdt, og da man var klar over, at der foruden de negativt ladede elektroner ogsaa maatte findes positive elektriske ladninger inden for atomet, maatte man først og fremmest spørge, hvordan disse ladninger var fordelt.

Vejen til at faa svar paa saadanne spørgsmaal aabnedes af den store opdagelse, at visse grundstoffer uden ydre paavirkning udsender gennemtrængende straaler. I kender jo alle den æventyrlige historie om, hvordan Madame Curie fra uranerts udvandt radium, der selv i smaa mængder er en saa kraftig straalekilde. Straalingen fra radium og de andre radioaktive stoffer har en sammensat karakter, og vi plejer at skelne imellem alfa-, beta- og gammastraaler. Medens gamma-straalerne viste sig at have lignende natur som røntgenstraalerne og betastraalerne at være elektroner med store hastigheder, fandt man, at alfa-straalerne er langt tungere partikler, der hver medfører to positive enhedsladninger og har en masse omtrentlig fire gange saa stor som hydrogenatomet. Paa grund af denne forholdsvis store masse har alfa-partiklerne saa stor bevægelsesenergi, at hver enkelt af dem, naar den standses, kan frembringe synlige virkninger. Mange af Jer har maaske selv haft den ejendommelige oplevelse at kigge i et saakaldt scintillaskop, hvor man ser et helt fyrværkeri af lysglimt, som frembringes paa en zinksulfidskærm, hvor den træffes af alfa-straalerne fra et radiumpræparat.

Disse alfa-straaler kan trænge igennem tynde lag af stoffer, men til trods for, at de paa deres vej passerer gennem millioner af atomer, er deres baner i almindelighed meget nær retliniede. Sometider ser man imidlertid, at banen fremviser et skarpt knæk, som fortæller os, at partiklen er kommet ud for et stærkt sammenstød med noget, der er langt tungere end elektronen, der paa grund af deres lethed bare vil skubbes af vejen. Det var netop saadanne iagttagelser og betragtninger, der førte Rutherford, som har været foregangsmand paa hele dette omraade, til opdagelsen af atomkernen, hvori næsten hele atomets masse er samlet, og som samtidig er sæde for den positive elektriske ladning i atomet.

Opdagelsen af atomkernen gav os et meget simpelt billede af atomet, der paa en maade kan sammenlignes med et lille solsystem, hvori et antal lette elektroner er bundet til den tunge kerne i afstande, der langt overgaar kernens eller elektronens egen størrelse. Antallet af positive enhedsladninger paa kernen og dermed det lige saa store antal elektroner omkring denne har vist sig at være netop det samme som det tal, det saakaldte atomnummer, som angiver det paagældende grundstofs plads i det kendte system, hvori stofferne ordnes efter den ejendommelige periodiske optræden af ligheder i deres kemiske egenskaber. Den omstændighed, at disse egenskaber bestemmes af atomnummeret, fører ogsaa til en simpel forstaaelse af, at et grundstof kan optræde i forskellige afarter, som man kalder isotoper, og hvis atomkerner har samme elektriske ladning, men forskellig masse. En af de mest kendte isotoper er det tunge hydrogen, som indgaar som bestanddel i det tunge vand.

Aarsagen til grundstoffernes bestandighed ligger i, at atomkernen ikke ændres under sædvanlige kemiske omdannelser, hvor det blot er elektronerne, der omplaceres eller endda under omstændigheder helt fjernes fra kernen. Som Rutherford viste, kan dog ogsaa selve atomkernen ændres ved særlig stærk paavirkning, f. eks. ved bombardement med hurtige alfa-partikler. Derved blev for første gang muligheden af at omdanne eet grundstof til et andet paavist, og en udvikling begyndte, der skulle faa de største følger i mange retninger.

Ikke alene har man gennem studiet af atomkerneomdannelser lært kernernes byggestene at kende, men gennem en rent fantastisk udvikling af tekniske hjælpemidler har man efterhaanden opnaaet et saadant kendskab til mange forskellige kernesønderdelinger, at det til slut, som alle ved, er blevet muligt i stor stil at frigøre den energi, der er bundet i atomkernerne, og som er millioner gange større end den, der kan vindes ved selv de voldsomste kemiske reaktioner.

I denne korte fortælling om lidt af det, vi har lært ved udforskning af atomernes verden, har vi hidtil kun talt om ting, der kan forstaas i simple billeder som dem, man benytter ved beskrivelser af, hvad man har set paa en rejse, men ligesom vi paa rejser i fremmede lande lærer andre skikke at kende og derved bliver klar over mange fordomme, hvormed vi drog ud, er man i atomfysikken blevet belært om en utilstrækkelighed af de principper, som fysikerne hidtil saa fast havde stolet paa.

Som jeg allerede har nævnt, minder et atom med sin atomkerne og elektroner i mange henseender om et solsystem, men en nærmere betragtning viser, at vi langtfra kan forstaa atomets egenskaber og navnlig dets særlige bestandighed ud fra de principper, hvormed Newton forklarede planeternes bevægelse omkring solen. Vel maa vi antage, at planeterne i lange tider har bevæget sig meget nær i samme baner, men dette skyldes kun at vort solsystem i saa ringe grad er blevet forstyrret af paavirkninger udefra. Skulle imidlertid engang en klode fra verdensrummet komme i nærheden af os, maa vi være forberedt paa, at baade jordens bane og aarets længde fra da af ville blive ændret. Helt anderledes forholder det sig imidlertid med atomets elektronsystem, der, saa længe kernen ikke splittes, efter enhver forstyrrelse vil reorganiseres saaledes, at den oprindelige tilstand igen vender tilbage.

En nøgle til forstaaelsen af de for atomerne gældende ejendommelige lovmæssigheder har vi faaet igennem Plancks

opdagelse i det første aar af vort aarhundrede af det saakaldte virkningskvantum, der er udtryk for en begrænsning af selve de mekaniske processers delelighed. Det er dog slet ikke muligt at forklare en saadan begrænsning ved hjælp af vore sædvanlige fysiske begreber, der, hvor udstrakt deres anvendelsesomraade end er, har vist sig at være idealisationer, der kun gælder, naar de virkninger, det drejer sig om, er store i forhold til det enkelte kvantum, og som svigter, naar vi søger at anvende dem til at gøre rede for de nye erfaringer fra atomernes verden.

Nogle af Jer har maaske allerede hørt om de overraskende vanskeligheder, som spørgsmaalet om elektronernes natur har stillet os overfor. Paa den ene side maa vi jo betragte elektronen som en partikel, idet maalinger af en elektrons masse og elektriske ladning altid giver samme resultater. Paa den anden side er man ved beskrivelsen af andre af elektronernes egenskaber henvist til at bruge bølgebilledei i lighed med dem, der har vist sig uundværlige for beskrivelse af lysets forplantning. Overraskelser af samme art har vi ogsaa mødt ved spørgsmaalet om lysets natur, idet bølgebillederne er ganske utilstrækkelige for at forklare love, der gælder for den maade, hvorpaa atomerne optager og udsender lysenergi, og for hvis redegørelse man har maattet gribe til at beskrive lyset som sammensat af enkelte lyskvanter eller fotoner med partikelkarakter.

En saadan situation, der hidtil var helt ukendt i fysikken, maatte jo til at begynde med virke ganske forvirrende, men efterhaanden indsaa man, at naar man benyttede de modstridende billeder, var det aldrig for at beskrive eet og samme fænomen, men kun for at gøre rede for erfaringer, der var vundet under forskellige forsøgsbetingelser, der udelukkede hinanden. Saadanne erfaringer staar derfor i et forhold til hverandre, som man betegner som k o m p l e m e n t æ r t for at understrege, at de, til trods for at de ikke kan forenes i eet enkelt anskueligt billede, hver især giver Udtryk for lige vigtige sider af de samlede oplysninger, der overhovedet kan vindes.

I denne forbindelse maa vi først og fremmest gøre os klart, at vi for at beskrive en bestemt forsøgsopstilling og de derved vundne iagttagelsesresultater altid vil være henvist til at benytte begreber udtrykt i den sprogbrug, som menneskene har udviklet for at finde sig til rette ved dagliglivets begivenheder. Dette betyder jo ikke andet end, at det ved ethvert fysisk forsøg maa dreje sig om noget, hvorom vi kan fortælle andre, hvad vi har gjort, og hvad vi har lært. Helt anderledes stiller spørgsmaalet sig imidlertid, naar vi skal sammenfatte resultater, vundet under forskellige forsøgsbetingelser.

Kun saa længe, det drejer sig om fænomener, hvor vi kan se helt bort fra den vekselvirkning, der altid maa være til stede mellem genstanden for undersøgelsen og de apparater, der benyttes ved dens iagttagelse, som f. eks. naar vi bestemmer en planets sted i dens bane til forskellige tider, kan vi tale om genstandens selvstændige opførsel uafhængig af vor iagttagelse. Naar vi derimod har at gøre med en elektron under forhold, hvor virkningskvantet spiller en afgørende rolle, kan vi slet ikke længere skelne skarpt imellem, hvad vi kunne kalde elektronens egen opførsel og dens vekselvirkning med de maaleinstrumenter, der indgaar i den forsøgsopstilling, hvorunder fænomenet fremkommer. Under saadanne forhold er derfor forudsætningen for paa sædvanlig maade at sammenligne resultater vundet under forskellige forsøgsbetingelser ikke til stede. Det drejer sig her ingenlunde om vanskeligheder af teknisk art, men om et principielt punkt, der er betinget af selve vilkaarene for erfaringernes beskrivelse.

Den komplementære sammenfatningsmaade er at betragte som en naturlig udvidelse af den sædvanlige fysiske beskrivelse, der ikke har vist sig rummelig nok til at omfatte de ejendommelige lovmæssigheder, der gælder i atomernes verden, og som er afgørende for egenskaberne af de stoffer, hvoraf alle vore redskaber, ja ogsaa vore egne legemer bestaar. I første øjeblik lyder det maaske underligt, at man skulle være nødt til at give afkald paa de krav om anskuelighed i naturbeskrivelsen, som man hidtil antog som helt selvfølgelige, men ved nærmere betragtning finder man, at vi ogsaa paa mange andre omraader har at gøre med forhold, der viser en nøje lighed med dem, vi har mødt ved undersøgelsen af atomernes egenskaber.

Vi behøver jo blot at tænke paa den komplementære maade, hvorpaa vi bruger saadanne to ord som „tanker" og „følelser" for at beskrive de situationer, som ethvert menneske daglig befinder sig i. Disse ord henviser jo netop til sider af vore indre oplevelser, der er lige væsentlige, men som udelukker hinanden i den forstand, at selv vore varmeste følelser fuldstændig mister deres art, naar vi forsøger at udrede dem ad den kolde logiske tankes vej. Lignende forhold gør sig gældende i samlivet med vore medmennesker, hvor intet af ordene „retfærdighed" og „kærlighed" kan undværes. Enhver pige eller dreng forstaar, hvor meget „fair play" betyder i leg eller sport, og at man ingen agtelse kan have for sig selv, uden naar man stræber efter at handle retsindigt mod andre, men samtidig maa vi gøre os klart, at brugen af retfærdighedsbegrebet i dets yderste konsekvens udelukker den kærlighed, som kaldes paa over for forældre, søskende og kammerater.

Naar man først har indlevet sig deri, kan man efterspore komplementaritetsforhold paa saa at sige alle kulturlivets omraader, og den omstændighed, at vi tvinges til at erkende saadanne forhold inden for et omraade som fysikken, der paa grund af dens anvendelser indtager saa stor en plads i skoleundervisningen, turde ogsaa komme til at øve en indflydelse paa de unges opdragelse. Selv om det i et kort foredrag ikke er muligt at give en klar og indgaaende fremstilling hverken af den store udvikling inden for atomfysikken eller af den belæring om vilkaarene for vor erkendelse, den har bragt, haaber jeg, at det er lykkedes mig i det mindste at give Jer et indtryk af problemernes betydning og rækkevidde.

TRANSLATION

Niels Bohr:

Atoms
and Human Knowledge

Upon request Professor Bohr has given us the manu-script for the talk he gave on the school radio yesterday for Danish and Norwegian gymnasium students

It is a great pleasure for me to have such an opportunity to come into contact with the young people in Denmark and Norway who through education in gymnasium prepare themselves for a career in the service of society. The title of my talk, "atoms and human knowledge", may sound very learned, but I shall try to tell a little about the insight into the world of atoms, gained by one of the most fantastic journeys of the human spirit into unknown reaches, where we have perpetually met great surprises and have had to overcome many difficulties. The experience we have gathered on this journey signifies a great advance in our mastery of nature, but at the same time we have gained a general human lesson which holds a message for everyone, both old and young.

The idea that all bodies we see or touch are built up of atoms, which by their tiny size evade direct observation, goes far back in time, and Greek philosophers such as Democritus already understood that in order to explain the stability of the elements it was necessary to assume a limit for the divisibility of substances. When, for example, ordinary water can be changed on vaporization into a completely different state and the steam can then in turn be condensed to water with exactly the same properties as before, this would not be possible to understand if we had to do with complete transformation of the substances, but the situation can be explained quite simply when we assume that what happens on vaporization is just that the tightly-packed small parts in the water are separated from each other so that they for a time can move more freely among one another, and that on condensation nothing more happens than that the molecules are again packed together and their mobility is limited.

As time passed, such ideas proved to be extremely fruitful and enlightening in many different fields, and especially at the beginning of the last century we gained through Dalton's work a simple explanation of the constant mass ratios in which elements combine into chemical compounds. We assume of course that in an element all atoms are identical while in a chemical compound there are small parts or molecules made up of atoms of various elements. For example, water is a combination of the elements *brint* and *ilt*, as they are called in Danish or *vandstof* and *surstof* in Norwegian, or, in terms which are used in so many other places and which therefore are also in the process of being introduced in Denmark: hydrogen and oxygen. A water molecule consists of two atoms of hydrogen and one atom of oxygen, but there is also another combination of these elements, hydrogen peroxide, whose molecules consist of two hydrogen atoms and two oxygen atoms, and on the formation of this substance the same amount of hydrogen will therefore combine with precisely double the amount of oxygen as on the formation of water.

Even though in many such ways the theory of atoms proved to be ever more indispensable and convincing, it was nevertheless the general view right up to our century that it would never be possible to give a direct proof of the existence of atoms, as it was assumed that the coarseness of our senses would exclude any demonstration of the presence and the effect of individual atoms. In the course of the development of physics, however, we have obtained more and more refined instruments to assist our senses, and we have for example long since been able, by means of the microscope, to study the fine details of the structure of bodies that we cannot see with the naked eye. I mention the microscope because, however surprising it may sound, it has made it possible for us today to count and weigh atoms. Still, this naturally does not mean that we in the microscope can see individual atoms and measure the distances between them. The smallest distance we can measure in this way is given by the wavelength of the light we use, and although light waves are extremely short, less than one thousandth of a millimetre, they are nevertheless exceedingly long as compared to the size of and distances between the atoms.

However, in an ordinary microscope we can observe small bodies which, although they consist of a very large number of atoms, nevertheless have so little weight that their movements, when they are suspended in water, will be affected by the impacts from the water molecules which continually hit them from all sides. Although the density of the small bodies is greater than that of water, they will not therefore fall to the bottom of the water vessel but swim about in the most irregular way, though in a manner such that at any time more of them will be nearer the bottom than at a greater height above it. This situation corresponds closely to the invisible movements of the molecules in the air which due to gravity also becomes thinner and thinner the further away we are from Earth. By comparing the way in which the density decreases in an upward direction in these two cases, it is in fact possible to determine the ratio between the weight of an air molecule and the weight of small bodies that can be weighed directly, and as soon as we know how much a single air molecule weighs, we can immediately calculate how many molecules are present in a given amount of air.

A completely different ingenious, and in principle simple, way of counting atoms has been found by examining the magnitude of the electric charges that small bodies can assume. Thus it is possible to keep a quite small electrified drop of oil floating in the air by subjecting it to an electric charge, which pulls it upwards with exactly the same force as gravity pulls it downwards. When the weight of the drop is known, its electric charge is thus given, and this now always proves to be equal to a whole number of a certain smallest unit. Such a limited divisibility of electricity itself was recognized already much earlier through the discovery of the simple laws that hold for the amounts of elements that are separated out by electrolysis of chemical compounds. For the same quantity of electricity many elements are namely separated out in quantities corresponding to the weights of their atoms, whose ratios are known from the composition of the chemical compounds, and this can of course only be interpreted to mean that the electrical charge that each atom carries with it in the electrolysis is the same for all such substances. Since we now know exactly, for example from the oil drop method, the absolute value of the unit charge, we can easily calculate the number of atoms separated out by electrolysis for a given quantity of electricity and the weighing of the separated amount thus provides us directly with the weight of the individual atom.

These great advances took place in the first decade of this century, but already before this the development was begun that would give us detailed information about the manner in which the bodies are built up of atoms, and even give us knowledge about the internal constitution of atoms. Here, the first step was the discovery of X-rays, which have the peculiar ability of penetration that among other things has provided medical science with new tools, because it is possible to study the bones and inner organs of the human body from the outside. X-rays are of a nature similar to that of light but have a wavelength which is many times shorter, and by studying the reflection of the rays from the layers formed by the atoms in a crystal, it has been possible to determine the

[65]

distance between the atoms and to form a detailed picture of the way in which the atoms are arranged in various crystals. Indeed, it has even been possible by means of X-rays to gain information about how the individual atoms are placed inside the molecules of chemical compounds.

Through a series of other discoveries in physics, however, we got still more powerful tools in our hands which would give us knowledge about the very components of the atoms. This was a milestone, since with the electron we became acquainted with a constituent common to the atoms of all substances. We were guided to this discovery through the study of the beautiful phenomena which appear in connection with electric discharges in containers with extremely rarefied air. In such containers we can produce the so-called cathode rays, and where these hit the walls of the container, X-rays arise. Since narrow bundles of cathode rays, which can be observed due to their luminous tracks in air, are deflected in an electric or in a magnetic field, one had to conclude that these rays consisted of electrically charged particles moving at high speed.

Very surprisingly, it turned out that the ratio between the charge and the mass of the cathode-ray particles was much higher than the ratio found in atoms separated out in electrolysis, and since charge always appears in the same units, one thus had to conclude that the mass of the particles was far smaller than the mass of the atoms, indeed more than a thousand times smaller than the mass of the lightest atom, the hydrogen atom. It is precisely the cathode-ray particles that we now call electrons, and since they can be produced by discharges in many different gases, it was understood that electrons must be present in the atoms of all elements. For a long time, however, it was not known what else was inside the atom and, as it was clear that in addition to the negatively charged electrons there must also be positive electric charges inside the atom, the question to be asked first of all was how these charges were distributed.

The path to an answer to such questions was opened by the great discovery that certain elements emit penetrating radiation even without external influence. You all know the fantastic story of how Madame Curie extracted radium, which even in small quantities is such a strong source of radiation, from uranium ore.[1] The radiation from radium and other radioactive substances is of a complex nature, and we usually distinguish between alpha, beta and gamma rays. Whereas the gamma rays proved to have a nature similar to X-rays, and beta rays to be electrons at high velocities, it was found that alpha rays are much heavier

---

[1] Bohr wrote a tribute to Marie Curie in connection with her visit to Copenhagen in October 1926. N. Bohr, manuscript, *Inventory of Relevant Manuscripts in the Niels Bohr Archive*, below, p. [524], folder 4.

particles, each of which carrying two positive unit charges and having a mass about four times that of the hydrogen atom. Because of this relatively large mass, alpha particles have so great a kinetic energy that every single one of them, as it is stopped, can produce visible effects. Many of you have perhaps had the strange experience of looking into a so-called scintilloscope where one can see a whole fireworks of light flashes produced on a zinc sulphide screen when it is hit by the alpha rays from a radium preparation.

These alpha rays can penetrate thin layers of material, but although they pass through millions of atoms on their way, their paths are usually almost straight lines. Sometimes, however, one sees that the path shows a sharp break, which tells us that the particle has had a violent collision with something much heavier than electrons which, due to their lightness, will just be pushed aside. Just such observations and considerations led Rutherford, who has been a pioneer in this entire field, to the discovery of the atomic nucleus where almost the entire mass of the atom is concentrated and which at the same time is the seat of the positive electric charge in the atom.

The discovery of the atomic nucleus gave us a very simple picture of the atom, which in a way can be compared to a small solar system where a number of light electrons are bound to the heavy nucleus at distances far surpassing the size of the nucleus or the electrons themselves. The number of positive unit charges on the nucleus, and hence the equal number of electrons around it, has turned out to be just the same as the number, the so-called atomic number, denoting this element's place in the familiar system where the elements are ordered according to the peculiar periodic appearance of similarities in their chemical properties. The circumstance that these properties are determined by the atomic number also leads to a simple understanding of the fact that an element can appear in different variants, called isotopes, whose atomic nuclei have the same electric charge but different mass. One of the best known isotopes is heavy hydrogen which is a constituent of heavy water.

The reason for the stability of the elements is that the atomic nucleus is not changed during ordinary chemical transformations in which it is only the electrons that are rearranged, or, under certain circumstances, are even entirely removed from the nucleus. As Rutherford showed, also the atomic nucleus itself, however, can be changed by especially strong influence, for example during bombardment with fast alpha particles. In this way the possibility of transforming one element into another was demonstrated for the first time, and a development began which would have the greatest consequences in many domains. Not only have we learned about the constituents of the nuclei through the study of the transformations of atomic nuclei, but through a quite fantastic development of technical tools we have gradually gained so much information

[67]

about many different nuclear disintegrations that it has finally, as everybody knows, become possible to release on a large scale the energy bound in atomic nuclei, which is millions of times greater than that which can be gained in even the most violent chemical reactions.

In this brief description about a little of what we have learnt from investigating the world of atoms, we have so far only spoken of matters that can be understood in terms of simple pictures such as those we use when describing what we have seen on a journey, but just as we on journeys in foreign countries learn about other customs and thus become aware of many prejudices that we took with us, we have been taught in atomic physics about the insufficiency of the principles on which physicists had hitherto relied so completely.

As I have mentioned already, an atom with its nucleus and electrons resembles in many respects a solar system, but a closer look shows that we are far from understanding the properties of the atom, and its special stability in particular, on the basis of the principles by which Newton explained the movements of the planets around the sun. We surely may assume that the planets for a long time have moved very nearly in the same orbits, but this is only because our solar system to such a small extent has been disturbed by outside influences. However, should at some time a body from outer space come close to us, we must be prepared that both the orbit of the Earth and the length of the year will be changed from then on. The situation is quite different, however, with regard to the electron system of the atom, which as long as the nucleus is not split will be reorganized after every disturbance, so that the original state returns again.

We have obtained a key to the understanding of the strange regularities valid for atoms through Planck's discovery in the first year of our century of the so-called quantum of action, which is an expression of a limitation of the divisibility of the mechanical processes themselves. It is, though, not at all possible to explain such a limitation by means of our usual physical concepts which, however broad their area of application might be, have proved to be idealizations which are only valid when the effects in question are large in relation to the individual quantum and which fail when we try to use them to explain the new experiences from the world of atoms.

Some of you have perhaps already heard about the surprising difficulties with which the question of the nature of electrons has confronted us. On the one hand we may of course regard the electron as a particle, as measurements of the mass and the electric charge of an electron always give the same results. On the other hand, when describing other properties of the electrons we have to use wave pictures similar to those that have proved indispensable for the description of the propagation of light. We have also met surprises of the same kind concerning the question of the nature of light, as the wave pictures are quite insufficient to explain laws that apply for the way in which atoms absorb and emit light energy, and to explain this we have had to resort to a description of light as composed of individual light quanta, or photons, behaving as particles.

Such a situation, which was hitherto quite unknown in physics, must initially seem quite confusing, but gradually it was realized that when the contradictory pictures were used, it was never to describe one and the same phenomenon, but only to explain experiences gained under different experimental conditions excluding one another. Such experiences are therefore in a relation to each other that we call *complementary* in order to emphasize that although they cannot be united in one single visualizable picture, each of them is an expression of equally important aspects of the complete information that altogether can be gained.

In this connection we must first of all be aware that in order to describe a given experimental arrangement and the observation results thereby obtained, we will always have to employ concepts expressed in the language usage that people have developed in order to cope with everyday events. This means nothing more than that every physical experiment must have to do with something about which we can tell others what we have done and what we have learnt. The question is quite another, however, when our purpose is to comprehend results obtained under different experimental conditions.

Only as long as we have to do with phenomena where we can entirely ignore the interaction which must always exist between the object of investigation and the apparatus used in its observation, as for example when we determine a planet's position in its orbit at different times, can we speak of the object's independent behaviour regardless of our observation. When, on the contrary, we have to do with an electron under conditions where the quantum of action plays a decisive role, we cannot any longer distinguish sharply between what we might call the electron's own behaviour and its interaction with the measuring instruments which are part of the experimental arrangement in which the phenomenon occurs. Therefore, under such circumstances the prerequisite for comparing in the usual manner the results obtained under different experimental conditions is no longer present. Here we do not at all have to do with difficulties of a technical kind but with a fundamental point which depends upon the very conditions for the description of the experiences.

The complementary mode of comprehension is to be regarded as a natural extension of the usual physical description, which has not proved sufficient to accommodate the strange regularities which apply in the world of atoms and which are decisive for the properties of the substances of which all our

[69]

tools, indeed also our own bodies, consist. At first it may sound surprising that we should be forced to renounce the requirements for visualizability in the description of nature that was hitherto assumed to be quite self-evident, but by closer consideration we find that also in many other domains we have to do with circumstances that show a close similarity to those we have met in the study of the properties of atoms.

We need only to think of the complementary way in which we use two such words as "thoughts" and "feelings" in order to describe the situations which every human being experiences every day. These words refer precisely to aspects of our inner experiences, each of which is equally important but excludes the other in the sense that even our warmest feelings completely lose their character when we try to clarify them along the path of cold logical thought. Similar conditions hold in our lives together with our fellow beings in which neither of the words "justice" and "love" can be dispensed with. Every girl or boy understands how much "fair play" means in games or sport and that one cannot have respect for oneself unless one tries to act honourably towards others, but at the same time we must be aware that the use of the concept of justice in its utmost consequence excludes the love which is called upon in relation to parents, sisters, brothers and comrades.

When first one has become familiar with this, it is possible to trace complementarity in so to speak all spheres of cultural life, and the circumstance that we are forced to recognize such situations within a field such as physics, which because of its applications occupies such a large place in the school curriculum, may also be expected to influence the upbringing of young people. Although in a short talk it is not possible to give a clear and detailed exposition of either the great development within atomic physics or the lesson regarding the conditions for human knowledge this has brought, I hope that I have at least succeeded in giving you an impression of the importance and extent of the problems.

Christian Møller and Ebbe Rasmussen, 1938.

4. FOREWORDS TO BOOKS

IX. FOREWORD

C. Møller and E. Rasmussen, "The World and the Atom",
Allen & Unwin, London 1940, p. 9

.

See Introduction to Part I, p. [13].

FOREWORD (1938)

Versions published in Danish and English

Danish: Forord

A   C. Møller and E. Rasmussen, "Atomer og andre Smaating", Hirschsprung, Copenhagen 1938, pp. 5–6

English: Foreword

B   C. Møller and E. Rasmussen, "The World and the Atom", Allen & Unwin, London 1940, p. 9

The two versions agree with each other

Foreword

I BELIEVE that this book will give a wide circle of readers an excellent opportunity of learning something about the wonderful developments in atomic physics in our own day. Our penetration into new worlds hitherto closed to men's eyes has brought to light a wealth of forces in nature of which we had no inkling. At the same time, through the new light thrown on our own position as observers of the world about us, we have won a fresh outlook upon deep problems of knowledge in general.

That the book results from the joint work of two physicists —the one belonging to the experimentalists, the other to the theoreticians—is by no means the least of the reasons why it is so well balanced an account of how the new knowledge has been achieved, and of how it has been successfully brought within the framework of ever-widening concepts. I believe, too, not only that each reader of this book will correctly perceive to how great an extent any advance has been based on a reciprocally fruitful linkage of research with technology, but also that he will have brought home to him very strongly the fact that such a far-reaching development as this in one part of the scientific field has been made possible only by the most intimate co-operation between scientists from, so to speak, all parts of the scientific world. The modest contribution made by the individual to this co-operative work finds its value only in being fitted into that structure whose deepest foundation in the end is the abiding impulse in every human being to seek order and harmony behind the manifold and the changing in the existing world.

PROFESSOR NIELS BOHR

9

X. FOREWORD

FORORD
G. Gamow, "Mr. Tompkins i Drømmeland",
Gyldendal, Copenhagen 1942, pp. 7–8

TEXT AND TRANSLATION

See Introduction to Part I, p. [14].

George Gamow, 1938.

*Vort Aarhundredes fysiske Videnskab kendetegnes ikke alene ved Ud-
forskningen af helt nye Erfaringsomraader, men ogsaa ved en gennemgribende Revision af Grundlaget for Anvendelsen af de mest elementære Begreber. Erkendelsen af Rum-Tids-Beskrivelsens Relativitet og Aarsagssæt-
ningens Begrænsning har ført til Udviklingen af mere omfattende Synspunkter, der har vist sig overordentlig frugtbare for Arbejdet paa at forøge
vort Indblik saavel i Universets Struktur som i Atomernes Egenskaber.*

*Paa dette rige Arbejdsfelt har faa ydet større Indsats end Professor
Gamow, der i 1928, da han som ganske ung for første Gang kom til København — hvor han siden ofte har været saa velkommen en Gæst — medbragte
den nye kvantemekaniske Forklaring af de radioaktive Atomers Sønderdeling. Den Udvikling, der derved sattes i Gang, førte i Løbet af faa Aar,
ikke mindst gennem hans egne Bidrag, til uanede Fremskridt og især til
Klarlæggelsen af Atomkerneomdannelsernes afgørende Betydning for den
uhyre Energiudvikling i Stjernernes Indre, der er bestemmende for det
umaadelige Spand af Aar, i hvilket vor Sol har fortsat sin for Livet paa
Jorden uundværlige Straaling.*

*Den Fantasirigdom, der er Baggrunden for Gamows Opdagelser, er hos
ham uadskilleligt forbundet med en sjælden Sans for Humor, der overalt,
hvor han har virket, har været en rig Kilde til Inspiration. Det er netop
denne Fantasi og Humor, der lyser ud af den foreliggende Bog, hvori Forfatteren — i Lighed med det Forbillede, som den engelske Matematiker
Charles Doddson, under Pseudonymet Lewis Carrol, har skabt med sine
berømte Æventyr om Alice i Vidunderland — med megen Finhed og men-*

[79]

neskelig Forstaaelse benytter en Sammenstilling mellem Drømmenes og Dagliglivets Verden til at lette Indlevelsen i de nye saa fremmedartede Forestillinger.

Man maa med største Glæde hilse Fremkomsten paa Dansk af denne Bog, der omkring i Verden har fundet begejstret Modtagelse saavel blandt Gamows egne Fagfæller som i langt videre Kredse. For mange vil den sikkert blive en fornøjelig Indførelse i den Tankeverden, som Fysikkens seneste Fremskridt har aabnet for os.

Niels Bohr

TRANSLATION

Physical science in our century is characterized not only by the investigation of entirely new areas of experience but also by a thorough revision of the basis for the application of the most elementary concepts. The recognition of the relativity of the space–time description and the limitation of the principle of causality has led to the development of more comprehensive viewpoints which have proved exceedingly successful for the work towards increasing our insight into both the structure of the universe and the properties of atoms.

In this rich field of work few have achieved more than Professor Gamow who in 1928, when quite young he came to Copenhagen for the first time – where since he has often been so welcome a guest – brought the new quantum-mechanical explanation of the disintegration of the radioactive atoms.[1] The development thus initiated led in the course of a few years, not least through his own contributions, to unimagined advances and, in particular, to the clarification of the decisive importance of the transformations of the atomic nuclei for the enormous generation of energy inside stars, which is determining for the immense span of years in which our sun has continued its radiation, indispensable for life on Earth.

The richness of imagination that is the background for Gamow's discoveries, is in him inextricably bound up with a rare sense of humour which has been a rich source of inspiration everywhere he has worked. It is precisely this imagination and humour that shine out from of the present book, in which the author – just as in the ideal which Charles Dodgson,[2] the English mathematician, under the pseudonym Lewis Carroll[3] has created with his famous tales about Alice in Wonderland – with great delicacy and human understanding uses a comparison between the world of dreams and that of everyday life to simplify the familiarization with the new and so strange ideas.

---

[1] For a historical account of Gamow's theory, see R.H. Stuewer, *Gamow's Theory of Alpha-Decay* in *The Israel Colloquium. Studies in History, Philosophy, and Sociology of Science*, Vol. 1: *The Kaleidoscope of Science* (ed. E. Ullmann–Margalit), Reidel, Dordrecht 1986, pp. 147–186.

[2] (1832–1898).

[3] Both surnames are misspelt in the Danish original.

One must welcome with greatest pleasure the appearance in the Danish language of this book which around in the world has found an enthusiastic reception, both among Gamow's fellow scientists and in wider circles. For many it will surely be an enjoyable introduction to the world of ideas that the latest advances of physics have opened for us.

Niels Bohr

XI. [FOR THE TENTH ANNIVERSARY OF THE JOURNAL "NUCLEONICS"]

Nucleonics **15** (September 1957) 89

See Introduction to Part I, p. [14].

Three Nobel Laureates Look into the Future

Harwell, England

During the next decade I forsee a great development of the nuclear power program in Great Britain. Already we have planned to install 5,000—6,000 Mw of capacity by 1965, and we have seen that the capacity of an individual two-reactor power station will jump from 300 to 500 Mw between 1960 and 1962. This increase in output of individual stations is likely to proceed fast and be limited only by what the electricity networks can stand. In Britain nuclear power is likely to achieve parity with oil and coal power by 1962 and thereafter to be cheaper.

Beyond this I can foresee new types of nuclear power stations coming into commission, operating at higher temperatures and efficiencies and over-all burnups.

During this period too I foresee research and development on fusion reactions expanding and moving toward the phase where a positive power gain is in sight. I do *not* foresee economic fusion reactors for this decade.

During this period several "demonstration nuclear ships" will be sailing the oceans and by the end of the decade parity with oil propulsion may be in sight.

Forecasts of developments in pure nuclear physics are well known to be virtually impossible. We can, however, foresee new types of accelerators reaching out to the 100-Bev energy region and that many new "strange particles" will be found. I doubt whether theory will catch up with the flood of new discoveries during the next decade.

J. D. Cockcroft

John D. Cockcroft, *director of the Atomic Energy Research Establishment at Harwell, recipient with E. T. S. Walton of 1951 physics award for pioneer work on transmutation of atomic nuclei.*

Goettingen, Germany

You have asked me, on the occasion of the tenth anniversary of NUCLEONICS, to give my opinion of the future development of my special field of work and the development of nuclear research in particular.

First, I must inform you, to my regret, that since 1945—that is, before the birth of NUCLEONICS—I have not been performing my own experimental work. After my not entirely voluntary eight months' stay, mostly in England, "at the pleasure of His Majesty," I was made president of the former Kaiser Wilhelm Institute, now the Max Planck Institute, a sort of general manager who is in charge of the forty institutions of the institute. However, I cannot pay very thorough attention to any of them, not even to those to the development of which NUCLEONICS is dedicated.

After I had demonstrated the fission of uranium and thorium with Mr. Strassmann at the end of 1938, we continued our radiochemical research and during all of the war years published our results. At the beginning of 1945 we had a table of exactly 100 fission products; the first American table, more detailed, appeared at the end of 1946.

Since then research has made great advances, and it is an unforgettable memory to have been able to be present when Glenn Seaborg discovered several atoms of element 101 in Berkeley at the end of November, 1955.

Out of artificially radioactive fission products, discovered only with the Geiger-Mueller counter, and other elements all the way up to transuranium ones have now come mass tables with which the chemist can work, just as if they were ordinary elements. I am convinced that with these the use of radioactive isotopes will be applied for peaceful purposes—medicine, agriculture, technology.

And how long will it take for the fission of uranium with its controlled chain reaction to be replaced as a source of energy by the fusion of hydrogen into helium, not, as I hope, for the production of hydrogen bombs, but for the production of energy? Then will the great problem of producing electricity be solved forever.

So let me wish NUCLEONICS for the next ten years a development as vigorous as it has had in the first ten years.

Otto Hahn

Otto Hahn, *president of the Max Planck Institute, recipient of 1944 award in chemistry for discovery of the fission of heavy nuclei.*

Copenhagen, Denmark

At the tenth anniversary of NUCLEONICS I want to offer my congratulations for the outstanding way in which this journal has contributed to keeping the world community of physicists and engineers abreast with the rapid development in a new field which holds out such great promises for science and technology. To those who have witnessed the development since Rutherford's discovery of the atomic nucleus in 1911, it has been a true adventure to follow the progress which step by step should so marvelously augment our knowledge of the atomic constitution of matter and entail consequences of such an immense practical importance. As regards the future, we reckon that in scientific respect, notwithstanding all surprises which may be revealed by continued research, it shall be possible to obtain an ever more harmonious synthesis of experience. From a human standpoint, it is equally clear that any increase of our knowledge means a greater responsibility, and we must trust that by growing understanding of the situation, humanity shall succeed in turning the great advance of science and technology to the benefit of all people.

Niels Bohr

Niels H. D. Bohr, *director of Copenhagen Institute for Theoretical Physics, recipient of 1922 physics award for development of atomic orbital theory.*

XII. FOREWORD

FORORD
J.R. Oppenheimer, "Naturvidenskab og Livsforståelse",
Gyldendal, Copenhagen 1960, pp. i–ii

TEXT AND TRANSLATION

See Introduction to Part I, p. [14].

[85]

FORORD

Naturvidenskabens udvikling i dette århundrede, der har stillet menneskeheden over for så store problemer, har også givet os ny belæring om vor stilling som iagttagere af den natur, vi tilhører. Udforskningen af atomernes verden har været som en rejse på ubanet vej i ukendt land, hvor vi, foruden at indhøste rige erfaringer, gang på gang har mødt overraskelser, der har tvunget os til at tage de elementæreste erkendelsesproblemer op til fornyet undersøgelse. Studiet af naturen, der altid har været en prøvesten for vore begrebsdannelsers forudsætninger og rækkevidde, har derved i vor tid fået ny forbindelse med tilsyneladende fjerntliggende områder af det menneskelige åndsliv.

På baggrund af sin intime fortrolighed med den atomfysiske forskning, til hvilken han selv har bidraget på fremragende måde, og med et for en naturvidenskabsmand sjældent kendskab til åndslivets historie, giver forfatteren af den foreliggende bog et tankevækkende overblik over dette emne. Med fortrinlig fremstillingskunst skildrer Robert Oppenheimer problemernes gradvise afklaring og de perspektiver, som de åbner. Bogen har overalt vakt megen opmærksomhed og er udkommet på mange sprog. Det må hilses med glæde, at den nu foreligger i dansk oversættelse; og for vide kredse herhjemme vil læsningen sikkert blive en lærerig og inspirerende oplevelse.

Niels Bohr

TRANSLATION

FOREWORD

The development of science in this century, which has faced humanity with such great problems, has also given us new lessons about our position as observers of the nature to which we belong. The investigation of the world of atoms has been like a journey on an untrodden path in unfamiliar territory, where, apart from reaping rich experience, we have again and again met surprises which have forced us to take up the most elementary epistemological problems for renewed examination. The study of nature, which has always been a touchstone for the conditions and extent of the formation of our ideas, has thereby in our time obtained new connections with seemingly remote fields of human culture.

On the basis of his intimate familiarity with atomic physics research, to which he himself has contributed in an outstanding way, and with knowledge of the history of culture rare for a scientist, the author of the present book gives a thought-provoking survey of this topic. With eminent descriptive power, Robert Oppenheimer portrays the gradual clarification of the problems and the perspectives they open. The book has attracted great attention everywhere and is published in many languages. It must be greeted with pleasure that it is now available in Danish translation; and for wide circles here at home the reading will surely be an instructive and inspiring experience.

Niels Bohr

XIII. PREFACE TO THE 1961 REISSUE

"Atomic Theory and the Description of Nature",
Cambridge University Press, Cambridge 1961, p. vi

See Introduction to Part I, p. [15].

PREFACE TO THE 1961 REISSUE

I am indebted to the Cambridge University Press for the suggestion to re-issue this collection of essays, which for some time has been out of print.

The articles were written at a time when the programme of developing a comprehensive treatment of atomic problems, on the basis of Planck's original discovery of the universal quantum of action, had obtained a solid foundation by the establishment of a proper mathematical formalism.

As is well known, the discussion of the epistemological aspects of quantum physics continued during the following years, and general consent has still not been reached. In the course of this discussion the attitude advocated in the articles was further developed, and especially a more adequate terminology was introduced to express the radical departure from ordinary pictorial description and accustomed demands of physical explanation. A collection of essays from this phase of the discussion has recently been published under the title *Atomic Physics and Human Knowledge* (John Wiley and Sons, New York, 1958).

Even if the old articles, which are here reprinted, thus contain utterances which now may be formulated in a more precise manner, acquaintance with the early discussions might, however, be helpful for a full appreciation of the new situation in natural philosophy with which the modern development of physics has confronted us.

NIELS BOHR

COPENHAGEN
January 1961

XIV. FOREWORD BY THE DANISH EDITORIAL COMMITTEE

DEN DANSKE REDAKTIONSKOMITÉS FORORD
I.B. Cohen, "Fysikkens Gennembrud",
Gyldendals Kvantebøger, Copenhagen 1962, pp. 5–6

With H. Højgaard Jensen, A. Petersen and S. Sikjær.

TEXT AND TRANSLATION

See Introduction to Part I, p. [16].

DEN DANSKE
REDAKTIONSKOMITÉS FORORD

Den fysiske videnskab har fra de ældste tider udøvet en særlig tiltrækning – ikke blot på forskerne, der stræber efter at forøge vor viden om den natur, hvori vi lever, og på teknikerne som arbejder med den praktiske udnyttelse af sådan viden, men også på de mange andre der føler trang til at udvide deres synskreds. Den inspiration, der kan hentes ved beskæftigelsen med fysiske problemer, kan sammenlignes med den berigelse, som kunstens store værker byder os. Skønt vi i fysikken ikke som i kunsten har at gøre med stemninger født og fastholdt af det menneskelige sind, men med fænomener som naturen stiller til vor beskuelse, kræver indlevelsen i den orden, som skridt for skridt er blevet bragt for dagen, brug og opøvelse af evner, som er menneskehedens fælles eje.

I vor tid har fysikkens anvendelse ikke alene ændret rammerne for vor daglige tilværelse på gennemgribende måde, men åbningen af hidtil utilgængelige erfaringsområder og deres udforskning ved systematisk eksperimentering har tillige i uanet målestok forøget vor kundskab om naturen og givet vort verdensbillede en enhed, der overgår fortidens dristigste drømme. Denne udvikling, der i stedse højere grad sætter sit præg på hele vor kultur, stiller store krav om udbredelse af oplysning i de videste kredse om fysikkens historie og nuværende stade. Ikke mindst gælder det ungdommen, der skal bringe nye kræfter til arbejdet på vor videns vækst og frugtbargørelse.

En stor indsats for dette formål er i de sidste år gjort af en gruppe af fremragende amerikanske videnskabsmænd og pædagoger. Denne såkaldte *Physical Science Study Committee*

5

har udfoldet energiske bestræbelser for at reformere fysik-undervisningen og stimulere interessen for og forståelsen af den fysiske videnskab. Komiteens nye lærebogssystem, suppleret med dens mange storartede undervisningsfilm, må betragtes som et af efterkrigstidens vigtigste pædagogiske fremskridt. Hovedvægten er lagt på fysikkens metode og principper såvel som på en forståelse af dens kulturelle betydning.

Som et led i sine bestræbelser udsender komiteen en række mindre bøger til selvstudium og til supplering af skolens fysikundervisning. Hver bog behandler sit emne, og serien skal efterhånden omfatte emner fra alle fysikkens områder, fra atomernes verden til himmelrummets dybder, fra oldtidens fysik til den moderne. Disse bøger udarbejdes af fremtrædende specialister, som foruden indgående viden har evne til at skrive almenforståeligt. Naturligvis tilegnes sådanne fremstillinger kun gennem en indsats fra læserens side; til gengæld er den vejledning, der her gives, i sjælden grad kyndig og alsidig.

Serien har allerede fået en meget stor udbredelse og er oversat til mange sprog. I dansk oversættelse indgår udvalgte bind i *Gyldendals Kvantebøger*. Når man har valgt dette navn – skønt kvantefysikken ikke direkte berøres i alle bøgerne – er det ikke mindst på grund af den afgørende betydning, som opdagelsen af det universelle virkningskvantum har haft for indstillingen til naturvidenskabelige og erkendelsesteoretiske problemer. Det er vort håb, at bøgerne vil vise sig nyttige for alle, der føler trang til at orientere sig om den fysiske videnskabs fremskridt, og at de også herhjemme vil stimulere oplysningsarbejdet på dette og andre områder af videnskaben, hvis udvikling rummer rige løfter om fremme af velfærd og kultur.

Niels Bohr · H. Højgaard Jensen · Aage Petersen · Søren Sikjær

TRANSLATION

FOREWORD BY THE DANISH EDITORIAL COMMITTEE

Physical science has from ancient times exercised a special attraction – not only on the researchers who strive to increase our knowledge of that nature in which we live and on the technicians who work with the practical application of such knowledge – but also on the many others who feel the need to widen their horizon. The inspiration that can be found by working with physics problems may be compared to the enrichment offered to us by great works of art. Although in physics we do not, as in art, have to do with moods conceived and contained in the human mind, but with phenomena which nature displays for us to see, insight into the order which has been revealed step by step requires the use and exercise of abilities that are common to humanity.

In our time, the application of physics has not only changed the framework for our daily existence in a radical way, but the opening of previously inaccessible fields of experience and their investigation by systematic experimentation have, in addition, increased our knowledge about nature on an unimagined scale and given our world picture a unity that exceeds the most daring dreams of the past. This development, which to an ever increasing degree sets its mark on the whole of our culture, makes great demands on the spreading of information in the widest circles about the history and the present state of physics. This is true not least for young people, who are to bring new strength to the work for the growth and utilization of our knowledge.

A great effort towards this end has been made in recent years by a group of eminent American scientists and educators. This so-called *Physical Science Study Committee* has made vigorous efforts in order to reform physics teaching and to stimulate the interest in and understanding of physical science. The committee's new system of textbooks, supplemented by its many excellent educational films, must be regarded as one of the most important educational advances in the postwar period. The main emphasis is placed on the method and principles of physics as well as on an understanding of its cultural significance.

As a part of its efforts the committee is publishing a series of small books for self-study and as a supplement to physics teaching in school. Each book

[95]

is devoted to one topic, and the series will eventually comprise topics from all fields of physics, from the world of atoms to the depths of the skies, from the physics of ancient times to modern physics. These books are prepared by distinguished specialists, who apart from thorough knowledge also have the ability to write popularly. Naturally, such accounts can only be understood through an effort on the part of the reader; on the other hand, the guidance given here is competent and all-round to a rare degree.

The series has already achieved very great distribution and has been translated into many languages. Selected volumes in Danish translation are included in Gyldendals Kvantebøger (Gyldendal's Quantum Books). This name has been chosen – although quantum physics is not directly touched upon in all the books – not least because of the decisive significance that the discovery of the universal quantum of action has had for the attitude towards scientific and philosophical problems. It is our hope that the books will prove useful for everyone who wants to become familiar with the advance of physical science, and that they here at home, too, will stimulate educational work in this and other fields of science, the development of which holds rich promises for the advancement of welfare and culture.

Niels Bohr · H. Højgaard Jensen · Aage Petersen · Søren Sikjær

PART II

PEOPLE

INTRODUCTION

by

FINN AASERUD

During his lifetime Niels Bohr wrote many tributes and obituaries for a variety of predecessors and contemporaries. These writings vary considerably in size and scope and in the effort that Bohr put into them. Most of them may rightly be considered occasional writings, while others, notably extensive published articles on the work of his mentor Ernest Rutherford (1871–1937)[1] and the discussions with his colleague and intellectual sparring partner Albert Einstein (1879–1955),[2] can be counted among his most important writings.

In addition to the two major articles on Rutherford and Einstein, publications by Bohr devoted to eight further individuals were included in previous volumes of the Collected Works: Luigi Galvani (1737–1798),[3] Hans Christian

[1] N. Bohr, *The Rutherford Memorial Lecture 1958: Reminiscences of the Founder of Nuclear Science and of Some Developments Based on his Work*, Proc. Phys. Soc. **78** (1961) 1083–1115. Reproduced in Vol. 10 (with a previously unpublished reference list), pp. [381]–[420]. Reprinted in N. Bohr, *Essays 1958–1962 on Atomic Physics and Human Knowledge*, Interscience Publishers, New York 1963, which has been reissued as *The Philosophical Writings of Niels Bohr, Vol. III*, Ox Bow Press, Woodbridge, Connecticut 1987, pp. 30–73.

[2] N. Bohr, *Discussion with Einstein on Epistemological Problems in Atomic Physics* in *Albert Einstein: Philosopher–Scientist* (ed. P.A. Schilpp), The Library of Living Philosophers, Vol. VII, Evanston, Illinois 1949, pp. 201–241. Reproduced in Vol. 7, pp. [339]–[381]. Reprinted in *Atomic Physics and Human Knowledge*, John Wiley & Sons, New York 1958, which has been reissued as *The Philosophical Writings of Niels Bohr, Vol. II*, Ox Bow Press, Woodbridge, Connecticut 1987, pp. 32–66.

[3] N. Bohr, *Biology and Atomic Physics* in *Celebrazione del secondo centenario della nascita di Luigi Galvani*, Bologna – 18–21 ottobre 1937-XV: I. Rendiconto generale, Tipografia Luigi Parma 1937, pp. 68–78. Reproduced in Vol. 10, pp. [49]–[62]. Reprinted in *Atomic Physics and Human Knowledge*, ref. 2, pp. 13–22.

Ørsted (1777–1851),[4] James Clerk Maxwell (1831–1879),[5] Harald Høffding (1843–1931),[6] Johannes Robert Rydberg (1854–1919),[7] Pieter Zeeman (1865–1943),[8] S.P.L. Sørensen (1868–1939)[9] and Werner Heisenberg (1901–1976).[10] Also reproduced in earlier volumes of the Collected Works are articles celebrating achievements of Vilhelm Thomsen (1842–1927),[11] Johannes Pedersen (1883–1977),[12] Niels Erik Nørlund (1885–1981),[13] Paul Bergsøe (1872–1963),[14] Jens Anton Christiansen (1888–1969),[15] Richard Courant (1888–

---

[4] N. Bohr, *Hans Christian Ørsted*, Fys. Tidsskr. **49** (1951) 6–20. Reproduced in Vol. 10, pp. [341]–[356] (Danish original), [357]–[369] (English translation).

[5] N. Bohr, *Maxwell and Modern Theoretical Physics*, Nature (Suppl.) **128** (1931) 691–692. Reproduced in Vol. 6, pp. [357]–[360].

[6] N. Bohr, *At Harald Høffding's 85th Birthday*, Berlingske Tidende, 10 March 1928; reproduced in Vol. 10, pp. [305]–[307] (Danish original), [308]–[309] (English translation). N. Bohr, *Tribute to the Memory of Harald Høffding*, Overs. Dan. Vidensk. Selsk. Virks. Juni 1931 – Mai 1932, pp. 131–136; reproduced in Vol. 10, pp. [311]–[318] (Danish original), [319]–[322] (English translation). N. Bohr, *Harald Høffding's 100th Birthday*, Overs. Dan. Vidensk. Selsk. Virks. Juni 1942 – Maj 1943, pp. 57–58; reproduced in Vol. 10, pp. [323]–[324] (Danish original), [325] (English translation).

[7] N. Bohr, *Rydberg's Discovery of the Spectral Laws*, Proceedings of the Rydberg Centennial Conference on Atomic Spectroscopy, Lunds Universitets Årsskrift. N.F. Avd. 2. Bd. 50. Nr 21 (1955) 15–21. Reproduced in Vol. 10, pp. [371]–[379].

[8] N. Bohr, *Zeeman Effect and Theory of Atomic Constitution* in *Zeeman, Verhandelingen*, Martinus Nijhoff, The Hague 1935, pp. 131–134. Reproduced in Vol. 10, pp. [335]–[340].

[9] N. Bohr, *Meeting on 20 October 1939*, Overs. Dan. Vidensk. Selsk. Virks. Juni 1939 – Maj 1940, pp. 25–26; reproduced in Vol. 11, pp. [379]–[381] (Danish original), [382]–[383] (English translation).

[10] N. Bohr, *Die Entstehung der Quantenmechanik* in *Werner Heisenberg und die Physik unserer Zeit*, Friedr. Vieweg & Sohn, Braunschweig 1961, pp. IX–XII. English version, *The Genesis of Quantum Mechanics*, in Bohr, *Essays 1958–1962*, ref. 1, pp. 74–78. The English version is reproduced in Vol. 10, pp. [421]–[428].

[11] N. Bohr, *Meeting on 30 January 1942*, Overs. Dan. Vidensk. Selsk. Virks. Juni 1941 – Maj 1942, pp. 33–34; Reproduced in Vol. 11, p. [393]–[394] (Danish original), [395]–[396] (English translation).

[12] N. Bohr, *Physical Science and the Study of Religions*, Studia Orientalia Ioanni Pedersen Septuagenario A.D. VII id. Nov. Anno MCMLIII, Ejnar Munksgaard, Copenhagen 1953, pp. 385–390. Reproduced in Vol. 10, pp. [273]–[280].

[13] N. Bohr, *On the Problem of Measurement in Atomic Science* in *Festskrift til N.E. Nørlund i Anledning hans 60 Års Fødselsdag den 26. Oktober 1945, Anden Del*, Ejnar Munksgaard, Copenhagen 1946, pp. 163–167. Reproduced in Vol. 11, pp. [655]–[661] (Danish original), [662]–[666] (English translation).

[14] N. Bohr, *Tenth and Eleventh Presentation of the H.C. Ørsted Medal*, Fys. Tidsskr. **57** (1959) 145, 158; reproduced in Vol. 11, p. [487]–[489] (Danish original), [490]–[491] (English translation).

[15] *Ibid.*

1972),[16] Alex Langseth (1895–1961)[17] and Kai Ulrik Linderstrøm-Lang (1896–1959),[18] which do not, however, delve into their work or careers.

The nineteen articles just referred to, five of which were presented in Volume 10 as constituting Bohr's "Historical Papers", make up only a minority of Bohr's publications celebrating particular individuals. For completeness, all these articles are presented briefly in this introduction, along with the fifty-eight further such writings reproduced below.

The writings are divided into four groups corresponding to the roles of the individuals in Bohr's life. The contributions in the "Autobiographical" section are presented in the order in which they were written. Next follows "Celebrations of Earlier Scientists", which is arranged in the order that the foreign and Danish subjects, respectively, were born. Each of the two most substantial sections, "Physics Teachers and Colleagues" and "Family and the Broader Danish Milieu", introduces the individuals in the order in which they entered Bohr's life, in effect using Bohr's biography as a point of reference.

Although some of the contributions are very brief, with little real content, they are included to attain completeness in the Collected Works with regard to Bohr's published writings. Their inclusion ensures as good a representation of Bohr's scientific and personal network as is possible on the basis of his publications.[19] The translation into English of the several Danish-language articles makes them available to an international audience for the first time.

1. AUTOBIOGRAPHICAL

When asked about his contributions to the world of knowledge, Niels Bohr always emphasized the role of cooperation in science over the efforts of individuals. Nevertheless, his long and brilliant career was bound to produce public and half-public autobiographical statements.

---

[16] N. Bohr, *Mathematics and Natural Philosophy*, The Scientific Monthly **82** (1956) 85–88; reproduced in Vol. 11, pp. [667]–[672].

[17] N. Bohr, *Ninth Presentation of the H.C. Ørsted Medal*, Fys. Tidsskr. **51** (1953) 65–67, 80; reproduced in Vol. 11, pp. [477]–[481] (Danish original), [482]–[485] (English translation).

[18] N. Bohr, *Eighth Presentation of the H.C. Ørsted Medal*, Fys. Tidsskr. **39** (1941) 175–177, 192–193; reproduced in Vol. 11, pp. [469]–[472] (Danish original), [473]–[475] (English translation).

[19] The many unpublished talks that Bohr gave to honour friends and colleagues are not included. The manuscripts for such talks in the BMSS are listed below in the *Inventory of Relevant Manuscripts in the Niels Bohr Archive*, pp. [523] ff. In this introduction, unpublished manuscripts adding to Bohr's published biographical statements are referred to as appropriate. Additional important manuscripts include tributes to Ernst Mach (*Inventory*, p. [526], folder 18), Georg Brandes (p. [528], folder 27), Tycho Brahe (p. [529], folder 33) and Max Born (p. [530], folder 42).

The first contribution is the autobiography Bohr wrote for the University of Copenhagen upon completing his doctorate there in 1911.[20] Upon being created a Knight of the Dannebrog – an order established by the Danish King Christian V in 1671 – nine years later, Bohr was asked to write another autobiography, which provides instructive testimony about the development of his career in the intervening eventful years.[21]

In 1922 Bohr received the Nobel Prize for Physics, which required that he contribute three different writings to be printed in the special publication issued by the Nobel Foundation. Two of these have been reproduced earlier in the Collected Works.[22] The third, which Bohr wrote for a section introducing the laureates' biographies, is reproduced here.[23]

After another short autobiography prepared in 1928 for the twenty-fifth anniversary of completing his gymnasium education[24] follows one of the rare newspaper interviews with Bohr, conducted for a major Copenhagen newspaper in connection with his fiftieth birthday on 7 October 1935.[25] More than any other publication by Bohr, the interview represents the immediacy of his spoken word in contrast to elaborate writings prepared by himself. As such, it includes several statements which he is known to have made in conversation, but which do not exist elsewhere in writing. In 1949 Bohr agreed to an interview with another major Copenhagen newspaper which testifies to his role as a statesman of science after World War II.[26]

[20] N. Bohr, [Autobiography] in *Festskrift udgivet af Kjøbenhavns Universitet i Anledning af Universitetets Aarsfest, November 1911*, Schultz, Copenhagen 1911, pp. 76–77. Reproduced and translated on pp. [133] ff.

[21] N. Bohr, [Autobiography for the Danish College of Arms], 23 November 1920. Transcribed and translated on pp. [137] ff.

[22] The Nobel Prize Lecture was published in N. Bohr, *Om Atomernes Bygning* in *Les Prix Nobel en 1921–1922*, P.A. Norstedt, Stockholm 1923, pp. 1–37 (paginated independently); reproduced in Vol. 4, pp. [425]–[465]. An English version, *The Structure of the Atom*, published in Nature **112** (1923), 29–44, is reproduced in Vol. 4, pp. [467]–[482]. Bohr's speech at the banquet is printed in *Les Prix Nobel en 1921–1922*, pp. 102–104 (Danish original), 104–106 (English translation). It is reproduced in a new English translation in Vol. 4, pp. [26]–[27].

[23] N. Bohr, *Niels Bohr* in *Les Prix Nobel en 1921–1922*, ref. 22, pp. 126–127. Reproduced on pp. [143] ff.

[24] N. Bohr, [Autobiography] in *Studenterne MCMIII: Personalhistoriske Oplysninger*, Berlingske Bogtrykkeri, Copenhagen 1928, p. 275. Reproduced and translated on pp. [147] ff.

[25] P. Vinding, *Conversation with Niels Bohr*, Berlingske Aftenavis, 2 October 1935. Reproduced and translated on pp. [151] ff.

[26] M. Bonnesen, *Niels Bohr on Jest and Earnestness in Science*, Politiken, 17 April 1949. Reproduced and translated on pp. [177] ff.

In 1952 Bohr was portrayed in a state-sponsored Danish film series introducing important personalities in contemporary Danish culture. Previous films had depicted, among others, the authors Johannes V. Jensen (1873–1950, Nobel Prize in literature 1944) and Martin Andersen Nexø (1869–1954) as well as the painter Jens Ferdinand Willumsen (1863–1958). As a complete manuscript for Bohr's monologue has not been found, it has been transcribed from the film and translated especially for the present volume. Even as a written text, it constitutes Bohr's most spirited account of his life.[27]

The last contribution of this section is Bohr's extensive autobiographical account written in connection with the awarding in 1956 of an honorary degree by Denmark's second university, which was established in 1928 and is located in the city of Århus.[28]

2. CELEBRATION OF EARLIER SCIENTISTS

The portraits of earlier scientists still hanging in his office at the Niels Bohr Institute testify to Bohr's sense of being part of a long-standing tradition in the history of science. To Bohr, his predecessors and their work served less as a subject of study in their own right than as a basis for understanding the present state of science.

The earliest scientist for whom there exists a published tribute by Bohr is Isaac Newton (1642–1727), who himself became famous for stating that he stood "on the shoulders of giants". The publication is based on a talk that Bohr gave in his capacity as President of the Royal Danish Academy of Sciences and Letters in connection with the Royal Society of London's celebration in 1946 of the 300th anniversary of Newton's birth.[29] Characteristically, Bohr presented Newton's work in the light of contemporary physics, against the background of which, he wrote, "we can perhaps more clearly than ever before appreciate the wisdom and caution Newton exhibited in his untiring efforts to obtain the proper balance between analysis and synthesis."[30]

---

[27] N. Bohr, [Portrait film], 1952. Transcribed and translated on pp. [197] ff.

[28] N. Bohr, *Autobiography of the Honorary Doctor*, Acta Jutlandica **28** (1956) 135–138. Reproduced and translated on pp. [207] ff.

[29] Fourteen years later, Bohr communicated another address to the Royal Society on behalf of the Royal Danish Academy in celebration of the Royal Society's 300th anniversary. N. Bohr, *Meeting on 14 October 1960*, Overs. Dan. Vidensk. Selsk. Virks. Juni 1960 – Maj 1961, pp. 39–41. Reproduced in Vol. 11, pp. [443]–[445].

[30] N. Bohr, *Newton's Principles and Modern Atomic Mechanics* in *The Royal Society Newton Tercentenary Celebrations 15–19 July 1946*, Cambridge University Press, Cambridge 1947, pp. 56–61. Reproduced on pp. [219] ff.

Niels Bohr and Ivan Supek, Zagreb, 1958.

Chronologically, the next scientist after Newton discussed by Bohr is Rudjer J. Bošković (1711–1787). The occasion was the celebration in 1958 of the 200th anniversary of the publication of Bošković's "Theoria Philosophiae Naturalis", held over a period of several days at the end of October in various places in Yugoslavia. Bohr was only able to come towards the end of the celebrations, speaking at the Bošković Institute of Physics in Zagreb on 29 October. Bohr's talk was tape-recorded and a transcript of the recording was sent to Bohr upon his return to Denmark.[31] In early May 1959, Bohr wrote to his Yugoslavian colleague Ivan Supek (1915–1997) that the tape-recording could not be used as a basis for publication and that he intended "to send you a more concentrated account of the address which dealt with the relationship between the viewpoints of natural philosophy at Bošković's time and modern physical developments." The manuscript was ready less than a month later, after Supek had visited Bohr in Copenhagen. When submitting the manuscript, Bohr wrote to his colleague: "I have with the kind help of Rosenfeld tried to write a brief foreword to the Bošković volume, as you suggested." In addition to being a physicist, Bohr's colleague Léon Rosenfeld was also a historian of science who, among his many other activities, took the initiative to publish the Niels Bohr Collected Works. In his brief account, Bohr considered Bošković's attempt at developing "a systematic account of the properties of matter on the basis of interactions of mass points through central forces" as an important step in the history of physics connecting, for example, the contributions of Maxwell to those of Newton.[32] It was also in connection with the visit in 1958 to Yugoslavia that Bohr agreed to contribute a greeting to a Yugoslavian newspaper in celebration of the National Day of the Republic of Yugoslavia, in which he paid tribute to Bošković among others.[33]

Bohr to Supek,
9 May 59
English
Full text on p. [516]

Bohr to Supek,
1 June 59
English
Full text on p. [517]

More than twenty years before he wrote his tribute to Bošković, Bohr had been in Bologna celebrating the 200th anniversary of the birth of Luigi Galvani. His speech on that occasion, reproduced in Volume 10, placed Galvani in the context of the history of science and presented Bohr's views on the relationship between biology and atomic physics.[34]

Two years before taking part in the celebration of Bošković, Bohr had made a substantially more extensive visit to Yugoslavia together with his wife Margrethe and their youngest son Ernest and his wife. The main occasion was the celebration in Belgrade in 1956 of the centenary of Nikola Tesla's (1856–1943) birth. Bohr's main contribution was an elaboration of his major lecture, "On Atoms and Human Knowledge", given for the first time at the Royal Danish Academy of Sciences and Letters in October 1955.[35] In his

---

[31] For the transcript, see *Inventory* (ref. 19), p. [532], folder 55.

[32] N. Bohr, *R.J. Bošković* in *Actes du symposium international R.J. Bošković, 1958*, Académie Serbe des sciences, Académie Yougoslave des sciences et des arts, Académie Slovène des sciences et des arts, Belgrade, Zagreb, Ljubljana 1959, pp. 27–28. Reproduced on pp. [231] ff.

[33] N. Bohr, [Greeting], Slovenski Poročevalec, Ljubljana. Reproduced from the newspaper and from Bohr's manuscript in English on pp. [227] ff.

[34] Bohr, *Biology and Atomic Physics*, ref. 3.

[35] See Vol. 7, p. [397]. The talk was reproduced in the proceedings of the Tesla conference,

brief speech celebrating Tesla, Bohr praised the Yugoslavian scientist for his "most ingenious invention of polyphase transmission of electric power and his researches on the marvellous phenomena of high frequency electric oscillations" and expressed his admiration for Tesla's ability to "exert ... influence in countries then more advanced in scientific and industrial developments". Bohr's 1956 visit to Yugoslavia was also part of his international campaign for an "open world".[36] He thus ended his speech about Tesla by expressing the "sincere hope for fruitful collaboration with the people of Yugoslavia, so rich in heroic and cultural traditions."[37]

In 1931 Bohr took part in a celebration in honour of James Clerk Maxwell. Even more than in his tribute to Newton, Bohr used the opportunity to discuss fundamental questions in contemporary physics.[38]

The earliest of his Danish predecessors to whom Bohr devoted an entire article was Vitus Bering (1681–1741), the Dane who, as an explorer in the service of the Russian czars, in 1728 established that the Asian and American continents were divided by what came to be known as the Bering Strait. Bohr had accepted the post as chairman of the reconstituted Bering Committee, originally led by the Danish explorer Knud Rasmussen (1879–1933).[39] The 200th anniversary of Bering's death in 1941 prompted the establishment of a Vitus Bering Medal to be awarded by the Danish Geographical Society as well as the decision to erect a large monument in Bering's memory in his home town of Horsens in Jutland, conceived by the sculptor Mogens Bøggild (1901–1987) and the architect Kaare Klint (1888–1954). An extract from Bohr's speech, given at the celebrations in Horsens on 19 December 1941 and published in a book on Bering the year after, was intended to serve as an inscription for the monument.[40] Otherwise, the speech is less a biography than a description of the celebration of the Danish explorer in the midst of the German occupation.[41] On

---

Centenary of the Birth of Nikola Tesla 1856–1956, Nikola Tesla Museum, Belgrade 1959, pp. 106–115.

[36] See Vol. 11, Part I.

[37] N. Bohr, [Tribute to Tesla] in *Centenary*, ref. 35, pp. 46–47. Reproduced on pp. [235] ff.

[38] Bohr, *Maxwell and Modern Theoretical Physics*, ref. 5.

[39] Bohr had previously served on a committee with Rasmussen in the 1920s preparing the relocation and expansion of the Danish National Museum. See Vol. 10, figure caption, p. [250]. He had furthermore given a speech on the occasion of Rasmussen's 50th birthday in 1929. N. Bohr, manuscript, *Inventory* (ref. 19), p. [524], folder 6.

[40] See p. [244], ref. 1.

[41] N. Bohr, [Tribute to Bering] in *Vitus Bering 1741–1941*, H. Hagerup, Copenhagen 1942, pp. 49–53. Reproduced and translated on pp. [237] ff.

13 May 1942 Bering was also celebrated in Copenhagen, on which occasion Bohr gave a slightly revised version of the talk which was transmitted on the Danish national radio.[42]

Bohr continued as Chairman of the Bering Committee, yet the monument for Bering was never to materialize. By 1957, the project had been given up, and instead a park in Bering's memory was inaugurated in Horsens, adorned with two of the twelve cannons from Bering's expedition found and rescued in the meantime by the Russians. Bohr and his wife took part in the celebrations together with the Danish Prime Minister Hans Christian Hansen (1906–1960) and other notables.

Among earlier Danish scientists, Bohr had a particularly strong affinity to Hans Christian Ørsted, internationally renowned for his discovery of electromagnetism in 1820. Bohr presented a succinct biography of Ørsted at the University of Copenhagen in 1951, when the Ørsted-founded Society for the Dissemination of Natural Science celebrated the centennial of Ørsted's death.[43] The published version of Bohr's speech is reproduced and translated in Volume 10.[44]

In 1955 Bohr celebrated the eminent philologist Johan Nicolai Madvig (1804–1886) who, like his older contemporary Ørsted, was a leading public figure in Denmark. Madvig served as President of the Royal Danish Academy of Sciences and Letters for sixteen years and acted for a time as Chairman of the Board for the Carlsberg Foundation.[45] Bohr's tribute, co-authored with orientalist Johannes Pedersen, is in the form of a foreword to a book on Madvig, published jointly by the Danish Academy, of which Bohr was President,[46] and the Carlsberg Foundation, from which Pedersen retired as Chairman in 1955 after twenty-two years of service.[47]

---

[42] N. Bohr, *Tale ved Statsradiofoniens Mindefest for Vitus Bering den 13. Maj 1942*, manuscript, *Inventory* (ref. 19), p. [528], folder 29.

[43] Bohr's involvement in this Society is briefly discussed in Volume 11, pp. [352]–[354].

[44] Bohr, *Hans Christian Ørsted*, ref. 4.

[45] The Carlsberg Foundation was established in 1876. See Vol. 11, p. [339].

[46] See Vol. 11, pp. [338]–[348].

[47] N. Bohr and J. Pedersen, *Foreword* in *Johan Nicolai Madvig: Et mindeskrift*, Royal Danish Academy of Sciences and Letters and the Carlsberg Foundation, Copenhagen 1955, p. vii. Reproduced and translated on pp. [247] ff. The book was printed by the firm Bianco Luno, which had served as a printing house for the Danish Academy since 1837. Bohr briefly celebrated the 125th anniversary of this cooperation in N. Bohr, *Forord* in A. Lomholt, *Lærdoms Mosaik: Samlinger til Det kongelige danske Videnskabernes Selskabs Historie*, Bianco Luno, Copenhagen 1962, pp. 5–6.

3. PHYSICS TEACHERS AND COLLEAGUES

Upon completing his doctorate at the University of Copenhagen in 1911, Bohr continued his studies in England with a stipend from the Carlsberg Foundation. His first choice was the Cavendish Laboratory at the University of Cambridge, where he could study under Joseph John Thomson (1856–1940), renowned for the discovery of the electron. Bohr, who had great admiration for Thomson as a scientist and teacher, accepted the invitation in 1926 from the journal "Nature" to write a congratulatory note in connection with Thomson's seventieth birthday. Placing himself in time between Thomson and the latest generation of physicists, Bohr observed that "it is difficult for scientists of the younger generation, who are working on the new land to which Thomson has opened the gates, fully to realize the magnitude of the task with which the pioneers were confronted."[48] Bohr also wrote a brief message for a congress in 1956 celebrating the 100th anniversary of Thomson's birth.[49]

After spending two terms, or about half a year, in Cambridge, Bohr moved to Manchester at the invitation of the physics professor at the University there, Ernest Rutherford. Bohr has recalled that he had been greatly impressed by a lecture Rutherford gave at an annual dinner event at the Cavendish in October 1911.[50] It was only the following month, however, that Bohr got a chance to meet Rutherford personally. This took place at the home of James Lorrain Smith (1862–1932), Professor of Pathology at the University of Manchester.[51] Lorrain Smith had been a close friend of Bohr's father, at whose laboratory he had spent time in the early 1890s, and he also knew Rutherford well. Some fifteen years younger than Thomson, Rutherford represented a fresh approach to experimental physics. It was in Manchester that Bohr took his first steps towards his revolutionary atomic theory of the nuclear atom – which had recently been proposed by Rutherford on the basis of experiments – applying the quantum of action introduced by Max Planck in 1900.[52] The Manchester laboratory was to serve as a model when he set up his own institute in

[48] N. Bohr, *Sir J.J. Thomson's Seventieth Birthday*, Nature **118** (1926) 879. Reproduced on pp. [251] ff.

[49] N. Bohr, [Greeting], carbon copy, *Inventory* (ref. 19), p. [531], folder 49.

[50] Bohr, *Reminiscences*, ref. 1, p. [383].

[51] N. Bohr, *Notes on my Relations with Rutherford*, manuscript, *Inventory* (ref. 19), p. [526], folder 14.

[52] Bohr gave a strongly appreciative speech for Planck when the latter visited Copenhagen in October 1930. N. Bohr, *Rede bei einem Fest für M. Planck*, manuscript, *Inventory* (ref. 19), p. [524], folder 7.

Copenhagen ten years later, and Rutherford would remain a father figure and role model throughout Bohr's life.[53]

As already noted, Bohr wrote an extensive Memorial Lecture celebrating Rutherford and his work.[54] That was by no means all. There are six more publications, two of which appeared in Rutherford's lifetime.

It was important to Bohr that Rutherford should be one of the very first to see his new Institute for Theoretical Physics at Copenhagen University, and Rutherford's first visit to Denmark in September 1920 was timed with this in mind. However, the opening of the institute was postponed due to unforeseen circumstances. Bohr did make use of the occasion of Rutherford's visit to introduce his mentor to the Danish public in a substantial popular article in a Copenhagen newspaper. Here he noted in particular Rutherford's recent discovery of artificially induced radioactivity at the Cavendish Laboratory, where Rutherford had succeeded Thomson as director the year before.[55] Bohr's next item on Rutherford appeared in the supplement to the issue of "Nature" containing his tribute to Thomson. The occasion was the fiftieth anniversary of the Cavendish Laboratory.[56]

Rutherford visited Copenhagen again in September 1932, the "Miraculous Year" of nuclear physics that saw a revolution in Rutherford's field of research. Since Bohr had just moved into the Carlsberg Mansion, Rutherford could now see both Bohr's institute and his new residence. Bohr gave several speeches for Rutherford during his visit. Two items – part of a talk in appreciation of Rutherford's lectures on 14 September and an entire speech at the festivities at the Science Society of Denmark (Danmarks naturvidenskabelige Samfund) the following day – appeared in a book published as a tribute after Rutherford's death.[57] In the second lecture, Bohr noted that he would refrain from recounting any of the many "stories which circulate among the pupils in his laboratory

---

[53] Altogether thirty-five letters of the rich correspondence between Bohr and Rutherford are reproduced in Vol. 2, pp. [576]–[598] (25 letters), Vol. 3, pp. [682]–[684] (1), Vol. 5, pp. [486]–[490] (4), Vol. 8, pp. [783]–[788] (4) and Vol. 9, pp. [651]–[652] (1).

[54] Bohr, *Reminiscences*, ref. 1.

[55] N. Bohr, *Professor Sir Ernest Rutherford and his Significance for the Recent Development of Physics*, Politiken, 18 September 1920. Reproduced and translated on pp. [253] ff.

[56] N. Bohr, *Sir Ernest Rutherford, O.M., P.R.S.*, Nature (Suppl.) **118** (18 Dec 1926) 51–52. Reproduced on pp. [263] ff.

[57] N. Bohr, [Two speeches for Rutherford, 1932] in A.S. Eve, *Rutherford: Being the Life and Letters of the Rt Hon. Lord Rutherford, O.M.*, Cambridge University Press, Cambridge 1939, pp. 361–363. Reproduced on pp. [267] ff. According to its preface the book was prepared in order that "Rutherford may reveal himself, just as he was, in lectures, books, papers, speeches, portraits, letters, and casual talk."

Niels Bohr and Ernest Rutherford, Cambridge, 1923.

… because Lord Rutherford last time reproached me for telling stories against him." One such story was a colleague's observation that "he had seen many young men working in the Cavendish Laboratory but nobody who could swear at his apparatus like Rutherford", as Bohr would later formulate it in lectures in 1954 and 1958.[58] In his brief speech at the dinner party at the Carlsberg Mansion welcoming Lord and Lady Rutherford as the first guests at his and his wife's new residence, Bohr stated that he owed more to Rutherford than "perhaps … any other man living."[59] As a whole, Bohr's lectures for Rutherford in Copenhagen show the close personal bonds between the two.

Rutherford died on 19 October 1937. Bohr learned about the sad event while attending the Galvani commemoration in Bologna. He immediately gave an

[58] N. Bohr, *Greater International Cooperation is Needed for Peace and Survival* in *Atomic Energy in Industry: Minutes of 3rd Conference October 13–15, 1954*, National Industrial Conference Board, Inc., New York 1955, pp. 21–26, on p. 22. Reproduced in Vol. 11, pp. [561]–[570], on p. [566]. Bohr repeated the story in *Reminiscences*, ref. 1, p. [383].

[59] N. Bohr, manuscript, *Inventory* (ref. 19), p. [525], folder 9. The published extracts, ref. 57, reproduced below have been edited and abbreviated in relation to the manuscripts for Bohr's talks contained in these folders.

impromptu speech in Rutherford's memory, which is reproduced in Italian in the proceedings from the conference.[60] The speech was subsequently published in English in "Nature",[61] which a month and a half later also carried a more elaborate obituary for Rutherford written by Bohr.[62]

In 1947 his French colleague Frédéric Joliot-Curie (1900–1958) invited Bohr to give a talk at the commemoration in Paris of the tenth anniversary of Rutherford's death. The event was organized by the World Federation of Scientific Workers, of which Joliot was President. Bohr expressed a strong desire to take part in the commemoration, but due to the pressure of work decided at the last moment to be represented a younger colleague, Christian Møller (1904–1980), who read the tribute. In a key sentence, Bohr presented Rutherford's general view of the scientific endeavour as a model for his own present efforts in the political arena: "International scientific co-operation was to Rutherford not only a way to progress in Science, but also a means of promoting that mutual understanding between nations which is vital for the harmonious development of human culture."[63]

The substantial Memorial Lecture that Bohr gave for Rutherford in 1958 was first published in an expanded version in 1961 in the Proceedings of the Physical Society,[64] which was in turn reprinted in the proceedings of a conference held in September 1961 celebrating the fiftieth anniversary of Rutherford's discovery in Manchester of the atomic nucleus.[65] At the latter conference, Bohr joined four other students of Rutherford from half a century earlier – Edward Neville da Costa Andrade (1887–1971), James Chadwick (1891–1974), Charles Darwin (1887–1962) and Ernest Marsden (1889–1970) – to speak in its Commemorative Session. In his lecture, which was published in the conference proceedings and is reproduced below, Bohr expressed his purpose as "stress[ing] the point which formed the main theme for the [Memorial]

[60] *Parole pronunciate dal prof. Niels Bohr in morte di E. Rutherford* in *Celebrazione*, ref. 3, pp. 427–428.

[61] N. Bohr, [Obituary for Rutherford], Nature **140** (1937) 752–753. Reproduced on pp. [271] ff.

[62] N. Bohr, [Obituary for Rutherford], Nature (Suppl.) **140** (1937) 1048–1049. Reproduced on pp. [273] ff. Bohr also gave a memorial speech for Rutherford at the Danish Physical Society on 22 November 1937, the first few pages of which exist in manuscript form. See N. Bohr, *Foredrag i Fysisk Forening*, manuscript, *Inventory* (ref. 19), p. [526], folder 15.

[63] N. Bohr, [Tribute to Rutherford], in *Hommage à Lord Rutherford, sept huit et neuf novembre MCMXLVII*, La fédération mondiale des travailleurs scientifiques, Paris [1948], pp. 15–16. Reproduced on pp. [275] ff.

[64] Bohr, *Reminiscences*, ref. 1.

[65] *Rutherford at Manchester* (ed. J.B. Birks), Heywood, London 1962, pp. 114–167, which, incidentally, omits the next-to-the-last paragraph of the original version.

Lecture", namely that "Rutherford's discovery [of the atomic nucleus] from the very beginning presented physicists with the challenge to establish a comprehensive explanation of the experiences, accumulated through the centuries, about the physical and chemical properties of matter."[66]

During his first stay with Rutherford in Manchester Bohr developed friendships with several colleagues in physics, notably the experimentalist Henry Moseley (1887–1915). Moseley was an avid supporter of Bohr's theories, and his X-ray spectroscopical investigations constituted important early confirmation of Bohr's conception of the atom.[67] Moseley's death as a young man on the battlefield in World War I affected Bohr deeply. Although Bohr was not nominally the author of the obituary in "Nature", the person who was, Moseley's close friend E. Ray Lankester, informed his readers that the Danish scientist had enabled him "to add to this personal sketch a notice of Moseley's scientific work". For completeness, both Lankester's "personal sketch" and the subsequent "notice of Moseley's scientific work" are reproduced below.[68]

Immediately upon returning from his first stay with Rutherford in Manchester, Bohr accepted a position as teaching assistant at the University of Copenhagen. His place of work, however, was the Technical University (*Polyteknisk Læreanstalt*), which supplied minimal facilities to the university for experimental physics and where Bohr had done his undergraduate studies. The position was vacated by Martin Knudsen (1871–1949), an experimental physicist at the University of Copenhagen later internationally known for his work on highly dilute gases, now known as "Knudsen gases". Knudsen had just been appointed professor of physics at the University of Copenhagen as the successor of Bohr's teacher, Christian Christiansen (1843–1917),[69] who had reached retirement age in 1913.

While Bohr admired Knudsen's work greatly, his many tributes to him should also be seen in the light of his and Knudsen's respective obligations within the Danish scientific milieu, particularly in the Royal Danish Academy

[66] N. Bohr, *The General Significance of the Discovery of the Atomic Nucleus* in *Rutherford at Manchester*, ref. 65, pp. 43–44. Reproduced on pp. [279] ff.

[67] Vol. 2, pp. [544]–[547], contains two letters between Bohr and Moseley from November 1913.

[68] E.R. Lankester, *Henry Gwyn Jeffreys Moseley*, Phil. Mag. **31** (1916) 173–176. Reproduced on pp. [283] ff. The manuscript for the "notice" is in the BMSS. See *Inventory* (ref. 19), p. [524], folder 1. It is transcribed in Vol. 2, pp. [427]–[430].

[69] No tribute by Bohr to his teacher has survived. The two letters between Bohr and Christiansen reproduced in Vol. 2, pp. [493]–[497], shed some light on their relationship.

of Sciences and Letters. Knudsen became its Secretary in 1917, at the very first meeting attended by Bohr as a member,[70] and retained the post until 1945.[71]

Bohr's first known speech for Knudsen was given on 18 October 1939 in connection with Knudsen's retirement as Chairman of the Danish Society for the Dissemination of Natural Science,[72] a post that Bohr was now taking over.[73] Bohr's first published tribute was a newspaper article from 1941 marking the Danish experimentalist's seventieth birthday. In addition to giving a popular account of Knudsen's scientific contributions, Bohr emphasized that "in times like these we, here at home, will think with special gratitude of how much a man such as he has contributed to our country's standing". To Bohr, science in general, and its representative Knudsen in particular, was a rallying point during the German occupation of their native country.[74] On the occasion of Knudsen's seventieth birthday Bohr also gave a spirited speech at the Science Society of Denmark,[75] and the board of the Society for the Dissemination of Natural Science, of which Bohr was now Chairman, composed a "greeting in recognition of how profoundly indebted to you the members of the Society are for your almost 40 years of leadership of the Society's activity."[76] One year later the Academy celebrated the twenty-fifth anniversary of Knudsen's tenure as Secretary with a greeting by Bohr, who also handed Knudsen a gift on behalf of the Academy.[77]

Bohr's next published tribute arose from Knudsen's retirement in 1945 as Secretary of the Danish Academy at the age of seventy-four. Bohr's brief speech was published in the Proceedings of the Academy.[78] When Knudsen died four years later, Bohr wrote three separate obituaries for him – in a

---

[70] N. Bohr, manuscript 8 May 1942, *Inventory* (ref. 19), p. [528], folder 28.

[71] See ref. 78.

[72] N. Bohr, manuscript, *Inventory* (ref. 19), p. [527], folder 21.

[73] See Vol. 11, pp. [352]–[354].

[74] N. Bohr, *Professor Martin Knudsen*, Berlingske Aftenavis, 14 February 1941. Reproduced and translated on pp. [289] ff.

[75] N. Bohr, *Tale ved Festen for Martin Knudsen i Danmarks naturvidenskabelige Samfund den 10. Marts 1941*, manuscript, *Inventory*, (ref. 19), p. [527], folder 23.

[76] N. Bohr, N. Bjerrum, P.O. Pedersen and V. Poulsen, *Til Professor, Dr. phil. Martin Knudsen paa 70-Aars-Dagen den 15. Februar 1941* in *Selskabet for Naturlærens Udbredelse, 1939–1940*, J. Jørgensen, Copenhagen 1941, pp. III–IV; also published in Fys. Tidsskr. **39** (1941) 1–2. This brief publication adds nothing to Bohr's other published writings on Knudsen, and is therefore not included in the Bohr Collected Works.

[77] *Meeting on 8 May 1942*, Overs. Dan. Vidensk. Selsk. Virks. Juni 1941 – Maj 1942, p. 44.

[78] N. Bohr, *Meeting on 19 October 1945*, Overs. Dan. Vidensk. Selsk. Virks. Juni 1945 – Maj 1946, pp. 31–32. Reproduced and translated on pp. [295] ff.

prominent Danish newspaper,[79] in the leading Danish physics journal[80] and in the Proceedings of the Academy.[81]

Bohr's life-long friend and colleague, the Danish experimental physicist Hans Marius Hansen (1886–1956), had a distinguished career at the University of Copenhagen. He was appointed professor of physics in 1923 and, in 1928, director of the university's newly established Biophysical Laboratory. Between 1948 and 1956 he was Vice-Chancellor of the University.

When Bohr returned in 1912 from his first stay in Manchester, he had not yet connected his atomic theory to the empirical formulae for spectral lines, stemming from the field of experimental spectroscopy which thrived in the latter half of the nineteenth century. Bohr later wrote a tribute to the principal representative of this field in Scandinavia, J.R. Rydberg.[82] This connection forged between the spectral lines and Bohr's theory constituted an indispensable element in his path-breaking "trilogy" of scientific papers published from July to December 1913.[83]

As Bohr recalled, Hansen, who had recently returned from a research stay at Göttingen, a major centre for experimental work in atomic spectroscopy, "became a source of inspiration" in this regard. Hansen also assisted Bohr in his considerable teaching load, enabling him to take a leave of absence so that he could accept a temporary lectureship that Rutherford offered him in Manchester. Bohr occupied this post from October 1914 to the early summer of 1916.[84] He published his first appreciation of Hansen in 1946, on the occasion of Hansen's sixtieth birthday.[85] When Hansen died in 1956, Bohr wrote an obituary for him in a newspaper[86] and in the main Danish physics journal.[87]

[79] N. Bohr, *Professor Martin Knudsen Died Yesterday*, Politiken, 28 May 1949. Reproduced and translated on pp. [299] ff.

[80] N. Bohr, *Martin Knudsen 15.2.1871–27.5.1949*, Fys. Tidsskr. **47** (1949) 145–147. Reproduced and translated on pp. [303] ff.

[81] N. Bohr, [Obituary for Martin Knudsen], Overs. Dan. Vidensk. Selsk. Virks. Juni 1949 – Maj 1950, pp. 61–65. Reproduced and translated on pp. [313] ff.

[82] Bohr, *Rydberg's Discovery*, ref. 7.

[83] N. Bohr, *On the Constitution of Atoms and Molecules*, Parts I–III, Phil. Mag. **26** (1913) 1–25, 476–502, 857–875. Reproduced in Vol. 2, pp. [159]–[233].

[84] The Collected Works contain two letters from Bohr to Hansen: Vol. 2, p. [515]–[518] (1 letter) and Vol. 8, pp. [695]–[697] (1).

[85] N. Bohr, *A Personality in Danish Physics*, Politiken, 7 September 1946. Reproduced and translated on pp. [325] ff. The quotation is from this publication.

[86] N. Bohr, *A Shining Example for Us All*, Politiken, 14 June 1956. Reproduced and translated on pp. [331] ff.

[87] N. Bohr, [Obituary for H.M. Hansen], Fys. Tidsskr. **54** (1956) 97. Reproduced and translated on pp. [335] ff. Bohr also spoke in Hansen's memory at memorial occasions at the University

The tradition of experimental spectroscopy continued into the twentieth century, when it became a crucial empirical basis for the new quantum physics. One of the most important contributors to the field was Friedrich Paschen (1865–1947), for whose seventieth birthday Bohr wrote a tribute in 1935.[88] On the theoretical side another German, Arnold Sommerfeld (1868–1951), and his school of physicists in Munich made substantial improvements in Bohr's theory on the basis of new spectroscopical data.[89] In the confusion of the time, Sommerfeld said, half in jest, half in desperation, that Bohr's correspondence principle did away with problems like "a magic wand".[90] Bohr first spoke for Sommerfeld at the latter's visit to Copenhagen in September 1919, noting that he was "the first scientist from a foreign country who had come to us here in Scandinavia in these difficult times to tell us about your scientific results."[91] Bohr's published tribute to Sommerfeld was written for his sixtieth birthday in 1928.[92]

Shortly after his return from Manchester in 1916 Bohr was approached, without prior warning, by a young physicist from Holland who had travelled to Copenhagen in the hope of working with him to complete his doctor's degree.[93] The physicist, Hendrik Anthony Kramers (1894–1952), was a student from the University of Leiden. Bohr made it possible for him to stay in Copenhagen. After fulfilling the requirements for the Leiden doctorate in 1918, Kramers stayed on for another eight years and was Bohr's assistant at the crucial time when the Institute for Theoretical Physics at the University of Copenhagen was inaugurated in 1921. Kramers's dispersion theory, based on Bohr's correspondence principle, gave an important push to Heisenberg's development of quantum

of Copenhagen and the Royal Academy. N. Bohr, *Tale ved Mindehøjtideligheden på Universitetet den 21. juni 1956* and *Noter taget af A. Bohr ved N. Bohr's mindetale over H.M. Hansen i Videnskabernes Selskab*, manuscripts, *Inventory* (ref. 19), p. [531], folder 46.

[88] N. Bohr, *Friedrich Paschen on his Seventieth Birthday*, Naturwiss. **23** (1935) 73. Reproduced and translated on pp. [339] ff.

[89] Correspondence between Bohr and Sommerfeld is reproduced in Vol. 2, pp. [603]–[604] (2 letters), Vol. 3, pp. [686]–[692] (5), Vol. 4, pp. [740]–[742] (3) and Vol. 5, pp. [502]–[507] (2).

[90] Sommerfeld made this remark in the third edition of his seminal *Atombau und Spektrallinien*, Friedr. Vieweg & Sohn, Braunschweig 1922, p. 338.

[91] N. Bohr, *Tale holdt til Sommerfeld efter Foredraget i Fysisk Forening 22–9–1919*, manuscript, *Inventory* (ref. 19), p. [524], folder 2.

[92] N. Bohr, *Sommerfeld and the Theory of the Atom*, Naturwiss. **16** (1928) 1036. Reproduced and translated on pp. [343] ff.

[93] The letter to Bohr of 25 August 1916 in which Kramers introduced himself is reproduced in both Vol. 2, p. [537] and Vol. 3, p. [652]. Subsequent correspondence between Bohr and Kramers is printed in Vol. 3, pp. [652]–[662] (4 letters), Vol. 5, pp. [394]–[397] (1), Vol. 6, pp. [425]–[430] (2), Vol. 7, pp [461]–[462] (1), Vol. 8, pp. [726]–[739] (4) and Vol. 9, pp. [596]–[605] (3).

mechanics in 1925. Kramers took an active part in the subsequent development and refinement at Bohr's institute of this revolutionary theory before leaving Copenhagen in 1926 to take up a professorship in Utrecht. Heisenberg took over his positions as assistant and lecturer at Bohr's institute.

Kramers continued his close relations with Bohr and Denmark. When he died in 1952, Bohr wrote an obituary for him in the Danish press[94] and gave an address at a commemoration gathering at the University of Leiden arranged by the Dutch Physical Society, which was subsequently published in a major Dutch scientific journal.[95] Especially in the Dutch publication, Bohr presented Kramers's early scientific contributions in the context of the physics at the time and described him as a person and a friend. Bohr mentioned Kramers's tenure as the first Chairman of the Scientific Sub-Committee of the UN Atomic Energy Commission and his "untiring endeavours for promoting understanding about the necessity of common control of the formidable powers which through the development of atomic physics had come into the hand of man."

In his Danish obituary for Kramers, Bohr recalls that in the early 1920s

"… a group of younger theoretical physicists from various countries was gathered at the institute, among whom notably Pauli from Austria and Heisenberg from Germany distinguished themselves by their special talent and approach, which in many ways resembled Kramers's abilities."

Bohr's relationship with Heisenberg was special, both personally and scientifically. This comes to expression in particular in Bohr's New Years greeting to his younger colleague towards the end of 1928: "Rarely have I felt myself in more sincere harmony with any other human being, and I still rejoice, when I think back on our walks and discussions". Bohr's published tribute to Heisenberg, on the latter's sixtieth birthday, is printed in Volume 10.[96]

Bohr was very close to Wolfgang Pauli (1900–1958), as evidenced by the obituary reproduced in this volume. Appearing as a foreword to a memorial volume, it refers to Pauli's "extensive correspondence", which is unique indeed.[97] Pauli had a particularly deep interest in Bohr's views on physics and

Bohr to Heisenberg,
[Dec] 28
Danish
Full text, Vol. 6, p. [424]
Translation, Vol. 6, p. [24]

[94] N. Bohr, *On the Death of Henrik Anton Kramers*, Politiken, 27 April 1952. Reproduced and translated on pp. [347] ff.
[95] N. Bohr, *Hendrik Anthony Kramers †*, Ned. T. Natuurk. **18** (1952) 161–166. Reproduced on pp. [353] ff.
[96] Bohr, *Genesis*, ref. 10.
[97] The Collected Works include as many as sixty-four letters to and from Pauli, more than any other of Bohr's colleagues: Vol. 3, pp. [673]–[675] (2 letters), Vol. 5, pp. [408]–[482] (19), Vol. 6, pp. [329], [432]–[456] (14), Vol. 7, pp. [31]–[35], [251]–[252], [268]–[269], [270]–[271], [277]–[279], [463]–[501] (18), Vol. 9, pp. [606]–[608] (1) and Vol. 10, pp. [543]–[575] (10).

Margrethe and Niels Bohr (sitting in the middle) visited Toshio Takamine (leaning over between them) in his home in Japan in 1937. Bohr's son, Hans, is seen to the left of Takamine. Second from the right is Yoshio Nishina.

its epistemological implications, and throughout the years the two remained in close contact on these issues. Just as Bohr in one of his 1932 lectures in Copenhagen had refrained from telling stories about Rutherford, so he now did no more than suggest that "the anecdotes around [Pauli's] personality grew into a veritable legend." Bohr characterized Pauli as "the very conscience of the community of theoretical physicists" and as "a true friend who to many of us appeared as a solid rock in a turbulent sea."[98]

Toshio Takamine (1885–1959) was among the several foreign physicists who made research visits to Bohr's institute at about the time of its establishment. Takamine came in 1921 and 1925, staying altogether for approximately a year, working as an experimental physicist specializing in spectroscopy. In a memorial volume for Takamine, in which most of the contributions are in Japanese, Bohr wrote of him that, being an experimentalist, Takamine was an important part of "our institute designed for the promotion of intimate cooper-

[98] N. Bohr, *Foreword* in *Theoretical Physics in the Twentieth Century: A Memorial Volume to Wolfgang Pauli* (eds. M. Fierz and V. Weisskopf), Interscience, New York 1960, pp. 1–4. Reproduced on pp. [361] ff.

ation between experimentally and theoretically working physicists."[99] Together with H.M. Hansen and another young Danish physicist, Sven Werner (1898–1984), Takamine worked on the experimental effect named after the Dutch physicist Pieter Zeeman, to whom Bohr wrote a tribute published in Volume 10.[100] After visits to several physics research centres in Europe and the United States, Takamine returned to Japan where he continued to work as a physicist.

Albert Einstein and Niels Bohr are often portrayed as the two giants of twentieth century physics. Six years older than Bohr, Einstein was one of the first to recognize the importance of his Danish colleague's atomic theory, and the first nomination that Bohr made for the Nobel Prize was for Einstein. That was in 1920. Later that year the two met personally, first in Berlin and then in Copenhagen. They were both awarded the Nobel Prize for Physics in 1922 – Einstein the one for 1921 and Bohr the one for 1922.

Already in the early 1920s their correspondence shows great mutual admiration and affection.[101] During the latter half of the 1920s they began their debate on the interpretation and philosophical implications of quantum mechanics, which is rightly regarded as momentous in the history of physics. The main venue for the debate was the Solvay Conferences in Brussels in 1927 and 1930.[102] It is this debate between Bohr and Einstein that provides the focus for Bohr's important account of his relationship with Einstein first published in 1949 in a volume on Einstein as a philosopher celebrating his seventieth birthday.[103]

Bohr wrote three additional tributes to Einstein, the first of which was a radio message also celebrating Einstein's seventieth birthday in 1949.[104] It was part of a tribute by UNESCO, and, as Bohr wrote to his colleague Pierre Auger

---

[99] N. Bohr, *Recollections of Professor Takamine* in *Toshio Takamine and Spectroscopy*, Research Institute for Applied Optics, Tokyo 1964, pp. 384–386. Reproduced on pp. [367] ff.

[100] Bohr, *Zeeman Effect*, ref. 8.

[101] Einstein to Bohr, 2 May 1920 and Bohr to Einstein, 24 June 1920, Vol. 3, pp. [634]–[635]; Bohr to Einstein, 11 November 1922 and Einstein to Bohr, 11 January 1923, Vol. 4, pp. [685]–[686]. Subsequent correspondence between Bohr and Einstein is reproduced in Vol. 6, pp. [21]–[24], [418]–[421] (1 letter) and Vol. 7, pp. [281]–[282], [436]–[437] (2).

[102] Bohr has written an account of the development of these conferences. N. Bohr, *The Solvay Meetings and the Development of Quantum Physics* in *La théorie quantique des champs*, Douzième Conseil de physique tenu à l'Université Libre de Bruxelles du 9 au 14 octobre 1961, Interscience Publishers, New York 1962, pp. 13–36. Reproduced in Vol. 10, pp. [429]–[455].

[103] Bohr, *Discussion with Einstein*, ref. 2.

[104] N. Bohr, *The Internationalist*, UNESCO Courier **2** (No. 2, March 1949) 1, 7. Reproduced on pp. [369] ff. In addition to Bohr, Jacques Hadamard (1865–1963), French mathematician, and Arthur Compton (1892–1962), American physicist, greeted Einstein in the thirty-minute radio programme arranged and distributed by UNESCO.

(1899–1993), who coordinated the effort, "I have on the whole confined myself to an account of Einstein's great scientific achievements and their background." A few days later Bohr reported on the collaboration with the Danish national radio, which recorded the English, as well as a Danish, version of Bohr's talk.[105] Typically, he introduced "a few smaller corrections to the language."

Bohr to Auger,
26 Feb 49
English
Full text on p. [494]

Bohr's two last published tributes to Einstein were written in connection with the latter's death six years later. The first appeared in a Danish newspaper,[106] and the second, a translated and slightly rewritten version of the first, in "Scientific American".[107] Although brief and without reference to the philosophical discussion between them, the articles present a broad appreciation of Einstein's revolutionary change in our concepts of space and time as well as of his ability, as Bohr expressed it in the radio talk, to provide "new encouragement in tracing and combating the deeprooted prejudices and complacencies inherent in every national culture."

Bohr to Auger,
3 Mar 49
English
Full text on p. [495]

Bohr wrote obituaries for two physicists who joined the institute some years after it was established. The experimentalist Ebbe Rasmussen (1901–1959) started as a research assistant in 1928, contributing importantly to spectroscopic investigations. He subsequently took on substantial duties in connection with running the institute before accepting a professorship at the Royal Veterinary and Agricultural College in 1942.[108] Fritz Kalckar (1910–1938) was one of Bohr's closest collaborators after Bohr turned the main attention of the institute to nuclear physics in the 1930s.[109] Kalckar died from a cerebral haemmorrhage at the early age of twenty-seven.[110]

Strictly speaking, the prominent mathematician and philosopher Bertrand Russell (1872–1970) can scarcely be termed either a teacher or a colleague of Bohr. Nevertheless, they first met in Rutherford's laboratory in Manchester as early as about 1915 and corresponded in a number of connections.[111] Bohr

---

[105] The Danish version was published in Politiken on 15 March 1949.

[106] N. Bohr, *Obituary*, Børsen, 19 April 1955. Reproduced and translated on pp. [373] ff.

[107] N. Bohr, *Albert Einstein: 1879–1955*, Sci. Am. **192** (1955) 31. Reproduced on pp. [377] ff.

[108] N. Bohr, *Ebbe Kjeld Rasmussen: 12 April 1901 – 9 October 1959*, Fys. Tidsskr. **58** (1960) 1–2. Reproduced and translated on pp. [379] ff.

[109] Kalckar was one of the rare individuals who co-authored a paper with Bohr. N. Bohr and F. Kalckar, *On the Transmutation of Atomic Nuclei by Impact of Material Particles*, Part I: *General Theoretical Remarks*, Mat.–Fys. Medd. Dan. Vidensk. Selsk. **14**, No. 10 (1937). Reproduced in Vol. 9, pp. [223]–[264].

[110] N. Bohr, *Magister Fritz Kalckar*, Politiken, 7 January 1938. Reproduced on pp. [385] ff. Bohr gave a more personal speech at Kalckar's funeral. N. Bohr, manuscript, *Inventory* (ref. 19), p. [526], folder 16.

[111] The BSC includes nine letters from Russell to Bohr and eight from Bohr to Russell, none of which is reproduced in the Collected Works.

declined the invitation to contribute to a book published in honour of Russell's ninetieth birthday,[112] but did agree to write a note for the souvenir programme for a concert at the London Royal Festival Hall on 19 May 1962 in celebration of Russell's birthday the day before.[113] Bohr's brief tribute may serve as a transition to the next section dealing with his articles on contemporaries outside the field of physics.

4. FAMILY AND THE BROADER DANISH MILIEU

Bohr's grandfather on his mother's side, David Baruch Adler (1826–1878), a prominent figure in Danish finance and politics, left plans for developing his large country estate just north of Copenhagen into a children's home. He had worked on his plans in close cooperation with his English-born wife Jenny, née Raphael (1830–1902), and when she died in 1902, she left the realization of the plans to their children. The property was provided as a gift to the municipality of Copenhagen. The first group of children from the municipality moved in in 1908. The Adler family, including Bohr's mother Ellen (1860–1930), was deeply involved in the practical arrangements. Bohr wrote a brief foreword for the pamphlet celebrating the fiftieth anniversary of the children's home in 1958.[114] Neither a tribute nor an obituary, Bohr's brief note is an appreciation of the efforts of his ancestors. Nærumgaard children's home is still in use.

Ellen Bohr's one year older sister Hanna Adler (1859–1947) contributed importantly to realizing the children's home. Hanna was a strong presence in Niels Bohr's childhood and played an important role for him as long as she lived. She was a pioneer in Danish education, establishing the first co-educational school in Copenhagen in 1893 following her return from a study tour of the United States, the leading country in the field; co-education was only introduced as the main principle in Copenhagen municipal schools in 1946. In the publication celebrating the 100th anniversary of her birth, Bohr's foreword paints the life of his aunt with a broad brush.[115]

Bohr's father, Christian Bohr (1855–1911), died in 1911, shortly before his fifty-sixth birthday and only months before Niels defended his doctoral

---

[112] Eventually published as *Bertrand Russell: Philosopher of the Century* (ed. R. Schoenman), Allen & Unwin, London 1967.

[113] N. Bohr, [Tribute to Russell] in *Into the 10th Decade: Tribute to Bertrand Russell*, Malvern Press, London [1962]. Reproduced on pp. [389] ff.

[114] N. Bohr, [Foreword] in J. Lehmann, *Da Nærumgaard blev børnehjem i 1908*, Det Berlingske Bogtrykkeri, Copenhagen 1958, p. i. Reproduced and translated on pp. [393] ff.

[115] N. Bohr, *Foreword* in *Hanna Adler og hendes skole*, Gad, Copenhagen 1959, pp. 7–10. Reproduced and translated on pp. [397] ff.

Nærumgaard, 1905.

dissertation at the University of Copenhagen. In two publications, written more than four decades later and devoted mainly to other matters, Bohr remembered his father's scientific work, reproducing a substantial quotation from one of his writings[116] and referring to the small circle of intellectuals who met for discussion in his father's home before 1900.[117] These writings are reproduced in Volume 10, which also places the articles in context.[118] Bohr was also closely attached to other members of his father's circle and the generations following it, as can be seen from his unpublished and personal tributes to Christian Bohr's assistant and successor Valdemar Henriques (1864–1936) from 1936[119] and to the medical scientist Erik Warburg (1892–1969) from 1962.[120]

---

[116] N. Bohr, *Physical Science and the Problem of Life, 1957* in N. Bohr, *Atomic Physics and Human Knowledge*, John Wiley & Sons, New York 1958, pp. 94–101. Reproduced in Vol. 10, pp. [113]–[123], reference to Christian Bohr on pp. [117]–[118].

[117] Bohr, *Tribute*, ref. 6. The English translation of Bohr's reference to the meetings is in Vol. 10, p. [319].

[118] Vol. 10, pp. XXVIII, [5], [297]–[309].

[119] N. Bohr, *Mindeord om V. Henriques*, manuscript, *Inventory* (ref. 19), p. [525], folder 12.

[120] N. Bohr, [Talk at Warburg's 70th birthday], manuscript, *Inventory* (ref. 19), p. [534], folder 63.

The last of Bohr's published writings devoted to his family is of a quite different order of immediacy than the others, being dedicated to his eldest son Christian Bohr (1916–1934) who died only seventeen years old, weeks after completing his gymnasium education. It added to the tragedy that Christian drowned in an accident while sailing with his father and his father's friends on 2 July 1934. The body was found on the Swedish coast only towards the end of August, and on the twenty-sixth of that month, almost two months after Christian's death, Niels and Margrethe Bohr gathered family and friends in their home for a last farewell. Bohr's moving speech on this occasion was published privately, together with the subsequent announcement on Christian's eighteenth birthday on 25 November 1934 of an award established in his memory by Hanna Adler. The pamphlet was prepared for recipients of the award.[121]

Although not a family member, Kirstine Meyer, née Bjerrum (1861–1941), became a family friend. In 1892 she became one of the first two women, the other being Hanna Adler, to obtain a Danish master's degree in physics and in 1909 the first woman to receive a Danish doctorate in natural science. In 1898 she earned the gold medal of the Royal Danish Academy, just as Bohr would nine years later. Her paper dealt with a physics problem formulated by Christian Christiansen, the physics professor at the University of Copenhagen. Kirstine Meyer went on to teach physics in Danish schools, including thirty years (1900–1930) at the school established by Hanna Adler, and played an important role in establishing the physics curriculum within the Danish school system. She made important contributions to the history of physics, notably on the development of the temperature concept (which constituted the subject for her doctoral work) and the Danish pioneers Ole Rømer and Hans Christian Ørsted.[122]

Kirstine Meyer, who recognized Bohr's talents early on, attended his lectures upon his first return from Manchester in 1912 and supported his efforts to establish a new professorship at the University of Copenhagen in theoretical physics. Bohr in turn appreciated Meyer's work and personality, as evidenced in a letter from her in 1931:

Meyer to Bohr,
25 Oct 31
Danish
Full text on p. [511]
Translation on p. [511]

"You, who are young, have both helped me 'the old one' and have always, together with your wife, shown me great friendliness, and you, the famous

---

[121] N. Bohr, *Obituary for Christian Alfred Bohr: Born 25 November 1916 – Died 2 July 1934*, Private print, 1934. Reproduced and translated on pp. [407] ff.

[122] As part of her publication in 1920 of the *H.C. Ørsted Scientific Papers* at the centennial of Ørsted's discovery of electromagnetism, Meyer published her own book-length biography in English of the Danish scientist: K. Meyer, *Scientific Life and Works of H.C. Ørsted*, Andr. Fred. Høst & Søn, Copenhagen 1920.

one, have always shown interest in and given support to the humble effort that I have been able to offer."

On her seventy-fifth birthday, Bohr recalled their relationship, which had begun "right from my earliest childhood, when I first learnt from Aunt Hanne[123] to look up to you". Four years later, in a speech at the seventieth birthday of the Danish engineer Valdemar Poulsen (1869–1942), Bohr referred to Kirstine Meyer's historical expertise by stating that she had "penetrated into the world of thought of H.C. Ørsted."[124]

Bohr to Meyer, 11 Oct 36 Danish Full text on p. [512] Translation on p. [512]

The first of Bohr's two published tributes to Kirstine Meyer, printed in "Fysisk Tidsskrift" (which Kirstine Meyer had established in 1902), as well as in the press, was intended to celebrate her eightieth birthday, but became in effect an obituary when she died shortly before the occasion.[125] On her birthday she was to have been elected an honorary member of the Society for the Dissemination of Science, and as a replacement a memorial fund was established in her name by the Society. Bohr's announcement of this fund, published in Volume 11 as an illustration of his political and social involvements, constitutes in effect Bohr's second published tribute to Kirstine Meyer.[126]

Vilhelm Thomsen (1842–1927) was a world-renowned philologist at the University of Copenhagen, whom Bohr knew from an early age as one of the four intellectuals gathering in his father's home for discussion after meetings at the Royal Danish Academy. Thomsen served for eighteen years as President of the Academy (1909–1927). At the centennial anniversary of Thomsen's birth Bohr was President, and he used the occasion to say a few words about his predecessor. Since Bohr's contribution can better be described as an illustration of his tenure as President than as a biography of Thomsen, it is reproduced in Volume 11.[127]

The Danish philosopher, Harald Høffding, Thomsen's senior by a year, had also been one of the four "ph"'s (the physiologist Bohr, the philosopher Høffding, the physicist Christiansen and the philologist Thomsen) who retired to the Bohr home after the meetings at the Academy. Høffding was the first person to be chosen by the Academy to inhabit the Carlsberg Honorary Mansion, where

---

[123] Hanna Adler was known to her nephews and nieces as "Aunt Hanne".

[124] N. Bohr, *Festen i Anledning af Dr. Valdemar Poulsens 70-Aars Fødselsdag 23/11 1939*, Inventory (ref. 19), p. [527], folder 22.

[125] N. Bohr, *Kirstine Meyer, n. Bjerrum: 12 October 1861 – 28 September 1941*, Fys. Tidsskr. **39** (1941) 113–115. Reproduced and translated on pp. [425] ff.

[126] N. Bohr, *Memorial Evening for Kirstine Meyer in the Society for Dissemination of Natural Science*, Fys. Tidsskr. **40** (1942) 173–175. Reproduced in Vol. 11, pp. [493]–[496] (Danish original), [497]–[499] (English translation).

[127] Bohr, *Meeting*, ref. 11.

he lived from 1914 until his death in 1931. Bohr moved in with his family the following year. Høffding was Bohr's teacher in the obligatory philosophy course at the university and took an early interest in the philosophical implications of Bohr's physics, which continued throughout his long life. Occasionally the two met for discussion. Bohr wrote three tributes to Høffding, all of them reproduced in Volume 10.[128]

The first meeting of the Royal Academy with Bohr as President took place on 20 October 1939. He succeeded S.P.L. Sørensen (1868–1939), who had died less than a year after taking up the presidency. An outstanding capacity in his field, Sørensen is known for introducing the pH concept. In 1920 he became director of the Carlsberg Laboratory's Chemistry Division. Having been absent from the first meeting after Sørensen's death, Bohr took it as his first task as President to present a brief speech in Sørensen's memory.[129]

Specializing in historical aspects of Danish foreign policy, especially Danish–German relations, Aage Friis (1870–1949) was a scholar with political influence for most of the first half of the twentieth century. Bohr's obituary in the Danish press for Friis concentrates on his leadership, beginning in 1933 after Hitler had come to power in Germany, of the effort to help scientists and other intellectual workers arriving in Denmark as refugees.[130]

Bohr's concentration on this aspect of Friis's career reflects his own personal experience. They had had little contact before 1933, but soon after the Nazi take-over in Germany Bohr, Friis and a few others established the Danish Committee for the Support of Refugee Intellectual Workers. With Friis at the helm, the Committee – one of many similar ones established at about the same time in several countries – worked hard to alleviate the plight of the refugees, some of whom were physicists who spent time at Bohr's institute before moving elsewhere, mostly to Great Britain and the U.S.A.

Bohr's obituary for Friis from 1949 reflects their close cooperation and shared goals. "Much remains to be done", Bohr wrote,

> "... for the solution of the humanitarian task [of] ... acquiring reasonable living conditions for [the refugees] and for their integration into Danish

---

[128] Bohr, *At Harald Høffding's 85th Birthday*; Bohr, *Tribute to the Memory of Harald Høffding*; Bohr, *Harald Høffding's 100th Birthday*, ref. 6. As documented in Vol. 10, pp. [505]–[514], Bohr and Høffding carried on occasional correspondence.

[129] Bohr, *Meeting*, ref. 9. Just as Bohr's memorial words for Thomsen, the obituary for Sørensen is reproduced in Vol. 11, pp. [379]–[381] (Danish original), [382]–[383] (English translation), to illustrate Bohr's tenure as President of the Academy.

[130] N. Bohr, *He Stepped in Where Wrong had been Done: Obituary by Professor Niels Bohr*, Politiken, 7 October 1949. Reproduced and translated on pp. [433] ff.

society. ... That we feel certain that the task will be solved in the right way is due first and foremost to the spirit for which Aage Friis was so warm an advocate and champion."

Whereas Bohr and Friis were brought together by external circumstances relatively late in their lives, Bohr and Niels Janniksen Bjerrum (1879–1958) knew each other from childhood. Bjerrum's father, Jannik Petersen Bjerrum (1851–1920), was a prominent ophthalmologist at the University of Copenhagen and Kirstine Meyer's brother. In his speech at Bjerrum's funeral, Bohr recalled that while having learned early on from Kirstine Meyer and Hanna Adler about the high expectations of his namesake, he developed a close direct relationship with Bjerrum only after the two had completed their university education.[131] The two friends frequently went sailing together in "Chita", a boat that they shared, and both were present when Bohr's son Christian fell overboard. The correspondence between the two after the tragedy is a striking and moving testimony of their friendship.

Bohr to Bjerrum, 29 Dec 34
Danish
Full text on p. [496]
Translation on p. [497]
Bjerrum to Bohr, 2 Jan 35
Danish
Full text on p. [497]
Translation on p. [498]

Bjerrum was a chemist who upon gaining his doctorate from the University of Copenhagen in 1908 made extensive visits to research laboratories abroad. As assistant for Jean Perrin (1870–1942) in Paris in 1910, he applied Max Planck's quantum concept in order to explain the spectra of many-atomic gases three years before Bohr presented his revolutionary theory introducing the quantum concept in order to understand the atom itself. Bjerrum was one of the very few Danes able to discuss such questions with Bohr on the basis of personal research experience. Perhaps his best-known scientific contribution is his improvement in 1909 of the dissociation theory published by the Swedish scientist Svante Arrhenius (1859–1927) a quarter century earlier. Having lost the competition in 1908 for a professorship at the University of Copenhagen to his equally promising previous gymnasium classmate, Johannes Nikolaus Brønsted (1879–1947), Bjerrum accepted in 1914 the professorship of chemistry at the Royal Veterinary and Agricultural College, where he continued to make important contributions to research and where he was director during the difficult times of the German occupation of Denmark.

Bohr's first, and most extensive, tribute to Bjerrum was a fiftieth birthday notice. In this article, Bohr took care to mention the fiftieth birthday of Brønsted which had occurred less than a month before. Only a little earlier Bohr had been instrumental in helping Brønsted obtain his own Institute of Physical Chemistry adjacent to Bohr's institute with means from the Rockefeller-funded International Education Board. In his tribute, Bohr emphasized that the work

---

[131] N. Bohr, manuscript, *Inventory* (ref. 19), p. [532], folder 53.

of both chemists represented the recent merging of physics and chemistry on the basis of the quantum nature of the atom. He noted in particular Bjerrum's early use of the quantum concept and his resulting close contact with Bohr and other physicists.[132] Bohr's two subsequent publications on his friend and colleague are brief. The first is the foreword to an anthology of Bjerrum's scientific articles selected by a committee of his closest colleagues, with Bohr as Chairman, in celebration of Bjerrum's seventieth birthday.[133] The second is an obituary written for the Danish press.[134]

Ole Chievitz (1883–1946) was one of Bohr's closest friends. His father was the anatomy professor Johan Henrik Chievitz (1850–1901), and the family lived for several years in the Academy of Surgery when Christian Bohr and his family also lived there. Ole shared a desk with Niels at school for six years until they both completed their gymnasium education. Later in life Chievitz was part of the group, together with Bohr and Bjerrum, who went on long sailing

Chievitz to Bohr,
20 Apr 24
Danish
Full text on p. [502]
Translation on p. [504]

trips in the summer. As shown in a letter from 1924 reproduced below, the surgeon Chievitz was genuinely interested in the physics represented by Bohr and in how to apply it in the medical field. He is well known for volunteering in Finland as a medical doctor under the auspices of the Red Cross at the end of World War I during the Finnish war of independence with Russia. He served in Finland in a similar capacity during World War II.

Bohr's article is the published version of his moving speech at the memorial ceremony for Chievitz. It appeared promptly in the inter-Scandinavian journal "Ord och Bild" ("Word and Image") established in 1892.[135] Bohr's article on Chievitz reappeared ten years later when some of Chievitz's friends included it in a privately published book.[136]

The orientalist Johannes Pedersen, born the same year as Chievitz, became a close friend of Bohr, as is evidenced by his contribution to a memorial

---

[132] N. Bohr, *Professor Niels Bjerrum 50 Years*, Berlingske Tidende, 9 March 1929. Reproduced and translated on pp. [437] ff.

[133] N. Bohr, [Foreword] in *Niels Bjerrum: Selected Papers, edited by friends and coworkers on the occasion of his 70th birthday the 11th of March 1949*, Munksgaard, Copenhagen 1949, p. 3. Reproduced on pp. [443] ff.

[134] N. Bohr, *His Memory a Source of Courage and Strength*, Politiken, 1 October 1958. Reproduced and translated on pp. [445] ff.

[135] N. Bohr, *Speech at the Memorial Ceremony for Ole Chievitz 31 December 1946*, Ord och Bild **55** (1947) 49–53. Reproduced and translated on pp. [449] ff. In 1956, Bohr accepted the chairmanship of the committee responsible for the publication of "Ord och Bild" in Denmark in order to secure its continuation under difficult economical circumstances. In 1959, the editor of the journal, Johannes Lehmann (1896–1980), would be responsible for the publication commemorating the Nærumgaard children's home, Lehmann, *Da Nærumgaard blev børnehjem*, ref. 114.

[136] N. Bohr, *Ole Chievitz* in *Ole Chievitz*, Nordisk Boghandel, Copenhagen 1956, pp. 7–14.

volume for Bohr[137] and Bohr's tribute to Pedersen reproduced in Volume 10.[138] The two shared broad cultural and philosophical interests. Having previously been Chairman of the Carlsberg Foundation for twenty-two years, Pedersen succeeded Bohr as President of the Royal Danish Academy in 1962.

Kai Ulrik Linderstrøm-Lang spent his entire working career at the Carlsberg Laboratory, where he conducted pioneering research on protein chemistry. He succeeded S.P.L. Sørensen as the laboratory's director in 1938. Bohr spoke at the awarding of the eighth H.C. Ørsted Medal to Linderstrøm-Lang by the Society for the Dissemination of Natural Science, of which Bohr was Chairman. Bohr's speech is more a reflection of this event than a sketch of the recipient of the award. It is therefore reproduced in Volume 11 to illustrate Bohr's public involvement, together with his speeches given at two subsequent awards of the Ørsted Medal to Axel Langseth and (jointly) to Paul Bergsøe and J.A. Christiansen.[139]

Turning to Bohr's tributes to his friends and acquaintances outside the university, the oldest person in this group was Niels Laurids Møller (1859–1941), a Danish author known for his poetry, translations – particularly of classical writings – and his history of world literature. In early 1934 Bohr thanked Møller for sending him his translation into Danish of "On the Sublime" by the Greek writer Pseudo-Longinus of the first century, a work generally regarded as a major early contribution to literary criticism. Less than a year later, Bohr wrote to Møller again, thanking him for "your faithfulness towards Christian's memory". In his greeting on Møller's eightieth birthday, Bohr takes the Danish author and his work to illustrate that "art" and "science", while different, are expressions of common aspirations.[140]

> Bohr to Møller,
> 11 Mar 34
> Danish
> Full text on p. [513]
> Translation on p. [514]
>
> Bohr to Møller,
> 11 Dec 34
> Danish
> Full text on p. [514]
> Translation on p. [515]

Halfdan Hendriksen (1881–1961) was a successful businessman with special relations to Iceland who went on to make a career in Danish politics with a basis in the Conservative Party. Bohr came to know him well only after he became Director of the Carlsberg Breweries in 1946 and obtained living quarters adjacent to the Carlsberg Mansion. Bohr's contribution was printed in a book published on the occasion of Hendriksen's seventy-fifth birthday and is essentially a reprint of the speech he had given at the celebration of Hendriksen's

---

[137] J. Pedersen, *Niels Bohr and the Royal Danish Academy of Sciences and Letters* in *Niels Bohr: His life and work as seen by his friends and colleagues* (ed. S. Rozental), North-Holland, Amsterdam 1967, pp. 266–280.

[138] Bohr, *Physical Science and the Study of Religions*, ref. 12.

[139] Bohr, *Eighth Presentation*, ref. 18; Bohr, *Ninth Presentation*, ref. 17; Bohr, *Tenth and Eleventh Presentation*, ref. 14.

[140] N. Bohr, *Writer and Scientist* in *Festskrift til Niels Møller paa Firsaarsdagen 11. December 1939*, Munksgaard, Copenhagen 1939, pp. 80–81. Reproduced and translated on pp. [461] ff.

seventieth birthday. In his article Bohr emphasized Hendriksen's "undaunted handling" of his duties as Minister of Trade throughout the occupation period and his efforts at Carlsberg to "reestablish the old harmony" after the difficult war years.[141]

Jens Rosenkjær (1883–1976) was Bohr's two years older schoolmate at the Gammelholm Gymnasium. Having obtained a master's degree in chemistry at the University of Copenhagen, he became an active teacher, concentrating especially on the history of science, in the characteristically Danish folk high school system founded by the influential Nikolai Frederik Severin Grundtvig (1783–1872) in the previous century.[142] Rosenkjær was involved in the establishment of a national Danish radio broadcasting system which was founded in 1925, and in 1937 he was appointed Director of the newly established Section for Lectures and Cultural Broadcasts. He retained the position until 1948. Bohr worked closely with Rosenkjær in connection with his several radio lectures during this period, and it is this collaboration that Bohr refers to in his brief contribution to a volume celebrating Rosenkjær's seventieth birthday.[143]

Peter Freuchen (1886–1957) was a Danish explorer, adventurer and author. "Already in our student days", Bohr recalls, "my brother and I received ... a vivid impression of Freuchen's radiant spirit and his multi-faceted talent."[144] Freuchen made his first expedition to East Greenland in 1906, and later joined the explorer Knud Rasmussen on the first two Thule expeditions. After World War II Freuchen lived in the U.S.A. In a letter written at Christmas 1939, Bohr expressed great interest in Freuchen's writing. In his brief obituary, Bohr emphasized "the close life together with the polar Eskimoes and the impression of their unique culture, ... [which] opened those springs in his impressionable mind that gave him the inspiration for his rich authorship."

Bohr wrote his tribute for the Swedish ambassador to Denmark, Carl Fredrik Hugo Hamilton (1890–1977), in connection with the latter's retirement from his post in 1941 after having served in Denmark since 1934. Bohr used the occasion to emphasize how crucial it was in these difficult times "to maintain

Bohr to Freuchen,
[Christmas 39]
Danish
Full text on p. [505]
Translation on p. [505]

[141] N. Bohr, *My Neighbour* in *Halfdan Hendriksen: En dansk Købmand og Politiker*, Aschehoug, Copenhagen 1956, pp. 171–172. Reproduced and translated on pp. [467] ff.

[142] Bohr touches on the role of Grundtvig in his *Danish Culture: Some Introductory Reflections* in *Danmarks Kultur ved Aar 1940*, Det Danske Forlag, Copenhagen 1941–1943, Vol. 1, pp. 9–17. Reproduced in Vol. 10, pp. [251]–[261] (Danish original) and [262]–[272] (English translation).

[143] N. Bohr, *A Fruitful Lifework* in *Noter til en mand: Til Jens Rosenkjærs 70-aars dag* (eds. J. Bomholt and J. Jørgensen), Det Danske Forlag, Copenhagen 1953. Reproduced and translated on pp. [473] ff.

[144] N. Bohr, *Obituary* in *Bogen om Peter Freuchen* (eds. P. Freuchen, I. Freuchen and H. Larsen), Fremad, Copenhagen 1958, p. 180. Reproduced and translated on pp. [477] ff.

At the MGM studios, Hollywood, 1933. From left to right: Charles Christian and Sigrid Lauritsen; Margrethe Bohr; the Danish actor Nils Asther; Niels Bohr; Peter and Magdalena Freuchen.

cultural links between Sweden and Denmark, whose importance is greater at present than perhaps ever before".[145]

Bohr's acquaintance with Meyer Weisgal (1894–1977) arose from Weisgal's close connection with Chaim Weizmann (1874–1952), world-famous chemist and the first President of the state of Israel. As Bohr recalled in his article of 1954 on the rebuilding of Israel, reproduced in Volume 11, he became acquainted with Weizmann during his early years in Manchester, when Weizmann was chemistry professor there.[146] Later on, the younger Weisgal became Weizmann's right-hand man, and although his background was not in science, he played a major role in the inauguration in 1949 of the Weizmann Institute in Rehovot, Israel, of which he subsequently served as the day-to-day leader. Bohr

---

[145] N. Bohr, *Farewell to Sweden's Ambassador in Copenhagen*, Politiken, 15 November 1941. Reproduced and translated on pp. [481] ff.

[146] N. Bohr, *The Rebuilding of Israel: A Remarkable Kind of Adventure*, Israel **7** (No. 2, 1954) 14–17. Reproduced in Vol. 11, pp. [689]–[693] (Danish original) and [694]–[699] (English translation).

followed these developments closely and visited the Weizmann Institute more than once. He submitted his tribute to Weisgal in April 1960. The collection in which it appeared was published in 1966, four years after Bohr's death.[147]

With two exceptions Bohr's published writings about family members, colleagues outside the field of physics and other friends were all directed to Danes. They do not, however, give an exhaustive picture of his manifold connections within Danish society. The persons he wrote about were to some extent chosen by circumstance, such as a special occasion or special request. Especially towards the end of his life, Bohr was much sought after as a contributor to festive publications. Nevertheless, all of the persons about whom Bohr decided to write played an important part in his life. The publications testify to the broadness of Bohr's interests and of his personal network within Denmark.

---

[147] N. Bohr, [Tribute to Weisgal] in *Meyer Weisgal at Seventy* (ed. E. Victor), Weidenfeld and Nicolson, London 1966, pp. 173–174. Reproduced and translated on pp. [485] ff.

Drawing by Harald Jordan in the Danish newspaper "Dagbladet" of Niels Bohr defending his doctoral thesis, May 1911. 1. Niels Bohr; 2. Christian Christiansen, Bohr's dissertation adviser and second opponent; 3. Poul Heegaard, professor of mathematics at the University of Copenhagen and first opponent at the event.

1. AUTOBIOGRAPHICAL

I. [AUTOBIOGRAPHY]

"Festskrift udgivet af Kjøbenhavns Universitet
i Anledning af Universitetets Aarsfest, November 1911",
Schultz, Copenhagen 1911, pp. 76–77

TEXT AND TRANSLATION

See Introduction to Part II, p. [102].

[133]

FESTSKRIFT

UDGIVET AF

KJØBENHAVNS UNIVERSITET

I ANLEDNING AF

UNIVERSITETETS AARSFEST

NOVEMBER 1911

CHRISTIAN GRAM: BASEDOW'S SYGDOM

UNIVERSITETET I REKTORATSAARET 1910—1911

KJØBENHAVN

TRYKT I UNIVERSITETSBOGTRYKKERIET (J. H. SCHULTZ A/S)

1911

VI.

Jeg, Niels Henrik David Bohr, Søn af afdøde Professor, Dr. med. Christian Bohr og Hustru Ellen, født Adler, er født d. 7. Oktober 1885 i København. Jeg blev Student fra Gammelholms Skole 1903 og tog Philosophicum 1904. Fik Videnskabernes Selskabs Guldmedalje 1907 for Besvarelsen af en fysisk Prisopgave om Undersøgelse af Vædskestraalers Svingninger. Tog Magisterkonferens med Fysik som Hovedfag i Efteraaret 1909, samt forsvarede d. 13. Maj 1911 min for Erhvervelsen af den filosofiske Doktorgrad skrevne Afhandling: Studier over Metallernes Elektrontheori. Har af videnskabelige Arbejder, foruden den ovenfor nævnte Doktorafhandling, publiceret:

Determination of the Surface-Tension of Water by the Method of Jet Vibration (Philosophical Transactions of the Royal Society of London, Series A, Vol. 209, p. p. 281—317).

On the Determination of the Tension of a recently formed Water-Surface (Proceding of the Royal Society, A, Vol. 84, 1910).

TRANSLATION

VI.

I, *Niels Henrik David Bohr*, son of Professor Christian Bohr, MD, deceased,[1] and his wife Ellen, née Adler, was born on 7 October 1885 in Copenhagen. In 1903 I passed my matriculation examination[2] from Gammelholm School and took Philosophicum[3] in 1904. I gained the Royal Academy gold medal in 1907 for the response to a physics prize problem about the investigation of water jet vibrations. In the autumn of 1909 I took my master's degree with physics as the main subject,[4] and defended on 13 May 1911 my written thesis "Studies on the Electron Theory of Metals" to gain the doctor of philosophy degree.[5] In addition to the doctoral thesis mentioned above, I have published:

Determination of the Surface-Tension of Water by the Method of Jet Vibration (Philosophical Transactions of the Royal Society of London, Series A, Vol. 209, pp. 281–317).[6]

On the Determination of the Tension of a recently formed Water-Surface (Proceedings of the Royal Society, A, Vol. 84, 1910).[7]

---

[1] Bohr's father had died on 3 February 1911.

[2] Final examination from the gymnasium, preparing for the university.

[3] Obligatory university course on philosophy, in which Bohr followed the lectures of the Danish philosopher Harald Høffding.

[4] N. Bohr, *Account of the Application of the Electron Theory to Explain the Physical Properties of Metals*, Master's thesis, 1909. Translated in Vol. 1, pp. [131]–[161].

[5] Originally published in Danish by V. Thaning & Appel, Copenhagen 1911. Reproduced in Vol. 1, pp. [163]–[290] (Danish original) and [291]–[395] (English translation).

[6] Reproduced in Vol. 1, pp. [25]–[65].

[7] Reproduced in Vol. 1, pp. [79]–[89].

II. [AUTOBIOGRAPHY FOR THE DANISH COLLEGE OF ARMS], 23 NOVEMBER 1920

[SELVBIOGRAFI FOR ORDENSKAPITLET], 23. NOVEMBER 1920

TEXT AND TRANSLATION

See Introduction to Part II, p. [102].

[137]

København, 23. November 1920.

Til

Ordenshistoriografen.

Undertegnede, Niels Henrik David Bohr, Professor i Fysik ved København havns Universitet, der den 18. November d. A. af Hans Majestæt Kongen udnævntes til Ridder af Dannebrogsordenen, er født 7. Oktober 1885 som Søn af Christian Bohr, Professor i Fysiologi ved Universitetet, død 1911 og Hustru Ellen født Adler. Jeg blev Student fra Gammelholms Skole 1903 og begyndte samme Aar at studere Fysik ved Universitetet. 1907 modtog jeg Videnskabernes Selskabs Guldmedalje for en Afhandling om Bestemmelse af Vædskers Overfladespænding. 1909 tog jeg Magisterkonferencen i Fysik og forsvarede 1911 en Afhandling "Studier over Metallernes Elektronteori" for den filosofiske Doctorgrad. Samme Aar rejste jeg til England, hvor jeg først arbejdede paa det fysiske Laboratorium i Cambridge under Sir J.J. Thomsons Ledelse og dernæst i Manchester hos Sir Ernest Rutherford. I Tilknytning til denne Forskers store Opdagelser paabegyndte jeg en Række Arbejder over Atomernes Bygning,[1] om hvilket Emne mine videnskabelige Arbejder i de paafølgende Aar væsentlig har drejet sig. Sommeren 1912 vendte jeg tilbage til Danmark, hvor jeg i August blev gift med Frøken Margrethe Nørlund, Datter af Apotheker Nørlund i Slagelse og Hustru Sofie født Holm. 1913 blev jeg Docent i Fysik ved Københavns Universitet, og 1914 rejste jeg med Orlov fra denne Stilling atter til England for at virke som Docent i teoretisk Fysik ved Universitetet i Manchester og deltage i de videnskabelige Arbejder i det fysiske Laboratorium der. 1916 overtog jeg efter min Hjemkomst fra England den nyoprettede Stilling som Professor i teoretisk Fysik ved Københavns Universitet. Til denne Stilling var der oprindelig ikke knyttet noget Laboratorium, men paa Statens Bekostning og med Bidrag fra privat Side bygges for Tiden et Institut for teoretisk Fysik, der vil blive aabnet til Begyndelsen af Foraarssemestret 1921, og som vil blive Hjemstedet for Undervisningen og videnskabelige Arbejder indenfor denne Del af Videnskaben her i Landet. 1917 blev jeg Medlem af Det Kgl. Danske Videnskabernes Selskab. 1918 udnævntes jeg af Kong Haakon til Ridder af den norske Sct. Olavs Orden efter som tilkaldt

---

[1] See p. [114], ref. 83.

sagkyndig at have deltaget ved Besættelsen af Professoratet i Fysik ved Kristiania Universitet.[2] Samme Aar modtog jeg efter Indstilling fra det matematisk–naturvidenskabelige Fakultet i Kristiania Guldberg Medaljen for videnskabelige Arbejder.[3] I de sidste Aar har jeg foretaget forskellige videnskabelige Rejser, idet jeg efter Indbydelse har holdt Foredrag over Emner vedrørende Atomernes Bygning i Leiden, Berlin og Stockholm.

Ærbødigst
Niels Bohr

[2] Now the University of Oslo, established in 1811. The professor appointed was Lars Vegard (1880–1963), a prominent researcher on northern lights.

[3] Named after the Norwegian mathematician and physicist Cato Maximilian Guldberg (1836–1902) and given for achievements in "the mathematical and physical–chemical sciences". Bohr is the second of only three people who have received the award, the others being the Latvian and German physical chemist Wilhelm Ostwald (1853–1932) in 1908 and the British chemist Cyril Hinshelwood (1897–1967) in 1952.

TRANSLATION

Copenhagen, 23 November 1920.

To

The historiographer of the order of chivalry.

The undersigned, Niels Henrik David Bohr, professor of physics at the University of Copenhagen, upon whom 18 November this year His Majesty the King conferred a knighthood of the Order of the Dannebrog, was born 7 October 1885 as son of Christian Bohr, professor of physiology at the University, deceased 1911, and his wife Ellen, née Adler. I matriculated from Gammelholm School in 1903 and the same year started studying physics at the University. In 1907 I received the gold medal of the Danish Academy of Sciences and Letters for a thesis on the determination of the surface tension of liquids. In 1909 I took the examination for the master's degree in physics and in 1911 I defended a thesis "Studies on the Electron Theory of Metals" for the philosophical doctor's degree. I travelled to England in the same year, where at first I worked at the physical laboratory at Cambridge under the guidance of Sir J.J. Thomson and then at Manchester with Sir Ernest Rutherford. In connection with the great discoveries of this scientist I started on a series of papers on the constitution of atoms,[1] with which topic my scientific articles in the following years have predominantly been concerned. In the summer of 1912 I returned to Denmark, where in August I married Miss Margrethe Nørlund, daughter of Apothecary Nørlund in Slagelse and his wife Sophie, née Holm. In 1913 I was appointed lecturer in physics at the University of Copenhagen, and in 1914, on leave from this position, I again travelled to England to work as lecturer in theoretical physics at the University of Manchester and to take part in the scientific work at the physical laboratory there. In 1916 after my return from England I assumed the newly established position of professor of theoretical physics at the University of Copenhagen. Initially there was no laboratory connected with this position but at the expense of the State and with contributions from private quarters, an institute for theoretical physics is currently being built, which will be opened for the start of the spring term in 1921, and which will be the domicile of teaching and research within this domain of science in this country. In 1917 I became member of the Royal Academy of Sciences and Letters. In 1918 I was conferred a knighthood of the Norwegian St Olav Order by King Haakon of Norway after I had taken part as invited expert in the appointment of the professor of physics at the University of Kristiania.[2] The same year I was awarded the Guldberg Medal for scientific work[3] on the recommendation of the mathematical–scientific faculty at Kristiania. In recent years I have made several lecture tours, as on invitation I have given talks on topics concerning the constitution of atoms in Leiden, Berlin and Stockholm.

Most respectfully
Niels Bohr

III. NIELS BOHR

"Les Prix Nobel en 1921–1922",
P.A. Norstedt, Stockholm 1923, pp. 126–127

See Introduction to Part II, p. [102].

Niels Bohr

was born at Copenhagen on October 7, 1885, as the son of Christian Bohr, Professor of Physiology at Copenhagen University, and of his wife Ellen, née Adler. I matriculated at the Gammelholm Grammar School in 1903, proceeded to the Master's degree in Physics at Copenhagen University tn 1909, and took the Doctor's degree at the same university in 1911. In 1912 I married my wife, Margrethe, whose maiden name was Nørlund. In 1913—14 I held a lectureship in Physics at Copenhagen University and in 1914—16 a similar appointment at the Victoria University in Manchester. In 1916 I was appointed Professor of Theoretical Physics at Copenhagen University, and since 1920 I have been at the head of the Institute for Theoretical Physics then established at that university.

My interest in the study of Physics was awakened while I was still at school, largely owing to the influence of my father, and after matriculating I was fortunate enough to come under the guidance of Professor C. Christiansen, a profoundly original and highly endowed physicist. The announcement by the Academy of Sciences in Copenhagen of a prize to be awarded for the solution of a certain scientific problem, caused me to take up an experimental and theoretical investigation of the surface tension by means of oscillating fluid-jets. This work, which I carried out in my father's laboratory and for which I received the prize offered, was published in the Transactions of the Royal Society 1908. My subsequent studies, however, became more and more theoretical in character, my doctor's disputation being a purely theoretical piece of work on the explanation of the properties of the metals with the aid of the electron theory. In the autumn of 1911 I made a stay at Cambridge, where I profited by following the experimental work going on in the Cavendish Laboratory under Sir J. J. Thomson's guidance, at the same time as I pursued own theoretical studies. In the spring of 1912 I was at work in Professor Rutherford's laboratory in Manchester, where just in those years such an intensive scientific life and activity prevailed as a consequence of that investigator's fundamental inquiries into the radioactive phenomena. Having there carried out a theoretical piece of work on the absorption of α-rays which was published in the «Philosophical Magazine» 1913, I passed on to a study of the structure of atoms on the basis of Rutherford's discovery of the atom-nucleus. By introducing conceptions borrowed from the Quantum Theory as established by Planck, which had gradually come to occupy a prominent position in the science of theoretical physics, I endeavoured to work out and present a picture of atom-structure that might fitly serve as an elucidation of the physical and chemical properties of the elements. I have been engaged ever since with the study of this problem, efforts to develop the formal basis of the Quantum Theory having proceeded hand in hand with work on the development of our conceptions as to the structure of the atoms of the elements. Whereas my studies for several years were of a purely theoretical character, I have now, since the establishment of the Institute for Theoretical Physics at Copenhagen, found conditions for work in which the possibility exists of closely combining theoretical and experimental investigations, and I have also been fortunate enough to find excellent collaborators in both these departments of work.

IV. [AUTOBIOGRAPHY]

"Studenterne MCMIII:
Personalhistoriske Oplysninger",
Berlingske Bogtrykkeri, Copenhagen 1928, p. 275

TEXT AND TRANSLATION

See Introduction to Part II, p. [102].

[147]

STUDENTERNE
MCMIII

PERSONALHISTORISKE OPLYSNINGER

SAMLEDE AF

Anna Sophie Bering-Liisberg

Coelestem adspicit lucem

KØBENHAVN
TRYKT I DET BERLINGSKE BOGTRYKKERI AKTIESELSKAB
1928

[148]

NIELS BOHR
(Niels Henrik David Bohr)

Professor, København

Født i København 7. Oktober 1885, Søn af Christian Bohr og Hustru Ellen, f. Adler.

Udførte i min Studentertid paa det af min Fader ledede fysiologiske Laboratorium og foranlediget ved en Prisopgave, stillet af min afdøde Lærer, Professor i Fysik C. Christiansen, en eksperimentel fysisk Undersøgelse over Vandets Overfladespænding. Efter Afslutningen af mine Universitetsstudier med en Magisterkonferens i Fysik 1909 og Forsvaret af min Doktordisputats om Metallernes Elektronteori 1911, foretog jeg en eetaarig Studierejse til Cambridge og Manchester, hvor jeg arbejdede ved de fysiske Laboratorier, der lededes af Sir J. J. Thomson og Sir Ernest Rutherford. Under Indtrykket af de store Opdagelser vedrørende Atomernes Byggestene, som vi særlig skylder disse Forskere, paabegyndte jeg under denne Rejse de Undersøgelser over Atombygningen, hvormed jeg siden væsentlig har været beskæftiget. Det har navnlig været min Bestræbelse under Hensyntagen til de teoretiske Opdagelser, som vi først og fremmest skylder de tyske Forskere Planck og Einstein, at medvirke til Udviklingen af almindelige Synspunkter, der tillader en sammenfattende Behandling af det nye Erfaringsomraade, som Atomforskningen har aabnet for os. Efter min Hjemkomst blev jeg Docent i Fysik ved Københavns Universitet 1913, men rejste allerede Aaret efter med Orlov fra denne Stilling til Manchester for i to Aar at varetage en Stilling som Docent i teoretisk Fysik ved Universitetet der. 1916 overtog jeg det nyoprettede Professorat i teoretisk Fysik ved Københavns Universitet og har siden 1920 bestyret det til denne Stilling knyttede Institut paa Blegdamsvejen, der er indrettet saavel med teoretiske Undersøgelser for Øje som til Udførelsen af Eksperimentalundersøgelser i Tilknytning til teoretiske Studier.

Gift 1912 med Margrethe.

TRANSLATION

NIELS BOHR
(Niels Henrik David Bohr)

Professor, Copenhagen

Born in Copenhagen on 7 October 1885, son of Christian Bohr and his wife Ellen, n. Adler.

Made an experimental physics investigation of the surface tension of water in my undergraduate years at the physiological laboratory led by my father and prompted by a prize problem posed by my teacher, C. Christiansen, professor of physics, now deceased. After completion of my University studies with a master's degree in physics in 1909 and defence of my doctoral thesis on the electron theory of metals in 1911, I made a one-year study trip to Cambridge and Manchester, where I worked at the physical laboratories led by Sir J.J. Thomson and Sir Ernest Rutherford. Inspired by the great discoveries regarding the building blocks of atoms, which we owe especially to these scientists, I began during this trip the investigations of the structure of the atom with which I have primarily been concerned since. It has in particular been my endeavour, with due regard to the theoretical discoveries that we owe first and foremost to the German scientists Planck and Einstein, to contribute to the development of the general points of view which allow a comprehensive treatment of the new field of experience that atomic research has opened for us. After my return home I became a physics lecturer at the University of Copenhagen in 1913, but left already the year after, with leave of absence from this position, for Manchester to occupy a position as reader in theoretical physics at the university there for two years. In 1916 I took over the newly established professorship in theoretical physics at the University of Copenhagen and have since 1920 been director of the Institute at Blegdamsvej connected with this position. The Institute is furnished with a view to theoretical investigations as well as to carrying out experimental investigations in connection with theoretical studies.

Married to Margrethe in 1912.

V. CONVERSATION WITH NIELS BOHR

SAMTALE MED NIELS BOHR
Berlingske Aftenavis, 2 October 1935

Interview by P. Vinding

TEXT AND TRANSLATION

See Introduction to Part II, p. [102].

[151]

Berlingske Aftenavis 2.—10.—35.

SAMTALE MED NIELS BOHR

Ved Dr. techn. Povl Vinding.

Paa Mandag fylder den berømte Videnskabsmand, Atomforskeren, Professor Niels Bohr 50 Aar. I denne Anledning har Professor Niels Bohr for en enkelt Gangs Skyld indvilliget i at udtale sig, og nedenfor bringer vi et Interview ved Dr. techn. Povl Vinding.

FINDES der noget Sted i København, hvor Fortiden taler mere levende til en end Egnen omkring Søndermarken. Kan man passere Frederiksberg Bakke uden at se for sig den Karosse, som en Februardag i Aaret 1800 førte en fortvivlet Mand, den landsforviste P. A. Heiberg ud i Landflygtigheden, kun fulgt af Kamma Rahbeks Handske. Er Portene til Slotsgaarden aabne, ser man saa ikke for sig Frederik den Sjettes Ekvipager svinge op for Trappen, og kan man færdes paa de stille Veje ved Pilealléen uden at se for sig Knud Lyne Rahbek og Bakkehusets unge Venner som Poul Martin Møller, hos hvem Digtergave og Erkendelsestrang forenedes paa saa harmonisk Maade. Et Stykke nede paa Solbjergets Sydskrænt inde i Hjertet af Gamle Carlsberg ligger, omgivet af den skønneste Park af sjældne Træer, den Hovedbygning, som J. C. Jacobsen fuldførte i 1854 og som er betegnet som den skønneste Privatbolig i Landet fyldt med Kunstskatte, som den var

og er. Her mødtes Hall, Carl Ploug, Japetus Steenstrup, Orla Lehmann, H. N. Clausen, Bille, Professor Rasmus Nielsen og H. C. Andersen, og her dyrkede man Veltalenheden som en Kunst. Ude i den skønne doriske Peristyl bag Vinterhaven staar Statuer af Homér, Perikles og Scipio sammen med Statuer af H. C. Ørsted, Thorvaldsen og andre, som J. C. Jacobsen satte højt, og nu ogsaa af Harald Høffding og af J. C. Jacobsen selv. Stiftelsen af Carlsbergfondet var viet Mindet om H. C. Ørsted.

Men herude taler ikke alene Fortiden, ogsaa Fremtidens stærke Stemme taler sit tydelige Sprog. I J. C. Jacobsens Stuer bor nu Professor Niels Bohr, og her hvor Byens Larm knapt naar ind, afdækkes et Kunstværk uendelig meget større og rigere end noget andet Kunstværk, det nye Verdensbillede, som Niels Bohrs geniale Atomforskning har givet Menneskeheden Mulighed for at skue ind i og som vil gribe ind i hele den menneskelige Erkendelse. Den 7. Oktober fylder Professor Bohr 50 og det vil, ikke mindst i Fremtiden, føles som en Glæde for alle danske, at Sverige allerede i 1922 tildelte ham Nobelprisen og at hans Fædreland har kunnet give ham saadanne Arbejdsvilkaar, at hans videnskabelige Storværk har kunnet udføres her, saaledes at København blev et Verdenscentrum for Atomforskningen.

HVEM vakte Deres Interesse for Naturforskningen, spørger jeg Professor Bohr, der modtager mig i sit Arbejdsværelse, det tidligere „Captainens Værelse", hvor tre store Malerier, et af gamle C. Christiansen, et af Chr. Bohr og et af Harald Høffding straks fanger ens Blik.

— Det er ikke helt let at svare paa, jeg er vokset op i et Hjem med et rigt aandeligt Liv, hvor videnskabelige Diskussioner hørte til Dagens Orden, ja for min Far var der knapt noget skarpt Skel mellem hans egentlige videnskabelige Arbejde og hans levende Interesse for alle Menneskelivets Problemer. Saa har jeg haft den Lykke tidlig at komme i Berøring med en Fysiker som Professor Christiansen og modtage et stærkt Indtryk af hans dybe Forstaaelse af videnskabelig Forsknings prøvende Karakter. Efter sit hele Væsen var jo Christiansen meget ængstelig for at paavirke andre i bestemt Retning, men ved Smaabemærkninger, der afslørede hans Tankerigdom, kunde han fremdrage helt nye Sider af gamle Problemer paa mærkelig inspirerende Maade. Ogsaa Harald Høffding skylder jeg meget. Foruden ved sit store Overblik betog han især ved den Sindets Aabenhed, som han bevarede op i sin høje Alderdom, og som fremfor alt ytrede sig i hans Beredthed til at optage enhver ny Belæring, som maatte fremkomme paa et af Videnskabens Felter, til fordomsfri Bearbejdelse. Christiansen, Høffding, Vilhelm Thomsen og min Far dannede jo en lille Kreds, som holdt regelmæssige Sammenkomster hver fjortende Dag, og efterhaanden som vi Søskende voksede til, fik vi Lov til at lytte til

Professor Niels Bohr i sit Arbejdsværelse i Æresboligen.

Samtalerne, naar Møderne holdtes i vort Hjem. Kredsens Medlemmer, der hver for sig var saa udprægede og dog saa forskelligartede Personligheder, værdsatte i høj Grad hverandre, og ikke mindst mærkede man bag de hyppige Smaafægtninger mellem Christiansen, der følte sig uløseligt knyttet til sit Barndomshjems religiøse Traditioner, og Høffding, hvis filosofiske Udvikling havde ført til et Brud med disse, ikke alene det gamle Venskab, men ogsaa den dybe gensidige Forstaaelse, der forbandt dem. Christiansens lune

Kritik var da ogsaa i første Linie rettet mod enhver Art af overfladisk Filosofering. Om en ikke ringe Del af den filosofiske Litteratur, og ikke mindst den, der tilsyneladende kan se saa objektiv og stringent ud, gælder det jo ogsaa, at den ikke naar ind til de dybere Problemer, der bunder i Udtryksmidlernes Begrænsning. For Spøg kunde man næsten fristes til at sige, at jo mindre man kender Problemernes Vanskeligheder, desto lettere har man ved at skrive almén-filosofisk derom. Det er maaske heller ikke noget Tilfælde, at en Mand

Professor Niels Bohr og Frue i Æresboligens Have.

[154]

Professor Niels Bohr i den doriske Søjle hal. I Baggrunden Har. Høffdings Buste.

som Sokrates, hvis hele Liv, saaledes som Platons Kunst har lært os det at kende, var en aldrig afsluttet Kamp for gennem en Prøvelse af Udtryksmidlernes Bæreevne at harmonisere Indhold og Form, ikke har efterladt sig en skreven Linie.

* * *

KAN den moderne Fysik nogen Sinde komme til at betyde noget for andre end Videnskabsmænd?

— Ja, jeg tror, at den Begrebsanalyse, den har ført til, vil sætte sig almene Spor, omend maaske først i afgørende Grad, naar vor Tids store fysiske Opdagelser inddrages i Undervisningen i Skolerne. Den nye kraftige Paamindelse om alle tilsyneladende nok saa fundamentale Begrebsformers Begrænsning vil da maaske ogsaa paa andre Felter mane til Forsigtighed til ikke at bruge Ordene længere, end de rækker. Saa snart vi kommer udenfor de rent matematiske Formalismers snævre Anvendelsesomraade, vil vi altid være henvist til at bruge et Ordmaleri, hvor Ordene ikke kan være skarpt definerede. De Vanskeligheder, som Licentiaten hos Poul Møller kæmper med, vil vi altid støde paa, saa snart vi bevæger os udenfor Dagliglivets Rammer. Kun ved Begrebernes Almindeliggørelse skabes der Mulighed for at indordne en større Sammenhæng. Det var jo netop Erkendelsen af Relativiteten af et tilsyneladende saa fundamentalt Begreb som Samtidigheden, som muliggjorde det for Einstein ikke alene at afrunde Newtons Værk, men at skabe helt nye

[155]

Udsyn vedrørende Sammenhængen mellem Fænomener, der hidtil syntes uden nogen indbyrdes Forbindelse. Det er interessant at tænke paa, at denne fundamentale Erkendelse ikke er mere end en Menneskealder gammel, og, tilføjer Bohr med et Smil, at de Filosoffer, som saa heftigt angreb Einstein, trods al Lærdom, ikke paa disse Punkter var trængt dybere ned end de primitive Folkeslag, der for første Gang formede Ord som „før" og „efter". Det er særegent for vor Tid og dens videnskabelige Virksomhed, at man er klar over, at hvis ens Forudsætninger ikke stadig haves for Øje og stadig uddybes, saa bliver de til Fordomme, som let skjuler den større Sammenhæng. Dette gælder baade det allermeste abstrakte, vòre Tanker om vor egen Tankevirksomhed, og det allermest konkrete, Iagttagelserne af den livløse Natur. Inden for Fysikken er det jo imidlertid ikke blot Relativitetsteorien, men ogsaa de seneste Berigelser med helt nye Erfaringsomraader inden for Atomteorien, der har afdækket afgørende Brist i det Grundlag, hvorpaa vi troede trygt at kunne stole. Atter her har vi oplevet, at tilsyneladende uovervindelige Modsætninger forsvinder ved den fortsatte Udforskning af Forudsætningerne for vore Anskuelsesformer.

* * *

HVAD vil det sige, at Aarsagsloven ikke gælder i den moderne Fysik?

Ja, det er maaske ikke saa let at forklare i faa Ord, men Forholdet er jo det, at den store Udvidelse af vort Erfaringsomraade havde afsløret de simple mekaniske Forestillingers Utilstrækkelighed og stillet Problemerne i en ny Belysning. Ligesom Relativitetsteorien førte til en Revision af vort Tids- og Rumbegreb, førte Kvanteteorien til en Uddybelse af den gamle Diskussion om Aarsagssætningen. Kvanteteorien fremkom omkring Aarhundredeskiftet som et Udtryk for de mekaniske og elektromagnetiske Teoriers Begrænsning ved Anvendelsen over for Atomfænomener, og Udgangspunktet var Plancks berømte Opdagelse af det saakaldte Virkningskvantum, der i Atomprocesserne indfører et Træk af Diskontinuitet, der var ganske ukendt i den klassiske Fysik. Det har nu vist sig, at Virkningskvantet i stedse stigende Grad er uundværligt for at kunne indordne Erfaringerne angaaende Atomernes Egenskaber, men samtidig er Fysikken i stedse videregaaende Grad tvunget til at give Afkald paa en Beskrivelse af de enkelte Atomers Forhold i Rum og Tid udfra Aarsagssætningen. For at faa Plads til de nye Lovmæssigheder, man her lærer at kende, er man faktisk henvist til en væsentlig statistisk Beskrivelse, hvorefter man for de enkelte Atomprocesser kan tale om „frie Valg" fra Naturens Side mellem forskellige Muligheder, en Udtryksmaade, hvis Hensigt dog kun er at betone, hvor principiel en Forskel fra den klassiske Naturbeskrivelse, det her drejer sig om. Det er undertiden sagt, at Kvanteteorien giver op, idet den renoncerer paa Aarsagsbeskrivelse, men dette giver et misvisende Udtryk for Forholdene; der er ikke Tale om nogen vilkaarlig Opgiven, men tværtimod om en nyvunden Erkendelse af en principiel Begrænsning af Aarsagsbeskrivelsens Anvendelighed, som har ført til nye Landvindinger. For den Ingeniør f. Eks., der har indlevet sig i den klassiske Fysik og oplevet, hvorledes den kan lægges til Grund for Beregningen af en Staalbros Styrke og Svingninger, vil en ny Indstilling maaske falde lidt svært, men Forholdet er dog det, at hvis den klassiske Fysiks Love var gældende for de enkelte Atomer selv, kunde Stofferne, som vi kender dem, slet ikke være til, og der kunde hverken være Staal eller Staalbro.

* * *

ER der Grænser, Naturvidenskaben ikke kan overskride?

I og for sig er vor Evne til at samle og ordne Erfaringer naturligvis uden Grænse, hvis vi blot kan skaffe os egnede Udtryksmidler, de daglig anvendte strækker ikke altid til. I Atomfysikken befinder vi os saaledes i den ejendommelige Situation, at den Forsøgsanordning, som vilde være nødvendig for at anstille de for en sædvanlig Aarsagsbeskrivelse krævede Iagttagelser netop paa Grund af Virkningskvantets Eksistens væsentlig vilde ændre selve de Fænomener, der er Genstand for Undersøgelsen. Saaledes vil f. Sks. ethvert Forsøg paa at følge en Overgangsproces mellem stationære Tilstande i et Atom, altsaa en saadan hvorved enkelte Lyskvanter opstaar eller forsvinder, medføre ukontrollerbare Energiomsætninger mellem Atomet og Maaleinstrumenterne Det er denne Situation, der har givet

Anledning til Indførelse af Begrebet „Komplementaritet", som netop kendetegner den nødvendige Almindeliggørelse af Aarsagssætningen. Noget lignende gælder inden for den biologiske Forskning og i Psykologien. Man kan ikke undersøge sin egen Sindstilstand og Tankeverden uden at ændre den, og man kan ikke undersøge et levende Væsen helt uden at udslukke Livet. Derfor vil Striden mellem Vitalister og Mekanister inden for Biologien ogsaa være ørkesløs; den nye Udvikling af vor Naturerkendelse, saaledes som Atomteorien har ført den med sig, leder til en nærmere Begrundelse af den Indstilling over for Livets Problem, som Biologer intuitivt altid har indtaget, og for hvilken f. Eks. min Far ved forskellige Lejligheder har givet saa rammende Udtryk. Her træffer vi i sine sidste Konsekvenser det Træk, som giver vor Erkendelse sin ejendommelige Karakter, nemlig at vi selv samtidig er Deltagere i og Tilskuere til Tilværelsens store Spil.

Som jeg ved en tidligere Anledning har søgt at udtrykke det, kan man maaske sige, at det ikke er Erkendelsen af vor menneskelige Begrænsning, men Bestræbelserne paa at udforske denne Begrænsnings Natur, der præger vor Tid. Det vilde kun give os en fattig Forestilling om vore Muligheder, om vi vilde sammenligne vor Begrænsning med en uoverstigelig Mur, der paa bestemte Veje er sat for vore Tanker. Overalt aabner sig nye Udblik, og nye Sammenhæng lader sig erkende, men enhver Vej, som vi slaar ind paa, deler sig atter og atter og krummer sig, saa at vi hurtigt taber Retningen og før eller senere kommer tilbage til vort Udgangspunkt. Alligevel kan vi altid vende hjem med Udbytte, og for Rigdommen af det Indhold, som vi kan samle og ordne, ser vi ingen Grænser. Ud fra dybere og dybere udforskede Forudsætninger erkendes større og større Sammenhæng. Saaledes opfattet lever vi under et stadig rigere Indtryk af en evig og uendelig Harmoni; vel at forstaa, Harmonien selv lader sig kun ane, men aldrig gribe; ved ethvert Forsøg derpaa svinder den efter sit Væsens Natur ud af vore Hænder. Intet er fast, hver Tanke, ja selv hvert Ord er kun egnet til at understrege en Sammenhæng, der i sig selv aldrig kan fuldt beskrives, men altid uddybes. Saadan er jo nu engang Vilkaarene for menneskelig Tænken.

*   *   *

PROFESSOR Bohrs blide, velklingende Stemme kommer helt til sin Ret ved Samtale paa Tomandshaand; han gaar frem og tilbage i Værelset og rundt om sit Arbejdsbord, mens han taler; hvert Øjeblik tænder han en Cigar, som kort efter uvægerlig slukkes.

Til Tider virker hans knivskarpe Stringens og hans Omtale af de bundløse Afgrunde, Menneskets Erkendelse ikke kan undgaa at staa overfor, næsten isnende paa Tilhøreren; til andre Tider varmes man op af den Frodighed, hvormed hans Tankerigdom kaster Lys over Proble-mer, som Dagliglivets Mennesker knapt har anet, endsige vovet sig ind i; og man synes at kunne forestille sig den Tankekoncentration, han har sat ind paa Løsningen af tilsyneladende uløselige Opgaver, som Fremstillinger af den moderne Atomteori Gang paa Gang oplyser er løst paa den sindrigeste Maade af Bohr, ofte efter at adskillige af Verdens skarpsindigste Fysikere forgæves har arbejdet derpaa.

*   *   *

KAN de etiske Værdier begrundes ad videnskabelig Vej?

Den egentlige Videnskabs Rolle i Livet er jo nødvendigvis meget beskeden, og Rækkevidden af de videnskabelige Metoder i snævrere Forstand meget begrænset, men derfor ser jeg ikke nogen Anledning til paa noget som helst af Menneskelivets Omraader at gøre et skarpt Skel mellem saakaldte videnskabelige Analyser og enhver anden Bestræbelse paa at finde Sammenhæng og Skønhed. For mig staar det saadan, at enhver saakaldt religiøs Tænker eller Profet har været en Mand, som har anvendt en mere eller mindre intuitiv Indsigt i Menneskelivets Vilkaar paa en Maade, som vel dybest set ikke er forskellig fra Psykologers eller Erkendelsesteoretikeres, saaledes som det maaske kommer klarest frem i Buddha-Skikkelsen. Der er ingen Vej uden om en ærlig Fordybning i Problemerne, og at leve Livet i dets evige Spil mellem Sorger og Glæder; overalt gælder den almindelige menneskelige Belæring, som netop Videnskaben saa ofte skarpt har understreget, at enhver tilsyneladende Disharmoni kan bringes til at forsvinde ud fra et mere omfattende Synspunkt, uden at man dog derved kommer til Bunds og sikrer sig mod nye opdukkende Disharmonier, hvis Fjernelse kræver endnu videre Rammer. Lige saa lidt som Videnskabens Former har vel derfor de etiske Værdier en absolut Karakter, men ligesom Formerne tjener til at fastholde og overskue et vist Erfaringsmateriale, er Paamindelsen om de etiske Værdier, særlig i Opdragelsens Tjeneste, vort Middel til at minde om de Veje, ad hvilke forudgaaende Slægter har fundet, at den størst mulige Harmoni i Livet kan opnaas og om de Sammenhæng, paa hvilke hvert Brud maa smerteligt betales.

I og for sig har jo ogsaa Former og hævdvundne Regler en Værdi i sig selv, idet de saa at sige stiller os overfor en bunden Opgave, til hvis Løsning det ofte er lettere at bringe alle Sjælsevner til Anvendelse i Modsætning til Angreb paa friere Opgaver, hvor ingen Skranker minder os om at staa tilstrækkelig paa Vagt mod hver ensidig Udvikling. Især i Kunsten gælder det jo, at det er Formen, der giver Opgaven. Ved indviklede Versemaal vil en daarlig Digter mulig fordærve Meningen for at faa Rimet frem, men for en Goethe vilde enhver saadan Vanskelighed blot føre til en endnu større Fordybelse i Emnet.

* * *

HVILKEN Vægt tillægger De det internationale Arbejde indenfor Videnskaben?

Den allerstørste. Ikke alene fordi Fremskridtet naturligvis befordres med den voksende Størrelse af den Kreds, der tager aktiv Del i Arbejdet, men ogsaa fordi der indenfor de forskellige Lande, ja endda ved de enkelte Universiteter udvikles en karakteristisk Tradition, der vel medfører en særlig Agtpaagivenhed overfor visse Sider af Problemerne, men let fører til Ensidighed, naar Indstillingen ikke af og til korrigeres ved Berøringen med andre Traditioner. Dette har man maaske særlig Øjnene aabne for, naar man tilhører et lille Land, hvor man ikke saa let som i de store Lande fristes til at være sig selv nok. Iøvrigt har det jo været min egen Skæbne paa et vist Tidspunkt at skulle komme til at virke som et Mellemled mellem den Tradition paa den teoretiske Fysiks Omraade, som opererer med almindelige abstrakte Begrebsdannelser, og som særligt har været udviklet i Tyskland, og den engelske Retning indenfor Eksperimentalfysikken, hvis Ledetraad har været Kravet om at udvikle konkrete Billeder, ogsaa paa Omraader som Atomfysikkens, der falder udenfor Dagliglivets af den umiddelbare Sansning karakteriserede Erfaringer. Naar man selv har været med i en saadan Udvikling og véd lidt om, fra hvor forskellige Sider Kimene til! frugtbart Arbejde stammer, føres man næppe til at overvurdere den personlige Fortjeneste paa Videnskabens Virkefelt, men ser snarere i hvert Fremskridt en Frugt af den gensidige Støtte i Bestræbelserne paa det fælles Maal.

* * *

KOMMER der mange udenlandske Fysikere hertil for at studere hos Dem?

Universitetets Institut for teoretisk Fysik har jo i et ikke helt lille Omfang været Mødested for unge Fysikere, interesseret i Atomfysikkens Problemer, og vi har haft den Lykke blandt disse Gæster at tælle adskillige af de allermest fremragende Repræsentanter paa dette Forskningsomraade blandt den yngre Generation. Jeg kan slet ikke stærkt nok betone, hvad det har betydet for mig selv paa denne Maade at kunne bevare den nøjeste Forbindelse med den Ungdom, der ikke alene stadig fører friskt Mod og Kraft ind i Arbejdet, men som tillige ved egen ny Indsats har bevirket, at meget af det, som for blot faa Aar siden var løse Brudstykker og Forhaabninger om dybere Sammenhæng, nu fremtræder som en hel Bygning, der i Fasthed og Skønhed næppe staar tilbage for noget andet Værk paa Naturvidenskabens Grund. Vi maa herhjemme være meget taknemlig for, at der ved Oprettelsen af Rask-Ørstedfondet er blevet skabt Mulighed for ikke alene at indbyde ældre, højt fortjente Forskere til at berette om deres Resultater i vor Kreds, men tillige til at give unge, lovende udenlandske Videnskabsmænd Lejlighed til for en Tid

at arbejde iblandt os. I Forstaaelse af den Betydning, som det har været for mange unge Mennesker paa denne Maade at kunne uddanne sig og at stifte Bekendtskab med andre videnskabelige Traditioner har jo ogsaa Rockefeller-Fondet i de sidste Aar stillet en stor Del af sine rige Midler til Disposition for dette Formaal herhjemme som andetsteds. Selv kan jeg jo aldrig glemme, hvad det for mig betød efter Afslutningen af mine Studieaar herhjemme at faa Lejlighed til ved et Ophold i England at komme i personlig Berøring med det grundlæggende Arbejde paa den eksperimentelle Atomforsknings Omraade, ledet af de store Fysikere Thomson i Cambridge og Rutherford, der dengang arbejdede i Manchester, og som i de sidste 20 Aar, hvor han har virket som Thomsons Efterfølger i Cambridge gennem sine egne og sine Medarbejderes vidunderlige Opdagelser endnu engang har gjort denne dejlige gamle By med de store Traditioner fra Newtons Tid til et Verdenscentrum for fysisk Forskning.

En af mine største Glæder er det nære Venskabsforhold og den nøje Forbindelse, jeg stadig har bevaret med Rutherford, hvis enestaaende menneskelige Personlighed staar paa Højde med hans Forskergeni.

* * *

FINDER De dybtgaaende Forskelle i de forskellige Nationers Indstillinger til Videnskaben?

Nej, videnskabeligt Arbejde er vel

efter sin Art det Omraade, hvor Forskellen mellem Folkeslagenes Traditioner gør sig mindst gældende og paa mest tilfældig Maade. Medens Traditioner paa mange andre Omraader jo ofte kan bevares ubrudt gennem lange Tidsrum, har saakaldte videnskabelige Skoler indenfor de enkelte Lande som Regel kun en kortvarig Levetid, og efterfølges som oftest af nye Skoler af et ganske andet Præg, betinget af en indre Trang til Fornyelse eller af Paavirkning, udgaaede fra andre videnskabelige Centrer. Jeg tror derfor, at man, bortset fra ganske korte Perioder, ikke med Rette kan tale om en national præget Indstilling til Videnskaben. I det hele tror jeg, at de Synspunkter, man saa hyppigt ser fremsat om den nære Forbindelse mellem Folkenes aandelige Liv og deres racemæssige Særegenheder, ofte er af meget overfladisk og vildledende Karakter. Gennem min Virksomhed har jeg haft Anledning til at komme i Berøring med Videnskabsmænd af de mest forskellige Folkeslag, blandt andet har vi paa Instituttet i Aarenes Løb haft Besøg af et større Antal Japanere. I første Øjeblik kan det jo føles vanskeligt at komme disse paa nært Hold paa Grund af den store Forskel i Omgangsformerne. Deres velkendte Høflighed maa jo for os til at begynde med næsten staa som Udtryk for Uoprigtighed, men samtidig med at man gradvis lærer den dybe Forskel i den traditionelle Indstilling til saa

at sige alle menneskelige Foreteelser nærmere at kende, begynder man ganske langsomt at ane, at der under de fra vore saa forskellige Former kan findes en ikke ringere Menneskelighed og Harmoni i Livsindstillingen, og føres til at indse, at det saa at sige beror paa Tilfældigheder, at Japanerne ikke har en Tradition som vor, og vi ikke en som deres. Naar alt kommer til alt, beror det særegne menneskelige jo netop derpaa, at vi i Modsætning til Dyrene gennem de af Talen betingede Udtryksmidler føres til at uddanne Begreber og Traditioner, der ikke som de egentlig biologiske Artsejendommeligheder, nedarves direkte, men overføres fra Slægt til Slægt gennem Opdragelsen, for hvilken selve Tilegnelsen af Sproget danner Grundlaget. I den Forstand er et Barn jo en Skabning af Naturen, der blot forud for Planter og Dyr besidder den nødvendige organiske Udvikling til at tilegne sig en begrebsmæssig Tradition og derigennem blive et Medmenneske. Jeg tror, at den stærke Fremhæven af de racemæssige Forskelle mellem Folkeslagene og den Betydning, man saa ofte tillægger dem, rummer en Mangel paa Agtpaagivenhed overfor de Farer for Menneskene til at synke nedad mod Dyrets Stade, der altid lurer paa os.

* * *

HAR Naturvidenskaben gode Kaar i Danmark?

Jeg tror, at man er berettiget til at

sige, at der her i Landet er god Jord-
bund for videnskabelig Forskning i
alle dens Afskygninger. Gennem Tiderne har vi jo paa mange af Videnskabens Felter ydet en i Forhold til
vort Lands Størrelse betydelig Indsats, der ofte har været afgørende for
Udviklingen, og Videnskabsmændene
har hos Landets Styrelse i Almindelighed fundet den for deres Arbejde nødvendige Støtte og Forstaaelse. Ved
Carlsbergfondets Oprettelse skabtes
der endvidere nye Muligheder, hvis
Betydning ikke kan overvurderes, for
at opnaa Støtte til Iværksættelse og
Videreførelse af videnskabelige Arbejder af særlig paakrævet og kostbar
Karakter. Takket være dette Fond er
det lykkedes herhjemme paa mange
forskellige Omraader, saavel indenfor
Naturvidenskaben som de humanistiske Videnskaber paa effektiv Maade
at deltage i Løsningen af betydningsfulde Opgaver, som ellers vilde have
været forbeholdt videnskabelige Institutioner i større Lande med deres rigere Midler. Herude, hvor J. C. Jacobsen indrettede sit eget Hjem, fornemmer man overalt den dybe Kær-

lighed til Natur og Kunst, som var den
egentlige Baggrund for hans og hans
Slægts store Gaver til det danske
Samfund. Jeg er taknemlig for at leve
i disse Omgivelser, hvor saa meget,
snart et af de sjældne Træer i Haven,
snart Rummenes Arkitektur og snart
et af Kunstværkerne i Huset fortæller
om alt det Arbejde og alle de Tanker,
som han har nedlagt i Skabelsen af
dette skønne Sted. Her er jo ogsaa
for mig mange Minder om Harald
Høffding, hvis Venskab lige fra min
Ungdom betød saa meget for mig. Et
Foredrag, han holdt om Sokrates ikke
et Aar før han døde, indledte han med
en lille Anekdote, som ogsaa jeg hol-
der af at mindes. Det var en højt anset Sofist, som kom tilbage fra en
Rejse til Lilleasien og traf Sokrates
samtalende paa Gaden, som hans
Skik var. Staar du endnu der, Sokrates, sagde han overlegent, og siger det
samme om det samme. Ja, svarede Sokrates, der som bekendt ikke var særlig anset af sin Samtid, det gør jeg,
men du, som er saa klog, du siger sikkert aldrig det samme om det samme.
                                    Povl Vinding.

[162]

TRANSLATION

CONVERSATION WITH NIELS BOHR
By Povl Vinding, D.Tech.

Niels Bohr, famous scientist and atomic researcher, will be fifty years old on Monday. In this connection Professor Niels Bohr has as an exception agreed to talk, and below we bring an interview with Povl Vinding, D.Tech.

Is there anywhere in Copenhagen where the past talks to you more vividly than in the neighbourhood around Søndermarken.[1] Is it possible to go over Frederiksberg Hill without envisioning the coach that one day in February 1800 took a despairing man, the banished P.A. Heiberg, into exile, accompanied only by Kamma Rahbek's glove.[2] If the gates to the castle courtyard are open do you not see the carriage of Frederik VI[3] swing up in front of the steps and can you wander around the quiet roads near Pile Allé[4] without imagining Knud Lyhne Rahbek and Bakkehuset's young friends such as Poul Martin Møller,[5] in whom the gift of writing and the urge for knowledge were united so harmoniously. A short way down on the south slope of Solbjerget[6] in the heart of Gamle Carlsberg,[7] surrounded by the most beautiful park of rare trees, lies the mansion house,[8] which J.C. Jacobsen completed in 1854 and which is described as the most beautiful private residence in the country, full of

---

[1] Park near the Carlsberg Mansion.

[2] Peter Andreas Heiberg (1758–1841), author and social critic, was exiled from Denmark in 1800 after being sentenced for *lèse-majesté* and criticism of the absolutist state administration. Before his exile, Heiberg greatly appreciated the opportunity of staying with Karen Margrethe ("Kamma") Rahbek (1775–1829) and her husband, the author Knud Lyhne Rahbek (1760–1830), in "Bakkehuset" ("The House on the Hill"), where the couple had moved in 1798 and which then became an informal meeting place for many of the writers, artists and intellectuals of the time. When exiled, Heiberg is said to have taken Kamma's grey silk glove as a memento and kept it for the rest of his life.

[3] (1768–1839). King of Denmark and Norway from 1808 to 1814 and of Denmark until his death.

[4] Main road going past Søndermarken.

[5] (1794–1838), a teacher of Søren Kierkegaard (1813–1855). Møller's unfinished novel, *The Adventures of a Danish Student*, was one of Bohr's favourite books. See Vol. 10, p. XXXI.

[6] Literally, "The Sun Mountain".

[7] The brewery founded by Jacob Christian Jacobsen (1811–1887) in 1847.

[8] Upon Jacobsen's death this house became the Carlsberg Honorary Mansion, where Bohr lived from 1932 to 1962.

art treasures as it was and is. Here Hall,[9] Carl Ploug,[10] Japetus Steenstrup,[11] Orla Lehmann,[12] H.N. Clausen,[13] Bille,[14] Professor Rasmus Nielsen[15] and Hans Christian Andersen[16] met, and here eloquence was cultivated as a fine art. Out in the beautiful Doric peristyle behind the conservatory, statues of Homer, Pericles and Scipio stand together with statues of H.C. Ørsted, Thorvaldsen[17] and others whom J.C. Jacobsen valued greatly, and now too of Harald Høffding[18] and of J.C. Jacobsen himself. The establishment of the Carlsberg Foundation was dedicated to the memory of H.C. Ørsted.[19]

But out here not only the past speaks, the strong voice of the future also speaks its clear language. Now Professor Niels Bohr lives in J.C. Jacobsen's rooms, and here where the noise of the city scarcely penetrates, a work of art infinitely greater and richer than any other work of art is revealed, the new world picture which Niels Bohr's brilliant atomic research has given mankind the opportunity of looking into and which will have repercussions for the whole of human knowledge. On 7 October Professor Bohr will be 50 years old, and it will, not least in the future, be felt with joy by all Danes that Sweden already in 1922 awarded him the Nobel Prize[20] and that his native country has been able to give him such working conditions that his scientific masterwork could be carried out here, so that Copenhagen became a world centre for atomic research.

* * *

Who awoke your interest for natural science? I ask Professor Bohr, who receives me in his office, formerly "The Captain's Room"[21] where three large paintings, one of old C. Christiansen, one of C. Bohr and one of Harald Høffding, catch one's eye immediately.

---

[9] Carl Christian Hall (1812–1888), politician.

[10] Carl Parmo Ploug (1813–1894), author, journalist, politician.

[11] (1813–1894), zoologist.

[12] (1810–1879), politician.

[13] (1793–1877), theologian.

[14] Possibly Carl Steen Andersen Bille (1828–1898), newspaper editor and politician.

[15] (1809–1884), philosopher.

[16] (1805–1875), poet and author.

[17] Bertel Thorvaldsen (1770–1844), sculptor.

[18] Harald Høffding (1843–1931), Danish philosopher, the first resident of the Carlsberg Mansion.

[19] Hans Christian Ørsted (1777–1851), Danish physicist much admired by Bohr.

[20] See Vol. 4, pp. [25]–[30].

[21] The office of the brewer Jacobsen, who was popularly called "The Captain", because he was a captain in the civic guard.

Professor Niels Bohr in his office in the Honorary Mansion.

It is not so easy to give an answer. I grew up in a home with rich intellectual life where scientific discussions were the order of the day; indeed, for my father there was hardly any clear division between his actual scientific work and his lively interest in all problems regarding human life. And then I have had the good fortune at an early age to come into contact with a physicist such as Professor Christiansen and receive a strong impression of his deep understanding of the probing nature of scientific research. According to the whole of his nature Christiansen was of course very wary of influencing others in a definite direction, but with small remarks which revealed the richness of his thoughts he could point out completely new aspects of old problems in a peculiarly inspiring way. I also owe much to Harald Høffding. Besides his breadth of vision, he made an impression in particular with that openness of mind which he kept into his old age, and which above all expressed itself in his readiness to take up for unprejudiced consideration any new lesson which might appear in any field of science. Christiansen, Høffding, Vilhelm

[165]

Thomsen[22] and my father formed a small circle which held regular meetings every fortnight, and gradually, as we children grew up, we were allowed to listen to the conversations when the meetings were held in our home. The members of the circle, who each on their own were so remarkable and yet so different personalities, appreciated each other highly, and not least did one notice behind the frequent minor skirmishes between Christiansen, who felt inextricably tied to the religious traditions of his childhood home, and Høffding, whose philosophical development had led to a break with these, not only the old friendship, but also the deep mutual understanding that bound them together. Christiansen's amiable criticism was thus primarily directed towards any kind of superficial philosophizing. As regards a not inconsiderable part of the philosophical literature – and not least that which to all appearances may seem so objective and cogent – it also holds that it does not reach into the deeper problems which are rooted in the limitation of the means of expression. In jest one could almost be tempted to say that the less one knows about the difficulties of the problems, the easier it is to write about them in general philosophical terms. Thus it is perhaps not a coincidence that a man such as Socrates – whose whole life, as we have learnt to know from Plato's writings, was a never-ending battle to harmonize content and form through a test of the validity of the means of expression – has not left one written line behind.

* * *

Can modern physics ever come to mean anything for other people than scientists?

Yes, I think that the conceptual analysis it has led to will have a general impact, though perhaps only to a decisive extent when the great discoveries in physics in our time are included in education in the schools. The new strong reminder about the limitation of all the apparently quite fundamental conceptual forms will then perhaps also in other fields invite caution not to use words further than they reach. As soon as we come outside the narrow area of application of purely mathematical formalisms, we will always have to take recourse in using a word-painting where the words cannot be strictly defined. As soon as we leave the framework of everyday life, we will always encounter the difficulties with which Poul Møller's licentiate battles.[23] Only

---

[22] (1842–1927). Bohr spoke at a memorial meeting for Thomsen. N. Bohr, *Meeting on 30 January 1942*, Overs. Dan. Vidensk. Selsk. Virks. Juni 1941 – Maj 1942, pp. 32–34. Reproduced in Vol. 11, pp. [392]–[394] (Danish original) and [395]–[396] (English translation).

[23] Bohr refers to Møller, *Adventures*, ref. 5.

Professor Niels Bohr and his wife in the garden of the Honorary Mansion.

by generalization of concepts is it possible to arrange a greater coherence. It was precisely the recognition of the relativity of a seemingly so fundamental concept as simultaneity that made it possible for Einstein not only to round off Newton's edifice, but to create quite new perspectives regarding the coherence between phenomena that hitherto seemed to have no mutual connection. It is interesting to recall that this fundamental recognition is not more than one generation old, and, Bohr adds with a smile, that the philosophers who attacked

Einstein so violently, despite all learnedness had not penetrated more deeply into these points than the primitive peoples who for the first time formed words such as "before" and "after". It is characteristic for our time and its scientific activity that we have become aware that if our presuppositions are not always kept in mind and constantly deepened, then they turn into prejudices which easily conceal the greater coherence. This holds for both the most abstract of all, our thoughts about our own mental activity, and the most concrete of all, the observations of inanimate nature. Within physics it is not, however, the theory of relativity alone, but also the latest enrichments with quite new areas of experience within the atomic theory, that have uncovered decisive flaws in the foundation upon which we thought we could safely trust. Here, once again, we have experienced that seemingly insurmountable inconsistencies disappear with the continued exploration of the presuppositions for our forms of visualization.

* * *

What does it mean that the law of causality does not hold in modern physics?

Well, it is perhaps not so easy to explain in a few words, but the situation is that the great extension of our realm of experience had revealed the insufficiency of the simple mechanical ideas and placed the problems in a new light. Just as the theory of relativity led to a revision of our concepts of time and space, the quantum theory led to a deepening of the old discussion of the principle of causality. The quantum theory appeared around the turn of the century as an expression of the limitation of the theories of mechanics and electromagnetism when applied to atomic phenomena, and the point of departure was Planck's famous discovery of the so-called quantum of action, which introduces a feature of discontinuity into atomic processes, quite unknown in classical physics. It has now turned out that the quantum of action is to an ever increasing degree indispensable for arranging the experience regarding the properties of atoms. At the same time, however, physics is to an ever advanced degree forced to renounce a description of the situation of the individual atoms in space and time according to the principle of causality. In order to find room for the new regularities which one learns about here, one is in fact obliged to use an essentially statistical description, according to which one can speak of "free choice" on the part of nature between various possibilities, a mode of expression whose purpose, however, is only to emphasize how fundamental a difference from the classical description of nature that is involved here. It is sometimes said that the quantum theory acknowledges defeat, in that it renounces the causal description, but this gives a misleading impression of the

Professor Niels Bohr in the hall with Doric columns. In the background the bust of Harald Høffding.

situation; there is no question of any arbitrary renunciation but, on the contrary, of a newly gained recognition of a fundamental limitation of the applicability of the causal description, which has led to new conquests. For the engineer, for example, who is familiar with classical physics and has experienced how it can be used as a basis for the calculation of the strength and vibrations of a steel bridge, a change of attitude will perhaps cause some difficulty, but the situation is nevertheless that, if the laws of classical physics held for the individual atoms themselves, the substances as we know them could not exist at all, and there could neither be steel nor a steel bridge.

* * *

Are there limits science cannot transcend?

In itself, our ability to collect and order experience is of course unlimited, if only we can acquire suitable means of expression; those in daily use do not always suffice. In atomic physics we are thus in the peculiar position that the experimental arrangement that would be necessary to make the observations required for the usual causal description would, precisely because of the existence of the quantum of action, essentially alter the phenomena themselves that are subjected to investigation. Thus, for instance, any attempt to follow a transition process between stationary states in an atom, that is, one whereby the individual light quanta emerge or disappear, will cause uncontrollable energy transfers between the atom and the measuring instruments. It is this situation that has given rise to the introduction of the concept of "complementarity" which is precisely what characterizes the necessary generalization of the causality principle. Something similar holds in biological research and in psychology. One cannot investigate one's own state of mind and world of thought without altering it, and one cannot fully investigate a living organism without extinguishing life. Therefore, the conflict between vitalists and mechanists within biology will be to no avail; the new development of our knowledge of nature, such as the atomic theory has brought, leads to a more detailed justification of the attitude towards the problem of life which biologists have always intuitively held, and for which my father, for example, on various occasions has given such telling expression. Here we meet in its ultimate consequence the feature which gives our knowledge its peculiar character, namely that we ourselves are at the same time actors and spectators in the great drama of existence.

As I have tried to express it on an earlier occasion,[24] it can perhaps be said that it is not the knowledge of our human limitations, but the endeavours to investigate the nature of these limitations, which characterize our time. We would only get a poor idea of our possibilities if we were to compare our limitations with an insurmountable wall blocking our thoughts along certain paths. New vistas are opening everywhere and new connections appear, but each path taken divides again and again and winds, so that we soon lose our way and sooner or later return to our starting point. Nevertheless, we can always return home rewarded, and we see no end to the wealth of content that we can collect and order. From an ever deeper scrutiny of presuppositions, an ever greater interconnection is recognized. We are thus subject to a constantly growing impression of an eternal, infinite harmony; the harmony itself can, of course, only be dimly perceived, never grasped; at any attempt to do so, it slips by its very nature through our fingers. Nothing is constant, each thought, indeed each word uttered, serves only to underline an interconnection that in itself can

---

[24] This paragraph is taken literally from N. Bohr, *Speech Given at the 25th Anniversary Reunion of the Student Graduation Class*, private print, 1953. Reproduced in Vol. 10, pp. [223]–[232] (Danish original) and [233]–[236] (English translation). Quotation on p. [235]. The translation has been slightly revised.

never be fully described, but always deepened. Such are, after all, the conditions for human thinking.

* * *

Professor Bohr's gentle melodious voice does itself full justice in private conversation; he walks to and fro in the room and around his desk as he speaks; he constantly lights a cigar, which invariably goes out shortly after.

At times his razor-sharp stringency and his mention of the bottomless abysses with which human knowledge cannot avoid being faced, have a nearly freezing effect on the listener, while at others you are warmed by the abundance whereby his richness of thought illuminates problems that people in their everyday life have hardly thought of, far less dared to examine, and you seem to be able to imagine the concentration of thought he has put into the solving of seemingly insoluble problems, which the accounts of the modern atomic theory again and again relate have been solved in the most ingenious way by Bohr, often after several of the world's cleverest physicists have worked on them in vain.

* * *

Can ethical values be substantiated by scientific means?

The role of science proper in life is of course necessarily very modest, and the range of scientific methods in a narrow sense is very limited, but still I do not see any reason, in any of the areas of human life whatsoever, to make a sharp divide between so-called scientific analyses and any other endeavour to find coherence and beauty. It seems to me that all so-called religious thinkers or prophets have been men who have applied a more or less intuitive insight into the conditions of human life in a way that fundamentally is probably not different from that of psychologists or epistemologists, as it perhaps appears most clearly in the figure of Buddha. There is no way around an honest absorption in the problems and to live life in its perpetual play between sorrows and joys. The general human lesson, which precisely science so often has sharply emphasized – that any apparent disharmony can be made to disappear on the basis of a more comprehensive point of view – holds true everywhere, although one does not thus get to the bottom of things and ensure oneself against the appearance of new disharmonies whose removal requires even wider frameworks. Just as little as the forms of science have an absolute character, can ethical values therefore be assumed to be absolute, but just as the forms serve to retain and survey a certain set of experimental results, the reminder about ethical values, especially in the service of upbringing, is our means of bringing to mind the paths along which previous generations have found that the greatest possible harmony in life can be achieved, and the interconnections whose every breach must be paid with suffering.

Truly, forms and time-honoured rules have a value of their own in that they so to say face us with a compulsory task for whose solution it is often easier to apply all mental faculties, in contrast to attacking freer problems, where no barriers remind us to be sufficiently on guard against any one-sided development. Especially in art it holds true, of course, that it is the form that defines the task. With complicated metrical forms a poor poet might corrupt the meaning to bring out the rhyme, but for a Goethe any such difficulty would only lead to an even deeper concentration on the topic.

* * *

What importance do you ascribe to international work in science?
The very greatest. Not only because progress is furthered, of course, by the increasing size of the circle that takes active part in the work, but also because inside various countries, indeed even at the individual universities, a characteristic tradition is developed which probably causes special attention being paid to certain aspects of the problems, but easily leads to one-sidedness if the attitude is not now and then adjusted by contact with other traditions. One is perhaps especially on the lookout for this when one belongs to a small country, where it is not so easy as in large countries to be tempted to be self-sufficient. Incidentally, it has been my own fate at one time to come to act as a link between the tradition in the field of theoretical physics, which operates with general abstract conceptualizations and which has been developed in Germany in particular, and the British school within experimental physics, whose guiding line has been the requirement of developing concrete pictures, also in fields such as atomic physics which lie outside the experience of everyday life characterized by direct perception.[25] When one has personally taken part in such a development and knows a little about from how diverse quarters the germs for fruitful work stem, one is hardly likely to overestimate personal merit in the domain of science, but rather sees in every advance a fruit of the mutual support in the endeavours for the common goal.

[25] Bohr refers here to the role of the Institute for Theoretical Physics, established for him at the University of Copenhagen after World War I. He laid out his vision for the institute in his inauguration speech on 3 March 1921, N. Bohr, *Dedication of the Institute for Theoretical Physics*, the manuscript for which is reproduced in Vol. 3, pp. [283]–[292] (Danish original) and [292]–[301] (English translation).

* * *

Do many physicists from abroad come here to study at your institute?

The University Institute for Theoretical Physics has of course to a not altogether small extent been the meeting place for young physicists interested in the problems of atomic physics, and we have had the good fortune of counting among these guests several of the most eminent representatives of the younger generation in this field of research. I cannot at all emphasize strongly enough what it has meant for me personally to be able to maintain in this way the closest contact with the young people who not only continually bring fresh courage and strength into the work, but who in addition, by their own new efforts, have brought about that much of what only a few years ago was loose fragments and hopes for a deeper coherence, now shows itself as a whole structure which in firmness and beauty hardly falls short of any other edifice founded on science. Here at home we should be very grateful that with the establishment of the Rask–Ørsted Foundation[26] an opportunity was created not only for inviting older scholars of great merit to tell of their results in our circle, but in addition for giving young promising scientists from abroad the opportunity to work among us for a time. In awareness of the importance it has had for many young people in this way to be able to educate themselves and to become acquainted with other scientific traditions, also the Rockefeller Foundation has in recent years made a large part of its rich funds available for this purpose, here at home as well as elsewhere. I myself can of course never forget what it meant to me, at the end of my years of study here at home, to have the opportunity through a stay in England to become personally involved with the fundamental work in the field of experimental atomic research led by the great physicists, Thomson in Cambridge and Rutherford, who then worked in Manchester and who during the last 20 years, while he has been active as Thomson's successor in Cambridge, has, through his own and his collaborators' wonderful discoveries, made this lovely old city, with the great traditions from Newton's time, into a world centre for physics research once more.

One of my greatest joys is the intimate friendship and the close links I have continually kept with Rutherford, whose unique human personality is on a level with his genius as scientist.

---

[26] Established in 1919.

* * *

Do you find fundamental differences in the attitudes to science of the various nations?

No, scientific work is probably by its nature the area in which the difference between the traditions of the various peoples is least in evidence and in the most incidental way. While traditions in many other fields can often be preserved uninterrupted through long periods of time, the so-called scientific schools inside the individual countries are usually only short-lived, and are most often followed by new schools of a quite different stamp, contingent upon an inner urge for renewal or upon influence emerging from other scientific centres. Therefore I think that, apart from quite short periods of time, one cannot rightly speak of a nationally marked attitude to science. Altogether, I think that the points of view one so often sees presented about the close link between the intellectual life of peoples and their racial characteristics are often of a very superficial and misleading character. Through my work I have had occasion to come into contact with scientists of the most diverse peoples; among others we have been visited at the Institute by a large number of Japanese over the years. At first it can of course seem difficult to get on close terms with them because of the great difference in manners. Their renowned politeness must initially almost appear to us as an expression of insincerity, but at the same time as one gradually becomes more familiar with the deep difference in the traditional attitude to, so to say, all human behaviour, one begins quite slowly to sense that beneath those manners so different from ours can be found a not less rich humanity and harmony in the attitude to life, and one is led to recognize that it is, so to say, a matter of chance that the Japanese do not have a tradition like ours, and we not one like theirs. All in all, the uniquely human is precisely due to the fact that we in contrast to animals are led, through the means of expression given by speech, to form concepts and traditions which are not, like the essentially biological peculiarities of species, inherited directly, but are transferred from generation to generation by upbringing, for which the very acquisition of language forms the basis. In this light, a child is of course a creation of nature which merely, in superiority to plants and animals, possesses the necessary organic development to be able to acquire a conceptual tradition and through this to become a fellow human being. I think that the great accentuation on the differences between peoples as regards race, and the significance so often attributed to them, reflects a lack of alertness towards the dangers for human beings of sinking down towards the level of animals – dangers that are always lying in wait for us.

* * *

Are there good conditions for science in Denmark?

I think that one is justified in saying that here in this country there is a good basis for scientific research in all its aspects. Throughout the ages we have after all in many fields of science contributed, in relation to the size of the country, a significant effort that has often been decisive for the development, and the scientists have in general found the support and understanding on the part of the national authorities that is necessary for their work. Furthermore, the establishment of the Carlsberg Foundation created new possibilities, whose importance cannot be overestimated, for obtaining support for the initiation and continuation of scientific work of particularly urgent and expensive character. Thanks to this foundation it has been possible in this country – in many different fields, within the sciences as well as the humanities – to take part in an efficient way in the solution of important tasks that otherwise would have been reserved for scientific institutions in larger countries with their richer means. Out here, where J.C. Jacobsen built his own home, one can sense everywhere the profound love of nature and art, which was the real background for his and his family's large gifts to Danish society. I am grateful to live in these surroundings, where so much – now one of the rare trees in the garden, now the interior architecture and now one of the works of art in the house – tells of all the work and all the thoughts he has put into the creation of this beautiful place. Here too of course, there are for me many memories of Harald Høffding, whose friendship right from my youth meant so much to me. He introduced a talk he gave on Socrates less than a year before he died with a small anecdote, which I too like to remember. There was a highly respected sophist who returned from a journey to Asia Minor and met Socrates discussing on the street, as was his wont. "Socrates, are you still standing there", he said superciliously, "saying the same about the same things". "Yes", replied Socrates, who, as is known, was not held in particular respect in his own time, "I am, but you who are so clever, you undoubtedly never say the same about the same things".

Poul Vinding.

VI. NIELS BOHR ON JEST AND EARNESTNESS IN SCIENCE

NIELS BOHR OM SPØG OG ALVOR I VIDENSKABEN
Politiken, 17 April 1949

Interview by M. Bonnesen

TEXT AND TRANSLATION

See Introduction to Part II, p. [102].

Særtryk af interview i Politiken den 17. april 1949.

Niels Bohr om spøg og alvor i videnskaben

I en samtale fremsætter professor Niels Bohr sit syn paa dansk videnskabs traditioner, stilling, arbejdsvilkaar og krav.

[178]

Niels Bohr, fotograferet ved sit skrivebord i arbejdsværelset paa
æresboligen paa Carlsberg.

NAAR den engelske matematiker G. H. Hardy samlede studenter, lærere og gæster i the common room paa Trinity College i Cambridge, plejede han at sige, for at samtalen ikke skulle blive for vidtsvævende:

„Nu vil vi stille direkte spørgsmaal og have korte, klare svar." Vendt mod Niels Bohr tilføjede han:

„Det gælder dog ikke Dem."

Hardy vidste, at Niels Bohrs videnskabelige veje havde ført ham frem til den erkendelse, at sproget ofte skiller der, hvor det burde forbinde. For nylig var Niels Bohr selv i sit radioforedrag til gymnasiasterne inde paa tanken om „de komplementære begreber", de tilsyneladende modsætninger — tanke og følelse, kærlighed og retfærdighed, subjektivitet og objektivitet, begreber, som ikke hver for sig, men bare sammen rummer muligheder for den harmoni, som ogsaa er videnskabens maal.

Der fortælles en lille historie fra Niels Bohrs laboratorium. Man drøftede disse komplementære tanker, og en af Niels Bohrs elever lancerede ordet *sandhed* og bad de andre finde modbegrebet. Der opstod et øjebliks stilhed, idet man kviede sig ved at bruge det nærliggende modsætningsord, som laa paa alles læber. Saa sagde Niels Bohr: *Klarhed.*

Man vil saaledes forstaa, at det korte, det alt for klare, det uforbeholdne, man kunne fristes til at sige det plakatagtige udtryk ikke er Niels Bohrs sag. En samtale med ham bevæger sig i hans arbejdsmetodes store omcirklende buer, før han tøvende, næsten kvalfuldt nærmer sig den kerne, som kan fæstnes til en gangbar sætning i en avis. Derfor kan resultatet selv af daglangt samvær med ham aldrig blive et interview, men et nølende forsøg paa en oversættelse af en tekst i hans tungemaal. Saaledes nøjes man sammen med Niels Bohr med at følge de drivende skyer og kondensere dem efter evne. Her er nogle draaber, samlede ved den ydre foranledning, at Videnskabernes Selskab har formaaet ham til for anden gang at modtage genvalg som Selskabets præsident for en femaarig periode.

Vore traditioner og vor indstilling som kulturnation

Det er paa denne baggrund, at *Niels Bohr*, ellers karrig med udtalelser til en offentlighed uden for videnskabsmændenes kres, har indvilget i en samtale paa saa rummelig basis, at den kan give anledning til at berøre baade aktuelle og almene emner.

Om videnskabens stilling her i landet, der i den seneste tid har indtaget en saa central stilling i offentlig debat, siger Niels Bohr:

— Skulle vi ikke forsøge at se det hele i et lidt større perspektiv. Maaske vil det være frugtbart at trække linjerne lidt bagud og faa fornemmelsen af vor kultur som en enhed og resultatet ogsaa af dem, der kom før os. Jeg tror, det er karakteristisk for os, at vi betragter os kulturmæssigt som et led i bestræbelserne for menneskehedens fremskridt. I den forstand føler vi os som verdensborgere, men samtidig er vi klare over, at vi kun er en del af et større fællesværk.

Jeg tænker altid med særlig glæde paa de rammende ord i H. C. Andersens digt: I Danmark er jeg født, der har jeg hjemme, der har jeg rod, *derfra* min verden gaar. I den ejendommelige indstilling, der ligger bag disse ord, samler sig jo vor skæbne — at vi er os selv bevidst og har kunnet bevare vor selvstændighed, men samtidig befrugtes af hele verden.

— Hvordan ser De paa den diskussion, der har været ført om videnskabens øjeblikkelige kaar her i landet?

En appel til hele befolkningen

— Uden at komme ind paa enkeltheder og specielle tilfælde, som dog ikke paa nyttig maade kan uddybes i et blad, vil jeg gerne sige nogle ting, som forekommer mig væsentlige. Hvad det gælder om er at skabe forhold, saa vore traditioner kan fortsætte paa en maade, der kan blive til opmuntring og nytte for hele befolkningen. Videnskaben griber i dag saa stærkt ind paa de mangfoldigste

omraader, paa teknikens udvikling, paa lægekunstens fremgang, ja videnskaben er ogsaa nødvendig for at opretholde levevilkaarene og derved fremme hele befolkningens velfærd, saadan som vi til tider har haft held til, endda paa forbilledlig maade.

Det drejer sig nu om at skabe saadanne muligheder for den videnskabelige forskning herhjemme, at vi kan bevare og udbygge vor stilling og lægge en saadan grund, at vi ikke stagnerer under en udvikling, som rummer saa store løfter. Det kan ikke stærkt nok fremhæves, at det trods alt er en ganske anden indsats end hidtil, som kræves, hvis vi skal hævde vor stilling mellem nationerne paa videnskabens omraade. En tilfredsstillende og rimelig løsning kan kun naas ved, at alle samfundets medlemmer, der jo til syvende og sidst gennem rigsdag og regering har ansvaret for udviklingen, faar forstaaelsen af, at det er vor pligt som kulturnation at sørge for, at muligheder ikke forspildes. Det er nødvendigt, at stadig flere forstaar, at forskningen maa indtage den plads, som tilkommer den baade ved dens praktiske betydning og ved den værdi, som kundskabssøgen og erkendelsestrang betyder for vor livsudfoldelse.

Statens pligter over for videnskaben

— Er den støtte, som ydes videnskaben, tilstrækkelig til at fæstne og bevare landets videnskabelige indsats?

— Vi har saa mange værdifulde eksempler og traditioner for støtte til dansk videnskab, og det har længe ligget mig paa sinde at sige noget herom. Det er naturligt for mig først at tale lidt om Carlsbergfondet og den tillid, som brygger Jacobsen viste Videnskabernes Selskab, da han stillede hele sin virksomhed til tjeneste for videnskaben og kulturen. Man bør ogsaa erindre den aand, hvori denne gave blev givet — i beundring for H. C. Ørsteds rige og mangesidede virksomhed. Men det er vigtigt ogsaa at huske, at det udtrykkeligt fremhæves i fondets statutter, at midlerne ikke kan benyttes til løsning af opgaver, som med rimelighed paahviler staten. Det er opmuntrende at tænke paa, at det her ikke drejer sig om et isoleret tilfælde af offervilje fra enkelte borgeres side. Ved siden af Carlsbergfondet, hvis betydning har været saa overvejende, findes jo nu bl. a. Otto Mønstedfondet, Laurits Andersenfondet, Michaelsenfondet og Egmont Petersenfondet. De har ikke bare været til støtte for videnskab og teknik, men ogsaa til gavn for den fremadstræbende ungdom og dens uddannelse. Jeg selv har særlig anledning til at nævne det store fond, som Ths. B. Thrige skabte til fremme af teknisk virksomhed, et formaal som er blevet opfattet saa liberalt, at fondet har ydet støtte ogsaa til omraader af videnskaben, der rummer løfter for anvendelse i samfundets tjeneste.

Ingen kan være i tvivl om betydningen af denne offervilje, men jeg vil gerne sige stærkt og klart, at disse bidrag aldrig kan blive andet end supplement til den indsats, staten maa gøre for videnskaben. Det har altid været og maa altid være hele samfundet, som skaber grundbetingelserne for vort videnskabelige liv, og det er nødvendigt i dag at gøre sig klart, at det ikke er tilstrækkeligt at vende tilbage til forholdene før krigen. Videnskaben maa i dag rejse nye krav paa mangfoldige felter, og jeg føler det som en pligt at sige, at vi maa forstaa, at det er afgørende for vor stilling som kulturnation, at disse krav imødekommes.

Arbejdsvilkaarene for den videnskabelige ungdom

— De nævnede ungdommen og dens uddannelse. Hvordan er betingelserne for de unge, der søger ind i videnskabeligt arbejde?

— Unge mennesker har gennem aarhundreder fundet støtte ikke mindst gennem det særlige liv, der har udfoldet sig omkring Regensen og de andre gamle kollegier. Den inspiration, som samværet med jævnaldrende har betydet, har været frugtbar for hele landets kulturliv. Den hastige udvikling i vor tid og den store stigning i de studerendes tal har imidlertid rejst nye krav. Det gælder nu om, at unge, der har vist evner og begejstring for videnskabelig forskning faar saadanne arbejdsvilkaar, at der ikke kan tvivles om, at der herhjemme er baggrund for virkeliggørelsen af deres forhaabninger. Det sørgeligste, som kunne hænde, ville være, at de unge følte sig tvunget til at resignere eller søge arbejdsmuligheder for bestandig andetsteds. For de unges begejstring er jo selve det livsblod, som stadig tilføres samfundet.

— Det siges, at de unge i stigende grad søger netop saadanne arbejdsvilkaar andetsteds.

— Vi bør ikke undervurdere betydningen af, at de unge en tid kommer ud og faar nye impulser Aar i udlandet er uadskilleligt forbundet netop med forholdene i et lille samfund som vort, og der er netop herhjemme en dybt rodfæstet tradition og forstaaelse for betydningen af det internationale samarbejde. Et vidnesbyrd herom er jo statens oprettelse af Rask-Ørstedfondet efter første verdenskrig, og det er interessant, at dette fond blev et forbillede med betydelige konsekvenser. Det var kendskabet til det samarbejde med udlandet, som Rask-Ørstedfondet muliggjorde, der gav stødet til, at Rockefellerfondet i aarene mellem krigene udfoldede den meget store virksomhed, hvorved lovende unge fra saa at sige et hvilket som helst land kunne søge belæring paa steder, hvor vigtige fremskridt fandt sted. Men forudsætningen for at frugtbargøre indhøstede erfaringer i hjemlandet er naturligvis, at de unge ved tilbagekomsten finder tilfredsstillende arbejdsvilkaar.

Samfundet maa øge sine bidrag

— Tænker De her særlig paa den praktiske anvendelse af videnskaben?

— Naturligvis tænker jeg ogsaa herpaa. Jeg haaber, at det paa dette omraade maa lykkes at bringe den rette arbejdsfordeling til veje mellem samfundets forskellige organisationer. Oprettelsen af Akademiet for de tekniske Videnskaber er et vigtigt led heri, og det er glædeligt, at Akademiets energiske virksomhed har faaet statens støtte ved oprettelsen af det tekniske forskningsraad Herigennem kan der bevilges understøttelser til opgaver af særlig aktuel karakter Men der er dog her kun tale om en første begyndelse. Skal vi tage del i den tekniske udvikling, som vore traditio-

ner og vor sagkundskab berettiger os til, maa de midler, samfundet anvender, øges i stadig stigende grad

Indstillingen i Videnskabernes Selskab

— Hvordan stiller Videnskabernes Selskab sig til spørgsmaalet om videnskabens praktiske anvendelse?

— Det er netop med henblik paa et saadant spørgsmaal, at jeg talte om naturlig arbejdsfordeling. I begyndelsen af selskabets 200aarige tilværelse var det ofte det naturlige forum for diskussion af spørgsmaal af praktisk natur. Men udviklingen har her som andetsteds ændret dette forhold. Blandt Selskabets medlemmer findes jo baade repræsentanter for de humanistiske videnskaber og for de naturvidenskabelige, og denne sammensætning har skabt en særlig tradition for forbindelsen mellem forskere fra de mest forskellige vidensomraader. Ved sin understregning af enhed i al kundskabsøgen, er der her skabt en tradition, der gennem tiderne har sat dybt præg paa dansk aandsliv. Selskabets opgave er for øjeblikket — foruden medlemmernes gensidige belæring og en udstrakt udgivervirksomhed — især at varetage den internationale forbindelse med tilsvarende selskaber verden over. Ved at sende repræsentanter til fælles møder deltager selskabet i organiseringen af forskningen paa de mange omraader, hvor disse forbindelser er af afgørende betydning.

Men denne arbejdsfordeling betyder ingenlunde, at der ikke blandt selskabets medlemmer nu som tidligere er mange af dem, som mest energisk arbejder for bedring af videnskabens kaar og mulighederne for dens anvendelse her i landet. Jeg be-

[183]

høver kun at minde om, at det var H. C. Ørsted, som i mange aar var Selskabets sekretær, der i sin tid tog initiativet til oprettelsen af polyteknisk læreanstalt, og det var en anden af Selskabets højt fortjente sekretærer, Chr. Abildgaard, som tog saa virksomt del i oprettelsen af Landbohøjskolen. Det var ogsaa et af Selskabets medlemmer, den som fysiker og tekniker lige fremragende P. O. Pedersen, der var sjælen i oprettelsen af Akademiet for de tekniske Videnskaber, og blev dets første præsident. I det hele taget kan det siges, at mange af selskabets medlemmer gennem deres tilknytning til universitetet eller andre institutioner f. eks. som ledere af laboratorierne har ansvar for at skaffe midler til undersøgelserne og til arbejdsvilkaar for medarbejderne. Man kan trygt sige, at den tid og de kræfter, der maa anvendes hertil ofte kan blive lige saa stor, som den der bliver tilbage til egen forskning.

Saadanne bestræbelser kan ofte rumme stor opmuntring ikke alene ved det liv, der kan skabes, men ogsaa ved den forstaaelse, man trods alle vanskeligheder gang paa gang møder. Selv kan jeg daarligt undlade, naar vi snakker herom, at give udtryk for taknemmelighed over, at man fra statsmyndighedernes side har vist forstaaelse for nødvendigheden af, at vi følger med i et omraade af videnskaben, hvor en saa rivende og løfterig udvikling finder sted som i atomfysiken.

Naar jeg har sagt dette, vil jeg dog gerne tilføje, at der aldrig kan blive tale om, at vi i vort lille samfund kan komme til at raade over saa mægtige hjælpemidler til atomfysisk forskning, som f. eks Amerika besidder. Hvad vi kan gøre er at sørge for, at vi ikke lades helt tilbage og navnlig skaber betingelser for at uddanne fysikere og teknikere, der kan løse de samfundsopgaver, som i nær fremtid ogsaa vil melde sig for os.

To slags sandheder og mange slags tvivl

— De nævnte før kundskabssøgen og meddelelsestrang. Kan De i den forbindelse sige noget om det bidrag til erkendelsen, som udviklingen af den fysiske videnskab har givet?

— Det er ikke let at skelne mellem forskellige omraader af forskningen Fælles for videnskabelig bestræbelse er det at udvide grænserne for vor viden og udvikle de synspunkter, hvorunder den lader sig sammenfatte Derfor drejer det sig altid om noget. der er fælles for alle. Det hænder imidlertid, at vi fra tid til anden paa

et eller andet felt, hvor vi stilles over for helt nye erfaringer, paa overraskende maade bliver belært om de almindelige forudsætninger for al kundskabssammenfatning, ja vi kunne maaske endda sige for selve den menneskelige erkendelse.

I den sidste menneskealder har vi netop ved udforskningen af atomernes verden mødt en saadan situation. Vi er stødt paa overraskelser, hvor erfaringerne i første øjeblik endda kunne synes at modsige hinanden, fordi deres forklaring synes at kræve indbyrdes modstridende billeder. Men ved en nøjere undersøgelse har det vist sig, at der ved anvendelse af saadanne billeder aldrig var tale om fænomener, der iagttoges under de samme forsøgsbetingelser. Billederne tog tværtimod sigte paa fænomener, der fremkom under forskellige forsøgsbetingelser, som — og dette er det afgørende — gensidig udelukkede hverandre Det hænger sammen med, at vi ved atomfænomener er afskaaret fra at skelne imellem, hvad vi kunne kalde de atomare objekters selvstændige opførsel og objekternes vekselvirkning med de maaleinstrumenter, der indgaar som nødvendige led i den forsøgsopstilling, hvorunder iagttagelserne fremkommer. Fænomenerne har derfor en ganske anden udelelighed end de sædvanlige fysiske fænomener, hvis forløb kan følges skridt for skridt paa den maade, vi kender fra dagliglivets erfaringer

Gennem fortvivlelse til spøg

Ja, jeg forstaar, at dette maa lyde underligt, men det har ogsaa kostet stor møje, før man blev klar over de ejendommelige forhold, vi her møder Fysikerne er jo kun almindelige mennesker, og ofte har vi i fortvivlelse maattet søge tilflugt og forfriskelse i spøg. Jeg tænker her paa den gamle historie, som vi gang paa gang har haft lejlighed til at mindes — om de to slags sandheder. Den ene slags er de udsagn, der er saa klare og simple, at det modsatte aabenbart ville være vrøvl. Den slags sandheder kaldt vi spøgefuldt for trivialiteter. Spiritualiteter derimod, de saakaldte „dybe sandheder", er udsagn, hvoraf det omvendte ogsaa er

en dyb sandhed Paa et nyt erfaringsomraade er udviklingen ofte den, at vi skridt for skridt fra kaos nærmer os orden. Men de mellemliggende stadier, hvor man tager sin tilflugt til dybe sandheder, er ofte de mest spændende, fordi indbildningskraften stadig opmuntres til at søge et fastere greb I atomfysiken har man netop gennem erkendelsen af, at en forklaring ved hjælp af billeder møder en absolut begrænsning efterhaanden opnaaet orden og udviklet metoder, hvorved vidtstrakte erfaringsomraader føjer sig harmonisk sammen.

Sandhed er noget man kan tvivle om

— De taler om to slags sandheder. Kan man overhovedet sige, hvad sandhed er?

— I første omgang kunne man i alt fald fristes til at sige, at sandhed er noget, man kan forsøge at tvivle om. Ved store anstrengelser kan man saa efterhaanden naa til erkendelse af, at i det mindste noget af ens tvivl var uberettiget.

— Hvad mente De med at stille sandhed og klarhed over for hinanden som modsætninger?

— Ja, saadan en talemaade stammer netop fra de spøgefulde bemærkninger, som vi en overgang trøstede os med. Det er naturligvis i og for sig ikke meningen at stille klarhed og sandhed op imod hinanden. Det var blot hensigten at understrege, at man ofte ved klarhed forstaar anvendelsen af tilvante forklaringsmaader; men de kan komme til kort, naar det drejer sig om at vinde plads for nye erfaringer, som slet ikke kan passe ind i de gamle former.

— Da De forleden dag i radioen til gymnasiasterne kom ind paa spørgsmaalet om forholdet mellem brugen af ord som retfærdighed og kærlighed, skulle saa ogsaa det optattes paa en saadan „spøgefuld" baggrund?

— Nej, det var ment helt anderledes og som rammeste alvor. Hvad jeg sagde, tog sigte paa den belæring, vi har vundet inden for atomfysiken om umuligheden af at gøre rede for erfaringerne ved hjælp af et enkelt anskueligt billede. Det forhold, vi her træffer, genfinder man — naar man først har faaet øjnene op derfor — paa saa at sige alle livets omraader. Det var derved, jeg kom ind paa at tale om, hvordan brugen af ord som tanker og følelser, der er lige uundværlige i de forhold, vi hver dag møder, strengt taget tager sigte paa situationer, som udelukker hverandre. I forhold til vore medmennesker gælder jo det samme om ordene retfærdighed og kærlighed. I det hele taget møder vi overalt det, vi i atomfysiken har kaldt *det komplementære forhold*. Det minder os paa den mest slaaende maade om den gamle visdom, at vi ved enhver bestræbelse paa at finde harmoni i tilværelsen aldrig maa glemme, at vi selv er baade skuespillere og tilskuere i livets drama.

Maaske er det et forfængeligt haab, men jeg selv stoler paa, at det vil blive af gennemgribende betydning for den fremtidige indstilling til kulturens problemer og for opdragelsen af de nye slægter, at vi inden for et erfaringsomraade, der ligger saa langt fra lidenskabernes virkefelt, saa eftertrykkeligt er blevet mindet om forudsætningerne for menneskelig erkendelse og om en livsvisdom, der ellers kunne synes at ligge naturvidenskaben fjern.

Merete Bonnesen.

TRANSLATION

Niels Bohr on jest and earnestness in science

In a conversation, Professor Bohr presents his views concerning the traditions, position, working conditions and demands of Danish science.

Niels Bohr, photographed at his desk in the office at the Honorary Mansion at Carlsberg.

When G.H. Hardy,[1] the English mathematician, gathered students, dons and guests in the common room in Trinity College at Cambridge, he used to say, in order to prevent the discussion from becoming too nebulous:

"Now we will ask direct questions and have short precise answers."

Turning to Niels Bohr, he added:

"This, however, does not apply to you."

Hardy knew that Niels Bohr's scientific paths had led him to recognize that language often separates there, where it should bind together. Recently Niels Bohr himself in his radio broadcast to gymnasium students[2] touched on the idea of "complementary concepts", the seemingly contradictory concepts of thought and emotion, love and justice, subjectivity and objectivity, concepts which do not each on their own, but only together, hold possibilities for the harmony which is also the goal of science.

There is a little story from Niels Bohr's laboratory. These complementary ideas were being discussed, and one of Bohr's pupils said the word *truth* and asked the others to find the opposite. There was a moment's silence, in that one was reluctant to use the obvious antonym which was on everyone's lips. Then Niels Bohr said: *clarity*.

You will thus grasp that the brief, the much too clear, the unqualified, you might be tempted to say the slogan-like, expression is not a matter for Niels Bohr. A conversation with him moves in the large circling arcs of his working method, until hesitatingly, almost agonizingly, he approaches the core that can assume definite form as a viable sentence in a newspaper. Therefore the result of even a whole day in his company can never be an interview, but a hesitating attempt at a translation of a text in his tongue. Thus together with Niels Bohr you are content to follow the drifting clouds and to condense them according to your ability. Here are some drops collected in connection with the Danish Academy having been able to persuade him to accept for the second time re-election as President of the Academy for a five-year period.[3]

Our traditions and our attitude as a civilized nation

It is on this background that *Niels Bohr*, otherwise sparing of statements to a public outside the circle of scientists, has agreed to a conversation on so broad

---

[1] Godfrey Harold Hardy (1877–1947).

[2] N. Bohr, *Atoms and Human Knowledge*, Berlingske Tidende, 2 April 1949. Reproduced and translated on pp. [57] ff.

[3] On Bohr's tenure as President of the Royal Danish Academy, see Vol. 11, pp. [338]–[348].

a basis as to give the opportunity of touching upon both current and general topics.

As regards the position of science in this country, which has recently assumed such a central role in public debate,[4] Niels Bohr says:

– Should we not try to see it all in a somewhat broader perspective. Perhaps it would be fruitful to look back a little and get a sense of our culture as a whole and as the result also of those who were here before us. I think it is characteristic for us to consider ourselves culturally as a link in the endeavours for the progress of humanity. In this sense we feel that we are world citizens, but at the same time we realize that we are only a part of a great common achievement.

I always think with special pleasure of the telling words in Hans Christian Andersen's poem: "In Denmark I was born, there is my home, there I belong, *from there* my world unfolds".[5] In the special attitude lying behind these words, our fate converges – that we are conscious of ourselves and have been able to keep our independence, while at the same time being inspired by the whole world.

– What is your view of the discussion that has been conducted about the present conditions for science in our country?

An appeal to the whole population

– Without discussing details or special cases, which cannot be usefully elaborated in a newspaper anyway, I should like to say a few things which to me seem essential. The important thing is to create conditions so that our traditions can continue in a way which can be of encouragement and benefit for the whole population. Science today has such a strong influence on the most diverse fields, on technological development, on the advance of medicine; indeed, science is also necessary to maintain living conditions and thus to further the welfare of the whole population, just as we have at times had success in doing, even in an exemplary way.

---

[4] Bonnesen refers in particular to the recent debate leading to the establishment of a Danish National Science Foundation. See Vol. 11, pp. [348]–[352], where the present interview is quoted on p. [349].

[5] This was a favourite quotation of Bohr's, which he also used, for example, in his seminal introduction to an encyclopedic work on Danish culture published during World War II. N. Bohr, *Danish Culture: Some Introductory Reflections* in *Danmarks Kultur ved Aar 1940*, Det Danske Forlag, Copenhagen 1941–1943, Vol. 1, pp. 9–17. Reproduced in Vol. 10, pp. [251]–[261] (Danish original) and [262]–[272] (English translation). The quotation from Andersen is on p. [265].

The point now is to create such opportunities for scientific research here at home that we can keep and develop our position and lay such a basis that we do not stagnate during a development holding such great promises. It cannot be emphasized strongly enough that after all a quite different effort than previously is required if we are going to uphold our position among nations in the field of science. A satisfactory and reasonable solution can only be reached if all members of society, who through the parliament and the government are ultimately responsible for the development, come to realize that it is our duty as a civilized nation to ensure that opportunities are not wasted. It is necessary that more and more people realize that research must assume the place it deserves both through its practical significance and through the value that the search for knowledge and the urge for understanding have for living our lives.

The duties of the State in relation to science

– Is the support provided to science sufficient to strengthen and maintain the scientific activity of the country?

– We have so many valuable examples and traditions for support of Danish science, and I have long wanted to say something about this. It is natural for me first to say a few words about the Carlsberg Foundation and the confidence that the brewer Jacobsen showed the Royal Danish Academy of Sciences and Letters when he placed his whole enterprise at the service of science and culture.[6] The spirit in which this gift was made must also be remembered – in admiration for H.C. Ørsted's rich and many-sided activity. But it is also important to recall that it is specifically emphasized in the statutes of the foundation that the funds cannot be used to solve problems which are reasonably the responsibility of the State. It is encouraging to consider that it is not here a question of an isolated case of selflessness on the part of individual citizens. Besides the Carlsberg Foundation, whose significance has been so prominent, there are now, among others, the Otto Mønsted Foundation, the Laurits Andersen Foundation, the Michaelsen Foundation and the Egmont Petersen Foundation. Not only have they supported science and technology, but they have also benefited ambitious young people and their education. I personally have special reason to mention the large foundation created by Thomas B. Thrige for the advancement of technological activity, a purpose which has been interpreted so liberally that the foundation has also supported fields of science holding promises for application in the service of society.[7]

---

[6] See Vol. 11, p. [339].
[7] See Vol. 11, p. [359].

Nobody can be in doubt about the importance of this philanthropy, but I should like to say loud and clear that these contributions can never be more than a supplement to the effort that the State must make for science. It has always been, and must always be, society as a whole that creates the basic conditions for our scientific life, and today it is necessary to realize that it is not sufficient to return to the situation as it was before the war. Today science must make new demands in many different fields, and I feel that it is my duty to say that we must understand that it is decisive for our position as a civilized nation that these demands are met.

Working conditions for the young generation of scientists

– You mentioned young people and their education. What are the conditions like for young people who want to work in science?

– Through centuries, young people have found support not least in the special life that was lived around "Regensen" and other old halls of residence.[8] The inspiration arising from the company of one's contemporaries has been fruitful for the cultural life of the whole country. However, the rapid development in our times and the large increase in the number of students have given rise to new demands. What matters now is that young people who have shown talent and enthusiasm for scientific research will get such working conditions that there can be no doubt that here at home there is a basis for the realization of their hopes. The saddest thing that could happen would be that young people felt forced to give up or to seek opportunities for work elsewhere. For the enthusiasm of young people is the lifeblood continually supplied to society.

– It is said that young people to an increasing extent seek just such working conditions elsewhere.

– We should not underestimate the importance of young people going abroad for a time and receiving new impulses. Spending some years abroad is inseparably connected precisely with the situation in a small society such as ours, and precisely here at home there is a deep-rooted tradition for and understanding of the importance of international cooperation. The establishment of the Rask–Ørsted Foundation by the State after the First World War gives testimony of this, and it is interesting that this foundation became an example with important consequences. It was the knowledge of cooperation with foreign countries made possible by the Rask–Ørsted Foundation that gave inspiration for the very great

---

[8] The student quarters of "Regensen", located in the centre of Copenhagen, have their origins in the beginning of the seventeenth century.

activity carried out by the Rockefeller Foundation during the years between the wars,[9] whereby promising young people from more or less any country could seek knowledge in places where important progress was taking place. But the precondition for making the harvested experience fruitful at home is, of course, that the young people find satisfactory working conditions when they return.

Society must increase its contributions

– Do you here think mainly of the practical application of science?

– Of course, I also think of this. I hope that in this area we shall succeed in achieving the proper division of labour between the various organizations in society. The establishment of the Academy of Technical Sciences[10] is an important step in this regard, and it is fortunate that the vigorous activities of the Academy have received support from the State through the establishment of the technical research council.[11] Thereby, support can be granted to tasks of particularly urgent character. But this is only a first beginning. If we are to take part in the technical development, as our traditions and our expertise entitle us to do, the funds that society uses must be increased to a steadily increasing extent.

The attitude of the Danish Academy

– What is the Danish Academy's attitude to the question of the practical application of science?

– It was precisely with such a question in mind that I spoke of a natural division of labour. At the beginning of the Academy's 200-year existence,[12] it was often the natural forum for discussion of questions of a practical nature. But here, as elsewhere, the development has changed this situation. Among the

---

[9] Bohr refers here first and foremost to the Rockefeller-funded International Education Board which was an important source of income for Bohr's institute in the 1920s. In the next decade, the Rockefeller Foundation took over the support of Rockefeller philanthropy for natural science, providing Bohr's institute with further substantial means.

[10] This Academy was established in 1937. See Vol. 11, p. [360].

[11] Bohr refers here to the Technical–Scientific Research Council (*Det Teknisk–Videnskabelige Forskningsråd*), established in 1946. See Vol. 11, p. [351].

[12] As President of the Academy, Bohr gave a spirited speech at its 200th anniversary. N. Bohr, *Meeting on 13 November 1942 on the 200th Anniversary of the Establishment of the Academy*, Overs. Dan. Vidensk. Selsk. Virks. Juni 1942 – Maj 1943, pp. 44–48. Reproduced in Vol. 11, pp. [401]–[405] (Danish original) and [411]–[414] (English translation).

members of the Academy there are representatives of both the human and the natural sciences, and this composition has created a special tradition for contact between people doing research in the most diverse fields of knowledge. With its emphasis on the unity of all search for knowledge a tradition has thus been created here, which through the ages has made a deep impression on Danish intellectual life. At the moment the task of the Academy – as well as the members' mutual learning from each other and an extensive publishing activity – is in particular to maintain the international connection with comparable academies all over the world. By sending representatives to joint meetings, the Academy takes part in the organization of research within the many fields where these connections are of decisive importance.

But this division of labour does not at all mean that among the members of the Academy there are not now, as previously, many who work most vigorously for the improvement of the situation for science and the possibilities for its application in this country. I need only mention that it was H.C. Ørsted, the Academy's Secretary for many years, who in his time took the initiative for the establishment of the Polytechnical Institute,[13] and it was another of the Academy's highly deserving Secretaries, Christian Abildgaard, who participated so effectively in the establishment of the Veterinary and Agricultural College.[14] It was also one of the Academy's members, equally outstanding as a physicist and as a technologist, P.O. Pedersen,[15] who was the driving force in the establishment of the Academy for Technical Sciences and was its first president. Generally, it can be said that through their affiliation to the university or other institutions, for example as the heads of laboratories, many of the Academy's members are responsible for obtaining funds for research and for the working conditions of their staff. One can safely say that the time and the energy required for this can often be just as great as that left for one's own research.

Such endeavours can often hold great encouragement, not only through the activity that can be created but also through the understanding one meets again and again despite all difficulties. As for myself, when we talk about this, I can hardly fail to express gratitude for the understanding shown by the state authorities for the necessity of keeping up in a realm of science where such a rapid and promising development is taking place as in atomic physics.

Having said this, I should nevertheless like to add that it will never be possible for us in our small society to be able to command such enormous

---

[13] Presently the Technical University of Denmark, established as the Polytechnical Institute in 1829. See Vol. 11, p. [352].

[14] At present the Royal Veterinary and Agricultural University. See Vol. 11, p. [412], ref. 18.

[15] Peder Oluf Pedersen (1874–1941), Danish engineer and physicist.

resources for research in atomic physics as are available, for example, in the United States. What we can do is to make sure that we are not entirely left behind and in particular that we create conditions for the education of physicists and technologists who can solve the tasks of society that in the near future will arise for us too.

Two kinds of truth and many kinds of doubt

– You mentioned earlier the search for knowledge and the urge to communicate. Can you say something in this connection about the contribution to human understanding that the development of physics has given?

– It is not easy to distinguish between different fields of research. It is common to scientific endeavour to extend the borders of our knowledge and to develop the viewpoints under which it can be synthesized. Therefore it is always a question about something shared by all. However, in one field or another where we are confronted with quite new experiences, it happens from time to time that we learn in a surprising way about the general conditions for any synthesis of learning – we might perhaps even say for human knowledge itself.

During the last generation we have met just such a situation through the investigation of the world of atoms. We have come across surprises where our experiences at the first instant even might seem to contradict each other because their explanation would seem to demand mutually contradictory pictures. But by closer examination it has appeared that in the application of such pictures there was never a question of phenomena observed under the same experimental conditions. On the contrary, the pictures concerned phenomena appearing under different experimental conditions which – and this is what is decisive – were mutually exclusive. This relates to the fact that in the case of atomic phenomena we are precluded from distinguishing between what we might call the independent behaviour of atomic objects and the objects' interaction with the measuring instruments constituting necessary parts of the experimental arrangement from which the observations arise. The phenomena therefore have quite another indivisibility than the usual physical phenomena whose course can be followed step by step in the manner we know from everyday experience.

Through despair to jest

Yes, I can see that this must sound strange, but it has also cost great effort until the peculiar circumstances we meet here became clear to us. Physicists are just ordinary people, and we have often in despair had to seek refuge and refreshment in jest. Here I have in mind the old story which we time and again have had occasion to remember – about the two kinds of truth. The one kind is those statements which are so clear and simple that the opposite would obviously be nonsense. That kind of truth we jokingly called trivialities. Spiritualities, on the other hand, the so-called "deep truths", are statements of which the opposite is also a deep truth. In a new field of experience, the development is often such that we approach order from chaos step by step. But the intermediate stages, where one takes refuge in deep truths, are often the most exciting, because the imagination is constantly encouraged to find a firmer foothold. In atomic physics, precisely by recognizing that an explanation by means of pictures will always encounter an absolute limit, one has gradually achieved order and developed methods whereby widespread fields of experience join harmoniously.

Truth is something to be doubted

– You speak of two kinds of truth. Is it at all possible to say what truth is?

– In the first instance you could in any case be tempted to say that truth is something you can try to doubt. With great effort you may eventually come to the realization that at least some of your doubts were unjustifiable.

– What did you mean by setting truth and clarity against each other as opposites?

– Yes, such a manner of speaking arises precisely from the playful remarks we consoled ourselves with for a while. Naturally, it is not really the intention to set clarity and truth against each other. The purpose was only to emphasize that by clarity one often understands the use of customary modes of explanation; but these can prove insufficient when there is a question of making room for new experiences which cannot fit at all into the old frameworks.

– In a radio broadcast for gymnasium students the other day[16] you touched upon the question of the relationship between the use of words such as justice and love. Should this too be understood on such a background of "jest"?

– No, it was meant quite differently and in full earnest. What I said con-

[16] See ref. 2.

cerned the lesson we have learned in atomic physics about the impossibility of explaining experiences by means of a single visualizable picture. The situation we meet here, when we first become aware of it, we find again in so to speak all areas of life. It was thereby that I came to speak about how the use of words such as thoughts and feelings, equally indispensable in the situations we meet every day, strictly speaking concern situations that are mutually exclusive. In relation to our fellow beings, the same holds for the words justice and love. On the whole, we meet everywhere what in atomic physics we call *complementary situations*. It reminds us in the most striking fashion of the old wisdom that in every endeavour to find harmony in our existence we must never forget that we are ourselves both actors and spectators in the drama of life.

Perhaps it is a vain hope, but I myself am certain that it will be of fundamental importance for the future attitude to the problems of civilization and for the bringing up of coming generations that, within a field of experience which lies so far from the sphere of passions, we have so emphatically been reminded of the conditions for human understanding and of a wisdom which might otherwise seem remote from natural science.

Merete Bonnesen.

VII. [PORTRAIT FILM]

Transcript of a film about Niels Bohr
recorded at the Carlsberg Mansion in 1952
by Dansk Kulturfilm

TRANSCRIPT AND TRANSLATION

See Introduction to Part II, p. [103].

Ja, nu har vi været rundt at se omgivelserne her på Carlsberg, hvor Jacob Christian Jacobsen indrettede så smukt et hjem midt i den store og blomstrende virksomhed, han selv havde skabt. Vi er nu i en stue, hvor den gamle brygger har arbejdet og siden ham Høffding, og hvor jeg har tilbragt en betydelig del af min tid i de sidste tyve år. Det er jo en stue med mange minder, der betyder meget for mig. Bag på væggen bagved ser De jo billedet af Høffding, ved siden af ham billedet af min far, fysiologen Christian Bohr, og ved siden af ham igen af fysikeren Christian Christiansen. Det var tre nære venner med mange fælles almindelige interesser, og sammen med den store sprogforsker Vilhelm Thomsen dannede de en lille gruppe, som igennem mange år mødtes regelmæssigt hver fjortende dag i hverandres hjem. Både min bror og jeg, der stod hinanden så nær hele livet igennem,[1] har ofte talt om de minder, vi havde fra den tid, hvor, da vi voksede lidt til, vi fik lov at lytte til diskussionerne og fik meget dybe indtryk deraf.

Mit eget liv har jo formet sig sådan, at jeg har haft den lykke at kunne beskæftige mig med spørgsmål, som meget tidligt jeg vandt interesse for og på områder af videnskaben, hvor vi har set en stor og rig udvikling. Det enkelte bidrag i sådan en udvikling må altid være meget beskedent. Videnskaben er et stort fælles menneskeligt værk, der gror ved mange bidrag af forskelligste art, gennem hvilke vore erfaringer gradvis øges, og begreberne afklares. Jeg studerede jo ved Københavns Universitet, hvor Christiansen blev min lærer; hans egne arbejder falder i forrige århundrede. Siden har vi lært meget nyt, men det var en forsker med meget vidt syn og dyb forståelse af videnskabens væsen. Det kom til at betyde så meget for mig, at jeg efter afslutningen af studiet i København, kom på en rejse til England, først til det berømte Cavendish Laboratorium i Cambridge, hvor den store fysiker Sir Joseph Thomson virkede, og derefter til Manchester, hvor Ernest Rutherford, en af de største skikkelser i videnskaben – den senere Lord Rutherford – netop på den tid drog til sig en stor skare af unge videnskabsmænd fra mange forskellige lande, der arbejdede under hans inspirerende ledelse.

Ja, det var meget store fremskridt i vore indblik i atomernes verden, som man i de år havde nået. Man havde jo forlængst forstået, at alle stoffer måtte være opbygget af atomer, men man tænkte sig, at vore sanser var alt for grove til, at vi nogensinde skulle få direkte oplysninger om de enkelte atomer. De er jo så små, at hvis man ville for eksempel forsøge at tælle atomerne i det yderste led af ens finger, og ville sætte hele menneskeslægten med dens tusinder af millioner til at tælle dag og nat, så ville det tage mange generationer, ja det

---

[1] Harald Bohr had died in 1951, the year before.

ville tage tusinder af år, før at man blev færdig med det. Men alligevel så er det jo gået sådan, at vi på grund af den vidunderlige udvikling af den fysiske eksperimentalteknik, eller rettere af den fysiske eksperimenterkunst, har kunnet studere virkninger af enkelte atomer og endda fået oplysninger om de enkelte atomers byggestene.

Ja, det var Thomson, som vi skylder, at man forstod, at i alle atomer indgår de såkaldte elektroner – partikler, der er meget lette i forhold til atomerne, og som er elektrisk ladede. Men lige før jeg kom over til Manchester, havde Rutherford gjort den skelsættende opdagelse, at elektronerne i atomet holdtes inde i dette ved tiltrækningen fra en kerne, som indeholdt næsten hele atomets masse, men alligevel var så lille i forhold til atomet, at dens størrelse sammenlignet med atomet er som denne pibe sammenlignet med hele Stor-københavn. Det er jo vidunderligt, at det alligevel har været muligt at undersøge atomkernernes egenskaber – finde ud af, hvordan de er bygget op. Ja, endda, som alle ved, i de seneste år er det lykkedes at frigøre en væsentlig del af den uhyre energimængde, der findes i atomkernen, hvorved vi har fået en sådan forøgelse af vort herredømme over naturens kræfter, der rummer meget store løfter, men jo også stiller menneskeheden overfor store problemer.

Den opgave, der imidlertid lå for i de år, var at se, hvordan man ud fra dette kendskab, som man havde fået til atombygningen, kunne forstå stoffernes sædvanlige fysiske og kemiske egenskaber, som man jo havde studeret, samlet erfaringer op i lange tider. Her viste det sig, at man kom ud for overraskende vanskeligheder. Men man fandt, at man havde fået en nøgle til at løse disse vanskeligheder ved opdagelsen af virkningskvantet, den store opdagelse, der skyldes den tyske fysiker Max Planck, og som fandt sted akkurat det første år af vort århundrede – en opdagelse, som vi mere og mere ser dannede indledningen til en hel ny epoke i naturvidenskabens historie. Det har for mig selv været en stor glæde i de senere år også at have lært Plancks skarpsindede og fine personlighed at kende.

I forståelse af betydningen af de opgaver, der trængte sig på inden for atomfysikken og af ønskeligheden af, at vi også herhjemme kunne være med til at bidrage til deres løsning blev, nogle år efter, der oprettet Institut for Teoretisk Fysik ved Københavns Universitet, først og fremmest med støtte fra staten, men jeg har også megen anledning til med taknemmelighed at tænke på den støtte, som blev vist både fra ældre – fra min fars kreds – og fra mine skolekammeraters kreds.[2]

---

[2] Thus, Bohr's schoolmate Aage Berlème (1886–1967) led a campaign to collect funds to pay for the site on which Bohr's institute would be built.

Vi har på Institutet haft den lykke gennem årene at have som medarbejdere mange dygtige og fremragende danske fysikere. Det har også været en meget stor glæde, at vi har haft et stort antal besøg af udenlandske videnskabsmænd – mellem dem nogle af de, der har bidraget allermest til udformningen af en virkelig atommekanik. Det er nu mennesker, som sidder i ledende stillinger mange steder rundt omkring i verden.

Når jeg brugte ordet atommekanik, så var det ikke meningen at bruge ordet på samme måde, som man bruger ordet mekanik, hvor man henviser til den store lærebygning, der hviler på de Newton'ske principper, der har vist sig så anvendelige på så store områder, og som især har kunnet gøre så nøje rede for himmellegemernes bevægelser. Men opdagelsen af virkningskvantet betød erkendelsen af, at denne lærebygning er en idealisation, der kun kan finde anvendelse på fænomener i stor målestok – på fænomener, hvor alle virkninger, der indgår, er store i forhold til det enkelte virkningskvantum. Når dette ikke længere er tilfældet, som ved de enkelte atomprocesser, så møder vi lovmæssigheder af en ganske anden art, hidtil helt ny i fysikken, og vi er endda henvist til at beskrive og sammenfatte erfaringer på anden måde. Det kommer af, at virkningskvantet fører med sig, at der vil altid være en væsentlig vekselvirkning imellem sådanne atomsystemer og de måleinstrumenter, ved hjælp af hvilke vi søger at få oplysninger om dem, og denne vekselvirkning kan vi ikke på nogen måde tage simpelt i betragtning. Hvis vi ville for eksempel prøve at finde systemernes reaktion på måleinstrumenterne, ja så måtte måleinstrumenterne selv gøres til genstand for undersøgelsen, og det betød, at så måtte vi indføre nye måleinstrumenter, og vi fik igen en ny ukontrollerbar vekselvirkning mellem disse og de gamle måleinstrumenter. Det hele ville bare være et indviklet spil, som slet ikke førte os videre. Under disse forhold, da må man forstå, at fænomener, der optræder under forskellige forsøgsbetingelser, ikke kan sammenlignes på sædvanlig måde – ikke kan sammenfattes i et enkelt billede – og at det endda kan se ud, som om de stred mod hverandre, hvis man ville forsøge derpå.

Fænomener, der optræder under forskellige forsøgsbetingelser, de kalder vi komplementære for at understrege, at de først taget tilsammen giver os alle de oplysninger, som vi kan få om atomerne. Denne komplementære beskrivelsesmåde giver en videre ramme, hvor vi får mulighed for at få ind disse nye lovmæssigheder som er af ren fundamental karakter, som faktisk bestemmer stoffernes egenskaber, de stoffer hvoraf vore redskaber er bygget op og de stoffer, hvoraf vore egne legemer er bygget op. Det har været en dyb belæring om vor situation som iagttagere af naturen, og det har været en belæring, der mindede om gammel visdom – for eksempel gammel kinesisk visdom, hvor man jo så stærkt understregede, at skulle man finde sig til rette i tilværelsen måtte man ikke glemme, at vi samtidig er tilskuerne og skuespillerne i livets spil.

Det har i de seneste år været en stor glæde og oplevelse for mig ud fra denne nye belysning af vor erkendelses grundlag, som atomfysikkens udvikling har givet os, at igen beskæftige mig med sådanne biologiske, psykologiske, filosofiske problemer, som lå den kreds, hvor jeg i min ungdom voksede op, så stærkt på sinde. Som et enkelt punkt, hvor vi finder forhold for iagttagelse og sammenfatning, som er forskellig fra det, som vi sædvanligt møder inden for naturvidenskaben: det møder vi, når vi forsøger at studere andre menneskers kulturer og sammenligner disse. Det er noget man meget nær kommer til at beskæftige sig med, når man deltager i det mellemfolkelige videnskabelige samarbejde – møder og kommer i nær berøring med mennesker fra langt bortliggende dele af verden. Når man her – og måske til sin overraskelse – lærer, at der under meget forskellige traditioner kan leves menneskeliv med værdighed, med gensidig respekt, og forstår, at det er medmennesker – med-mennesker som vi – ja, så har vi mistet nogle af de fordomme, der på godt og ondt gør os til medlemmer af en bestemt national kultur – og enhver sådan national kultur, den må indeholde naturligt et element af selvbehagelighed. Det skal ikke være ment som nogen kritik, men blot svarende til den glæde ved livet, som ethvert rask menneske føler. Men fordybelsen i sådanne pro-blemer, i kilderne til de misforståelser, der kommer op ved forbindelse imellem mennesker af forskellige kulturer, det tør man håbe kunne vise vej til bedre forståelse imellem folkeslagene, en forståelse, der jo nu om stunder er mere nødvendig end nogensinde, nu hvor videnskabens udvikling så givet rummer så store løfter for forbedring af menneskenes levevilkår, og som samtidig gør det mere nødvendigt end nogensinde at vi på vor lille klode finder frem til veje for fredeligt samarbejde på den fælles menneskekulturs udvikling og fremgang. Det er ikke mindst inden for det videnskabelige samarbejde, at man kan håbe, at sådanne bestræbelser vil fremmes, idet vi her måske mere end på noget andet område føler, at man på forskelligt grundlag stræber efter det samme mål – at forøge vor erfaring og uddybe vor erkendelse af den natur, hvori at vi selv er del.

[201]

TRANSLATION

Well, now we have been around to look at the surroundings here at Carlsberg where Jacob Christian Jacobsen established such a beautiful home in the middle of the large and flourishing enterprise he himself had created. We are now in a room where the old brewer worked, and after him Høffding, and where I have spent a considerable part of my time over the last twenty years. It is of course a room with many memories that mean much to me. On the wall behind me you see the picture of Høffding and next to him the picture of my father, the physiologist Christian Bohr, and next to him again the physicist Christian Christiansen. They were three close friends with many shared general interests, and together with the great philologist Vilhelm Thomsen they formed a small group which for many years met regularly every fortnight at each other's homes. Both my brother and I, who were so close to each other all our lives,[1] have often talked about the memories from that time when as we grew a little older we were allowed to listen to the discussions which made a very deep impression on us.

My own life has turned out in such a way that I have had the good fortune of being able to devote myself to questions which caught my interest very early and in realms of science where we have seen a great and rich development. The individual contribution to such a development must always be very modest. Science is a great common human construction which grows by means of many contributions of the most various kinds, through which our experience is gradually increased and the concepts are clarified. I studied at the University of Copenhagen where Christiansen was my teacher; his own papers are from the last century. Since then we have learnt much that is new, but he was a researcher with a very broad vision and a deep understanding of the nature of science. It came to mean so much for me that after the completion of my studies in Copenhagen I travelled to England, first to the famous Cavendish Laboratory in Cambridge, where the great physicist Sir Joseph Thomson worked, and then to Manchester where Ernest Rutherford, one of the greatest figures in science – later Lord Rutherford – just at that time attracted a large number of young scientists from many different countries, who worked under his inspiring guidance.

Indeed, very great advances in our insight into the world of atoms had been achieved in those years. One had of course long ago come to understand that all substances must be built up of atoms, but it was thought that our senses were much too coarse for us ever to be able to obtain direct information about the individual atoms. For they are so small that if, for example, one wanted to try to count the atoms in the outer joint of one's finger, and would ask the whole human race, with its thousands of millions, to count day and night, then it would take many generations, indeed it would take thousands of years, before the task was complete. But nevertheless it has turned out that due to the marvellous development of the experimental technique of physics, or rather of the experimental art of physics, we could study the effects of individual atoms and have even obtained information about the building blocks of the individual atoms.

Indeed, it is to Thomson that we owe the recognition that in all atoms there are so-called electrons – particles which are very light in relation to the atoms and which are electrically charged. But just before I came to Manchester, Rutherford had made the epoch-making discovery that the electrons in the atom were kept inside it by the attraction from a nucleus which contained almost the entire mass of the atom but was nevertheless so small in relation to the atom that its size compared with the atom is like this pipe compared with the whole of greater Copenhagen. It is marvellous that it has been possible, nevertheless, to investigate the properties of the atomic nuclei – to find out how they are constructed. Indeed, as everybody knows, it has even been possible in recent years to liberate a significant part of the immense amount of energy present in the atomic nucleus, whereby we have obtained such an increase in our mastery of the forces of nature, which holds very great promises, but of course also presents humanity with great problems.

The immediate task during those years, however, was to see how, on the basis of this knowledge which had been obtained about the constitution of atoms, one could understand the ordinary physical and chemical properties of substances, which one had studied, gathered experience about, for a long time. Here it turned out that surprising difficulties were encountered. But it was found that a key to resolving these difficulties had been given by the discovery of the quantum of action, the great discovery which is due to the German physicist Max Planck, and which took place in precisely the first year of our century – a discovery which we realize more and more formed the introduction to an entirely new epoch in the history of natural science. To me, personally, it has been a great pleasure in later years also to have become acquainted with Planck's discerning and fine personality.

In recognition of the importance of the tasks requiring attention within atomic physics and of the desirability that also we here at home could take part in contributing to their solution, the Institute of Theoretical Physics at the University of Copenhagen was established a few years later, first and foremost with support from the State, but I have also good reason to think with gratitude of the support shown both by the older generation – my father's circle – and by the circle of my schoolmates.[2]

Through the years we have had the good fortune at the Institute of having many able and excellent Danish physicists on the staff. It has also been a very great pleasure that we have had a large number of visits by scientists from abroad – among them some of those who have contributed most of all to the shaping of a genuine atomic mechanics. They are now people who occupy leading positions in many places around the world.

When I used the words atomic mechanics, it was not my intention to use the words in the same way as one uses the word mechanics, where one refers to the great edifice of knowledge resting on the Newtonian principles which have proven themselves so applicable in such large areas, and which in particular have been able to account so accurately for the movements of the heavenly bodies. But the discovery of the quantum of action meant the recognition that this edifice of knowledge is an idealization which can only be applied to phenomena on a large scale – to phenomena where all actions involved are large in relation to the individual quantum of action. When this is no longer the case, as with the individual atomic processes, then we encounter regularities of an entirely different kind, hitherto quite unknown in physics, and we are even obliged to describe and to arrange experiences in another way. This is because the quantum of action entails that there will always be a significant interaction between such atomic systems and the measuring instruments by means of which we seek to obtain information about them, and we cannot in any way take this interaction into account in a simple manner. If, for example, we were to try to find the reaction of the systems to the measuring instruments, then, indeed, the measuring instruments themselves should be made the object of the investigation, and that meant we then had to introduce new measuring instruments, and we obtained once more a new uncontrollable interaction between these and the old measuring instruments. It would all just be a complicated game which did not lead us any further at all. Under these circumstances, then one must understand that phenomena which occur under different experimental conditions cannot be compared in the usual manner – cannot be arranged in a single picture – and that it can even seem as though they contradicted each other if one would try to do so.

We call phenomena which occur under different experimental conditions complementary, in order to emphasize that only when taken together do they provide us with all the information we can obtain about the atoms. This complementary mode of description provides a wider framework where it is possible to place these new regularities of a purely fundamental character, which actually determine the properties of the substances – the substances of which our tools are made and the substances of which our own bodies are made. It has been a profound lesson about our situation as observers of nature, and it has been a lesson reminiscent of old wisdom – for example, old Chinese wisdom, where it was so strongly emphasized that if one should find one's way in existence, then one must not forget that we are at the same time spectators and actors in the drama of life.

In recent years it has been a great pleasure and experience for me to take up once more, on the basis of this new light on the foundation of our knowledge that the development of atomic physics has given us, such biological, psychological and philosophical problems which were so important for the circle in which I grew up during my youth. As a single point, where we find conditions for observation and synthesis different from what we usually meet within natural science: this we meet when we try to study the cultures of other peoples and compare them. That is something one experiences at very close hand when one takes part in international scientific cooperation – meets and comes into close contact with people from faraway parts of the world. When one here – and perhaps to one's surprise – learns that within very different traditions human lives can be lived with dignity – with mutual respect – and understands that it involves fellow human beings – human beings like ourselves – then indeed we have lost some of the prejudices that for better or for worse make us members of a certain national culture – and every such national culture must quite naturally contain an element of complacency. This is not intended as any criticism, but simply corresponds to the pleasure in life which every healthy person feels. But concentration on such problems, on the sources of the misunderstandings arising upon contact between people of different cultures, this, one dare hope, could show the way to better understanding between the peoples, an understanding which nowadays is more necessary than ever before, now that the development of science certainly holds such great promises for the improvement of people's living conditions and at the same time makes it more necessary than ever before that we on our small planet discover ways for peaceful cooperation on the development and progress of common human culture. Not least within scientific cooperation one may hope that such efforts will be advanced, as here, perhaps more than in any other field, we feel that all are striving, on different backgrounds, for the same goal: to increase our experience and deepen our knowledge of the nature of which we ourselves are part.

VIII. AUTOBIOGRAPHY OF THE HONORARY DOCTOR

SELVBIOGRAFI AF ÆRESDOKTOREN
Acta Jutlandica **28** (1956) 135–138

TEXT AND TRANSLATION

See Introduction to Part II, p. [103].

Selvbiografier, prisopgaver og prisbesvarelser

SELVBIOGRAFI AF ÆRESDOKTOREN

Jeg, *Niels Henrik David Bohr*, er født i København den 7. oktober 1885 som søn af professor i fysiologi ved Københavns universitet, Christian Bohr, og hustru Ellen, født Adler. Den 1. august 1912 ægtede jeg Margrethe, datter af apoteker Alfred Nørlund og hustru Sophie, født Holm; vi har fire sønner, Hans, Erik, Aage og Ernest, der alle er gift og selv har børn, medens vor ældste søn Christian som ung student kom af dage ved en sejlbådsulykke.

I 1903 tog jeg studentereksamen fra Gammelholms Latin- og Realskole og begyndte et fysikstudium ved Københavns Universitet, under hvilket jeg i 1907 erhvervede Videnskabernes Selskabs guldmedalje for en afhandling om vædske-strålers svingninger. I 1909 tog jeg magistergraden i fysik ved Københavns Universitet, og i 1911 blev jeg dr. phil. på en afhandling »Studier over Metallernes Elektronteori«.

I universitetsåret 1911–12 foretog jeg med støtte fra Carlsbergfondet en studie-rejse til Cambridge og Manchester, der på den tid var arnesteder for udforskningen af atomernes struktur. Netop i 1911, få måneder før min ankomst til Manchester, havde Professor Ernest Rutherford, den senere Lord Rutherford, ved sin opda-gelse af atomkernen skabt grundlaget for en helt ny udvikling af fysikken. Det blev min gode skæbne fra første færd at deltage i denne udvikling, og den for mig så værdifulde personlige forbindelse med Rutherford, der bevaredes lige til hans død i 1937, omfatter jeg med den dybeste taknemmelighed.

I Manchester gennemførte jeg en teoretisk undersøgelse over atompartiklers hastighedstab under deres gennemtrængning gennem stoffer og blev især grebet af de udsigter, som atomkernens opdagelse havde åbnet for en videregående opkla-ring af grundstoffernes egenskaber og slægtskabsforhold, der dog øjensynlig kun kunne virkeliggøres ved dybtgående afvigelser fra den klassiske fysiks på så mange områder frugtbare og fyldestgørende synspunkter.

Efter min hjemkomst til København fortsatte jeg disse undersøgelser og det blev mig mere og mere klart, at nøglen til fremskridt for forståelsen af atomernes egenskaber var givet gennem Plancks opdagelse i 1900 af det universelle virknings-kvantum, der, navnlig gennem Einsteins arbejder, havde vist sig så frugtbart for opklaringen af de overraskende træk, som energiudvekslingen ved atomare syste-mers bestråling frembyder.

I nogle i 1913 offentliggjorte arbejder lykkedes det mig således at vise, at det var muligt at tyde de empiriske love for grundstoffernes spektre og samtidig nå til oplysninger om elektronernes binding til atomkernen ud fra den antagelse, at

136

enhver ændring i atomets energi er knyttet til en elementær proces, hvorved atomet overføres fra en af dets såkaldte stationære tilstande til en anden.

I 1913 blev jeg ansat som docent i fysik ved Københavns Universitet og med orlov fra denne stilling opholdt jeg mig i universitetsårene 1914-15 og 1915-16 i England, hvor jeg som lektor ved Manchester universitetet genoptog det personlige samarbejde med Rutherford og navnlig beskæftigede mig med den videre udvikling af kvanteteorien for atomerne og for tydningen af de nye eksperimentelle fremskridt på atomfysikkens område.

I 1916 modtog jeg udnævnelse til professor i teoretisk fysik ved Københavns Universitet og virkede fra første begyndelse for oprettelsen af en til professoratet knyttet institution, hvor teoretiske og eksperimentelle undersøgelser kunne udføres i nøjeste forbindelse med hverandre. Med støtte fra Staten og Carlsbergfondet, såvel som ved bidrag fra private personers side, lykkedes det at gennemføre planerne, og i 1920 stod institutet for teoretisk fysik ved Københavns Universitet i sin første skikkelse rede til at byde danske og udenlandske fysikere muligheder for virksomt at deltage i undersøgelser indenfor det hastigt voksende arbejdsfelt.

På institutet, blandt hvis første danske medarbejdere især må nævnes den nylig afdøde og dybt savnede rektor ved Københavns Universitet, professor H. M. Hansen, den nuværende forskningchef ved Atomenergikommissionens forsøgsstation, professor J. C. Jacobsen, og professoren i fysik ved Aarhus Universitet, Sven Werner, udfoldede der sig fra begyndelsen et omfattende internationalt samarbejde.

For de teoretiske undersøgelsers vedkommende drejede det sig først og fremmest om den gradvise afklaring af de nye forestillinger, hvortil overordentlig værdifulde bidrag blev ydet af en række fremragende udenlandske fysikere, der deltog i arbejdet på institutet gennem kortere eller længere tid. Især må her nævnes de afgørende fremskridt og originale synspunkter, der skyldes fysikere som Hans Kramers, Oscar Klein, Wolfgang Pauli, Werner Heisenberg og Paul Dirac.

Hvad de eksperimentelle arbejder i de første år angik, blev navnlig George Hevesys tilknytning til institutet af største betydning. Jeg tænker ikke alene på opdagelsen af grundstoffet Hafnium, der i så høj grad bestyrkede den første orienterende forklaring af grundstoffernes periodiske system, men tillige på den af Hevesy udviklede isotop-indikatormetode, der skulle blive et for den fysiske og biologiske forskning uvurderligt hjælpemiddel.

For den så vigtige forbindelse mellem den teoretiske fysik og de matematiske videnskaber var det en lykkelig begivenhed, at det af Carlsbergfondet i 1932 oprettede Matematiske Institut ved Københavns Universitet blev anbragt i en med Institutet for teoretisk fysik sammenhængende bygning. Dette betød også for mig selv en kærkommen daglig forbindelse med min afdøde broder, Harald Bohr, der virkede som bestyrer for det Matematiske institut til sin død i 1953, og med hvem jeg fra barndommen var så nøje forbundet i fælles menneskelige og videnskabelige interesser.

[210]

Medens hovedemnet for arbejdet på institutet i den første halve snes år havde været studiet af den for stoffernes almindelige fysiske og kemiske egenskaber bestemmende elektronbinding til atomkernerne, samledes interessen i begyndelsen af trediverne om problemer vedrørende atomkernernes egen opbygning. Allerede i 1919 havde Rutherford påvist muligheden for omdannelse af atomkerner ved sammenstød med de energirige partikler, der spontant udsendes fra de radioaktive stoffer og derved indledt en udvikling, der skulle få så vidtrækkende følger. Denne udvikling tog især fart efter opdagelsen af neutronerne og af disses anvendelse til frembringelse af radioaktive isotoper. For fremskridt på dette nye forskningsområde blev det endvidere af afgørende betydning, at det viste sig muligt at frembringe atomkerneomdannelser ved atomioner, kunstigt accelererede til høj energi.

Betingelsen for herhjemme at deltage i den eksperimentelle atomkernefysik skabtes ved indretningen af et højspændingslaboratorium på institutet med støtte af Carlsberg- og Rockefellerfondet og især ved et af Thrigefondet skænket cyklotronanlæg. På dette tidspunkt udførtes også teoretiske undersøgelser på institutet, der tillod en forståelse af mange af de karakteristiske træk ved atomkerneprocesserne og som ikke mindst viste sig frugtbare ved forklaringen af den i 1939 opdagede fissionsproces, der skulle give grundlaget for atomenergiens udnyttelse til praktiske formål.

I årene inden krigen foretog jeg rejser til mange lande i forbindelse med det videnskabelige samarbejde på atomfysikkens område og bestræbelserne for at skabe arbejdsmuligheder for de talrige videnskabsmænd, der måtte forlade deres hjemland som følge af den politiske udvikling i Europa.

Under de første år af Danmarks besættelse kunne arbejdet på institutet fortsættes trods voksende vanskeligheder. Efter at have erfaret om tyske planer om min fængsling, lykkedes det mig imidlertid ved modstandsorganisationernes hjælp sammen med min hele familie at undvige til Sverige. Efter forud modtaget indbydelse fra den britiske regering bragtes jeg, sammen med min søn Aage, få dage efter til England, hvorfra vi som medlemmer af den britiske atomenergiorganisation nogle måneder senere rejste til U.S.A. for at deltage i de der igangsatte store projekter. Om nogle af mine oplevelser og bestræbelser i disse år har jeg berettet i 1950 i et åbent brev til De Forenede Nationer.

Efter krigen vendte jeg tilbage til København og fortsatte min virksomhed som professor ved Københavns Universitet indtil april i dette år, da jeg efter aldersbestemmelserne fratrådte denne stilling. Foreløbig fortsætter jeg dog som bestyrer for institutet, hvor der i årene siden krigen har fundet betydelige udvidelser sted såvel af de eksperimentelle hjælpemidler som af medarbejderstaben, og hvortil udenlandske fysikere atter har søgt i stigende omfang. Det er mig en dyb glæde at følge de stadige fremskridt på atomfysikkens område, der skyldes den yngre generation her som andet steds.

Ud over virksomheden på institutet har min egen interesse i de senere år særlig samlet sig om erkendelsesteoretiske spørgsmål rejst igennem atomfysikkens udvikling, der har krævet en dybtgående revision af grundlaget for naturbeskrivelsen. Min interesse for de biologiske og psykologiske problemer, som man derved føres ind på, stammer fra min første ungdom, hvor jeg lyttede til diskussionerne i min fars og hans venners kreds, blandt hvilke jeg især skulle få nærmere forbindelse med fysikeren Chr. Christiansen, der blev min lærer ved universitetet, og filosoffen Harald Høffding, med hvem jeg havde mange lærerige samtaler lige indtil hans sidste dage.

Foruden mine videnskabelige studier og pædagogiske virksomhed har jeg i årenes løb i stedse større grad deltaget i organisatoriske bestræbelser inden for den fysiske videnskab og dens anvendelser i lægevidenskaben og for praktiske samfundsopgaver. Siden 1939 har jeg virket som præsident for Det kongelige danske videnskabernes selskab, hvori jeg indvalgtes i 1918, og fra 1935 som præsident for Den danske kræftkomite, til hvis virksomhed jeg igennem en længere årrække havde været knyttet. Endvidere har jeg siden dannelsen af den af regeringen nedsatte danske atomenergikommission virket som dennes formand.

I 1922 modtog jeg Nobelprisen i fysik og er dr.h.c. ved flere udenlandske universiteter samt medlem af videnskabelige selskaber i forskellige lande. Den tilknytning til Aarhus Universitet, hvormed man har æret og glædet mig, har været mig særlig velkommen i forbindelse med de bestræbelser, der ligger os alle så dybt på sinde, om det nøjeste samarbejde mellem vore to danske universiteter.

TRANSLATION

AUTOBIOGRAPHY OF THE HONORARY DOCTOR

I, *Niels Henrik David Bohr*, was born in Copenhagen on 7 October 1885 as the son of Christian Bohr, professor in physiology at the University of Copenhagen, and his wife Ellen, née Adler. On 1 August 1912, I married Margrethe, daughter of Alfred Nørlund, apothecary, and his wife Sophie, née Holm; we have four sons: Hans, Erik, Aage and Ernest, who all are married and have children, while Christian, our eldest son, lost his life in a sailing accident as a young student.

In 1903 I passed my matriculation examination from Gammelholm Grammar School and started my physics studies at the University of Copenhagen, during which I received the gold medal of the Royal Danish Academy in 1907 for a paper about water jet vibrations. In 1909 I took my master's degree in physics at the University of Copenhagen, and in 1911 I got my D.Phil. Degree for the thesis "Studies on the Electron Theory of Metals".[1]

In the university year 1911–12 I went to study, with support from the Carlsberg Foundation, at Cambridge and Manchester, which at that time were seedbeds for the investigation of the structure of atoms. Just in 1911, a few months before my arrival in Manchester, Professor Ernest Rutherford, later Lord Rutherford, had created the foundation for a completely new development of physics with his discovery of the atomic nucleus. It became my good fortune to participate in this development from the very beginning, and I regard the, for me so valuable, personal ties with Rutherford, which were maintained right up to his death, with the deepest gratitude.

In Manchester I carried out a theoretical investigation concerning the decrease of velocity of atomic particles on passing through matter,[2] and was especially fascinated by the prospects that the discovery of the atomic nucleus had opened for a further clarification of the properties and relationships of the elements, which, however, could seemingly only be realized by radical

---

[1] Reproduced in Vol. 2, pp. [163]–[290] (Danish original), [291]–[395] (English translation).

[2] N. Bohr, *On the Theory of the Decrease of Velocity of Moving Electrified Particles on Passing through Matter*, Phil. Mag. **25** (1913) 10–31. Reproduced in Vol. 2, pp. [15]–[39], and Vol. 8, pp. [47]–[71].

departures from the conceptions of classical physics which are fruitful and complete in so many spheres.

After my return to Copenhagen I continued these investigations and it became increasingly obvious to me that the key to progress in the understanding of the properties of the atoms had been given through Planck's discovery of the universal quantum of action in 1900, which, especially through the contributions of Einstein, had proved to be so fruitful for the elucidation of the surprising features displayed by the energy exchange on the irradiation of atomic systems.

In some papers published in 1913 I thus succeeded in showing that it was possible to interpret the empirical laws for the spectra of the elements and at the same time obtain information about the binding of the electrons to the atomic nucleus on the basis of the assumption that any change in the energy of the atom is related to an elementary process whereby the atom is transferred from one of its so-called stationary states to another.[3]

In 1913 I was employed as a lecturer of physics at the University of Copenhagen and on leave from this position I stayed in England for the university years 1914–15 and 1915–16, where as a senior lecturer at the University of Manchester I resumed personal collaboration with Rutherford and in particular occupied myself with the further development of the quantum theory of atoms and of the interpretation of the new experimental advances in the field of atomic physics.

In 1916 I received the appointment of professor of theoretical physics at the University of Copenhagen and worked from the very beginning for the establishment of an institution linked to the professorship, where theoretical and experimental investigations could be carried out in the closest connection with each other. With support from the State and the Carlsberg Foundation as well as with contributions on the part of private persons, it was possible to effectuate the plans, and in 1920 the Institute for Theoretical Physics at the University of Copenhagen, in its original form, was ready to offer physicists from Denmark and abroad possibilities to take active part in research within the rapidly growing field of activity.

At the institute a wide-ranging international collaboration blossomed from the start. Among the first Danish staff should be mentioned in particular: Professor H.M. Hansen,[4] the recently deceased and sorely missed Vice-Chancellor

---

[3] N. Bohr, *On the Constitution of Atoms and Molecules*, Parts I–III, Phil. Mag. **26** (1913) 1–25, 476–502, 857–875. Reproduced in Vol. 2, pp. [159]–[233].
[4] Bohr published three tributes to Hansen. See pp. [325] ff., [331] ff., [335] ff.

of the University of Copenhagen; Professor J.C. Jacobsen,[5] at present research director at the experimental centre of the Atomic Energy Commission,[6] and Sven Werner,[7] professor of physics at the University of Aarhus.

As regards the theoretical research, this first and foremost involved the gradual clarification of the new ideas to which extremely valuable contributions were made by a number of excellent physicists from other countries, who participated in the work at the institute for shorter or longer periods of time. In particular should here be mentioned the decisive advances and original viewpoints due to physicists such as Hans Kramers,[8] Oscar Klein, Wolfgang Pauli,[9] Werner Heisenberg and Paul Dirac.

As regards the experimental contributions in the early years, the attachment of George Hevesy in particular to the institute was of the greatest importance. I am not only thinking of the discovery of the element hafnium, which so greatly strengthened the first preliminary explanation of the periodic system of the elements, but also of the isotope indicator method developed by Hevesy,[10] which was to become an invaluable tool for research in physics and biology.

It was a happy event for the important connection between theoretical physics and the mathematical sciences that the Mathematical Institute at the University of Copenhagen, established in 1932 by the Carlsberg Foundation, was placed in a building adjoining the Institute for Theoretical Physics. This meant also for myself welcome daily contact with my late brother Harald Bohr, who was director of the Mathematical Institute until his death in 1953 [1951], and with whom from childhood I have been so closely connected in common human and scientific interests.

While the main subject for work at the institute in the first decade had been the study of the binding of electrons to the atomic nuclei, which determines the general physical and chemical properties of substances, at the beginning of the thirties interest focused on problems regarding the constitution of atomic nuclei themselves. Already in 1919 Rutherford had demonstrated the possibility of transformation of atomic nuclei on collision with the highly energetic particles

[5] Jacob Christian Jacobsen (1895–1965), Danish experimental physicist who worked at Bohr's institute from its establishment.

[6] On Bohr's involvement with the Danish Atomic Energy Commission, see Vol. 11, pp. [363] ff.

[7] (1898–1984). See Vol. 11, p. [513], ref. 7.

[8] Bohr's two published tributes to Kramers are reproduced on pp. [347] ff., [353] ff.

[9] Bohr's published tribute to Pauli is reproduced on pp. [361] ff.

[10] The Hungarian-born physicist and chemist George de Hevesy (1885–1966) got the idea of using radioactive isotopes as tracers while working at Rutherford's laboratory from 1911 to 1912, during which period Bohr also arrived in Manchester. He developed his method further during his first stay at Bohr's institute from 1920 to 1926.

which are spontaneously emitted from the radioactive substances, and initiated thereby a development which should have far-reaching consequences. This development gained momentum especially after the discovery of neutrons and their application in the production of radioactive isotopes.[11] Furthermore, it was of decisive importance for progress in this new field of research that it proved possible to produce transformations of the atomic nucleus by means of atomic ions, artificially accelerated to high energy.

The prerequisite for participating in this country in experimental nuclear physics was created by the establishment of a high-voltage laboratory at the institute with support from the Carlsberg and Rockefeller Foundations and, in particular, by means of cyclotron equipment donated by the Thrige Foundation.[12] At that time theoretical research, too, was conducted at the institute, allowing an understanding of many of the characteristic features of nuclear processes and not least proving fruitful in the explanation of the fission process discovered in 1939,[13] which was to provide the basis for the exploitation of atomic energy for practical purposes.

In the years before the war I made journeys to many countries in connection with the scientific collaboration in the field of atomic physics and the endeavours to create opportunities for employment for the numerous scientists who had to leave their native country as a consequence of political developments in Europe.

During the first years of the occupation of Denmark, work at the Institute could continue despite growing difficulties. After having learned about the German plans for my imprisonment, I managed, however, to escape to Sweden with my whole family through the help of the resistance organizations. A few days later, after prior invitation from the British Government, I was brought together with my son Aage to England, from where we travelled to the U.S.A. a few months later as members of the British atomic energy organization to take part in the great projects initiated there. In 1950, I described some of my experiences and endeavours during these years in an open letter to the United Nations.[14]

[11] This discovery made it possible to make radioactive isotopes of virtually every element, which enabled Hevesy, during his second extended stay in Copenhagen from 1934 to 1943, to develop his tracer method into an important tool in several areas.
[12] Bohr describes the relationship with Thrige and the Thrige Foundation in N. Bohr, [Preface], ...fra Thrige 6, No. 1 (1953) 2–4. Reproduced in Vol. 11, pp. [555]–[558] (Danish original) and [559]–[560] (English translation).
[13] See R. Peierls, *Introduction*, Vol. 9, pp. [3]–[83], on [52]–[76].
[14] Bohr's activities for an "open world" is the subject of Vol. 11, Part I.

After the war I returned to Copenhagen and continued my work as professor at the University of Copenhagen until April this year, when I retired from this position because of the age clause. For the time being, however, I continue as director of the institute, where in the years since the war significant expansion as regards both the experimental facilities and the staff has taken place, and which physicists from abroad have again visited to an increasing extent. To me it is a deep pleasure to follow the continuing advances in the field of atomic physics which here and elsewhere are due to the younger generation.

In addition to work at the Institute, my own interest in recent years has focused especially on epistemological questions raised through the development of atomic physics, which has required a profound revision of the basis for the description of nature. My interest in problems of biology and psychology, into which one is thereby led, stems from my early youth when I listened to discussions in the circle of my father and his friends, among whom I should in particular become more closely acquainted with Christian Christiansen, the physicist, who was to be my teacher at the University, and Harald Høffding, the philosopher, with whom I had many enlightening conversations right up to his last days.[15]

Apart from my scientific studies and educational activity, through the years I have to an ever increasing extent taken part in organizational endeavours within physical science and its application in medical science and as regards practical tasks in society. Since 1939, I have been President of the Royal Danish Academy of Sciences and Letters, into which I was elected in 1918,[16] and since 1935, President of the Danish Cancer Committee with whose activities I have been connected for many years.[17] Furthermore, I have been chairman of the Danish Atomic Energy Commission since its establishment by the Government.[18]

In 1922, I received the Nobel Prize for physics,[19] and I am honorary doctor at several universities abroad as well as a member of learned societies in various countries. This connection to the University of Aarhus, whereby I have been honoured and pleased, has been especially welcome to me in relation to the endeavours, which mean so much to us all, to gain the closest possible cooperation between our two Danish universities.

---

[15] Bohr's relationship with Høffding is dealt with in Vol. 10, pp. [297] ff.

[16] See Vol. 11, pp. [338]–[348].

[17] See Vol. 11, pp. [373]–[375].

[18] See ref. 6.

[19] See Vol. 4, pp. [25]–[30].

2. CELEBRATION OF EARLIER SCIENTISTS

IX. NEWTON'S PRINCIPLES AND MODERN ATOMIC MECHANICS

"The Royal Society Newton Tercentenary Celebrations 15–19 July 1946",
Cambridge University Press, Cambridge 1947, pp. 56–61

See Introduction to Part II, p. [103].

NEWTON'S PRINCIPLES AND MODERN ATOMIC MECHANICS

By Professor N. Bohr, for. mem. r.s.

Every one of the delegates feels most privileged to be given the opportunity through the invitation of the Royal Society to attend the celebration of the tercentenary of the birth of the great genius to whom we owe so largely the foundation of all later work in science. When with veneration we contemplate Newton's great achievements, we do not know whether most to admire his vision in recognizing universal gravitation, or the mastership in disentangling the phenomena of nature, which led him to his fundamental discoveries in optics, or perhaps above all Newton's power in conceiving and formulating general principles and in creating the mathematical tools for their fruitful application. Truly it may be said that Newton's genius did not only bring order in all knowledge attainable at his time, but even led him to an amazing degree to anticipate later discoveries and developments.

The principles of mechanics enunciated by Newton, by which he created a model for any causal description of natural phenomena, served indeed as a basis for the subsequent development of physical science and have even, as is well known, afforded a main source of inspiration for philosophers attempting to set up ultimate categories of human thinking. In what follows I shall endeavour to discuss how far the revision of the foundation of the unambiguous application of even our most elementary concepts, like space-time co-ordination and causality, which the modern development of physics has necessitated, has thrown new light on this aspect of the situation, and especially to show how against this new background we can perhaps more clearly than ever before appreciate the wisdom and caution Newton exhibited in his untiring effort to obtain the proper balance between analysis and synthesis. For this purpose, I shall briefly recall some of the essential features of the development in physics since Newton's days.

One of the greatest advances bearing on fundamentals was surely the establishment in the last century of the electromagnetic theory. Of course, the knowledge of isolated electric and magnetic phenomena goes far back through the ages, but it was first at the beginning of that century that the deeper connexion between these phenomena came to light through the discovery of Ørsted and, above all, through the wonderful researches of Michael Faraday, about the fruits of which he first discoursed in the very room in which we are assembled. Faraday's fundamental discoveries and original concepts were soon made the foundation of the comprehensive edifice built up by Clerk Maxwell, in whom all the world recognizes so worthy a follower of Newton's great example.

56

In the beginning of his work Maxwell took his inspiration directly from the Newtonian concepts of mechanics; but, when his prediction as regards the propagation of electromagnetic radiation was so convincingly borne out by the investigations and discoveries of Hertz, and the self-consistency of the electromagnetic theory was gradually perceived, the question naturally came up of the possibility of basing all ideas of mechanics on electromagnetic field conceptions.

Whatever will be the eventual answer to such questions, which are still open and which are inseparable from the deepest problems of the constitution of matter, a far-reaching clarification of the foundation and scope of the framework for the description of physical phenomena has been obtained through the discovery of the finite velocity of propagation of all forces, resulting in the development of relativity theory. Not only do we owe to Einstein the clear recognition of the relative character of the concept of simultaneity and the inseparability of space-time co-ordination, but, through his conception of the relation between this co-ordination and universal gravitation, our whole world picture achieved a higher degree of unity and harmony than ever before. The power of the relativity theory in tracing regularities hitherto unnoticed, and the new inspiration it has given to philosophical thought by stressing the extent to which all appearance depends on the choice of viewpoint, can hardly be overestimated. Still, on this occasion, it is proper to recall, as Einstein himself has constantly stressed, how deeply the new development is in debt to Newton's piercing analysis of the paradoxes implied in the conceptions of absolute time and space by means of which he reached a synthesis of so wide a scope.

Quite new prospects as regards natural philosophy have, however, in our days been opened by the deepening of our insight in the atomistic interpretation of physical phenomena. Just as Newton trusted, his fundamental principles found here a wide field of application. In the first place, the mechanical theory of heat proved the adequate means of accounting for the universal laws of thermodynamics which rest, on the one hand, directly on the theorem of conservation of energy and, on the other, on the interpretation by probability considerations of the entropy concept. In this last connexion, it should be emphasized that in the statistical treatment of thermal phenomena we have not primarily to do with any renunciation on the causal mode of description of the motion of each atom, but simply with the proper mathematical methods of applying mechanical concepts to the behaviour of large numbers of particles. Even as regards the disclosure of the structure of individual atoms made possible by the marvellous development of the art of experimentation, it was again Newton's principles which served as a guide in interpreting the new evidence. Thus, it was reliance on these principles which led J. J. Thomson to the recognition of the electron as a universal constituent of matter and Rutherford to his discovery of the atomic nucleus, which went so far in completing our picture of atomic constitution. Few discoveries more than Rutherford's have, indeed, in a few years to such an extent deepened and widened our outlook and even brought about so great and unexpected a revolution in human resources.

Equally essential for this progress, however, has been the recognition of a fundamental feature of atomicity in the laws of nature, which goes far beyond the ancient doctrine of the limited divisibility of matter. In fact, the discovery by Planck of the quantum of action, based on such a sagacious analysis of the general laws governing thermal radiation, may truly be said to have inaugurated a new era in science by revealing that all concepts of mechanics and electrodynamics are idealizations, the unambiguous application of which is restricted to the account of processes in which the actions involved are large compared with the universal quantum. While this condition is, of course, amply fulfilled in all such phenomena as were open to investigation in Newton's time and long after, we have now in the atomic processes clear evidence of a feature of individuality which refuses all analysis in terms of classical mechanical and electromagnetic concepts. In particular, such concepts are entirely insufficient to explain the remarkable stability of atoms on which the specific physical and chemical properties of the elements depend, and which is even essential for the very existence of the solid bodies which are our indispensable tools in all physical measurements.

In view of the fact that even our knowledge of the properties of the atomic particles, like their mass and electric charge, is derived from measurements interpreted on classical mechanical lines, we have been confronted with the task of developing a mode of description which incorporates the existence of the quantum, but nevertheless forms a generalization of Newtonian mechanics in which such interpretations of measurements appear as unambiguous applications. From the outset it was clear that this theory would involve a renunciation in principle of the causal mode of description. In fact, from the very character of the problem, there can only be a question of considering the relative probabilities of occurrence of the different individual quantum processes which can take place under given conditions.

In the first attempts to face this problem, it was tried to preserve mechanical ideas so far as they did not directly contrast with the most conspicuous quantum effects. But, though it was possible in this way to bring a certain degree of order within a wide field of experience and, in particular, to reach a preliminary survey of the electron binding in atoms, capable of accounting for the remarkable relationship between the properties of the elements brought out by general chemical evidence, it soon became clear that a comprehensive account of atomic phenomena could only be reached by a consistent and more harmonious generalization of the classical physical theories which, like Newton's own work, had to be based on a thorough revision of the foundation of all mechanical concepts.

Decisive progress in this respect has been obtained by the concurring efforts of a number of the most eminent physicists of our day, some of those to whom we owe the boldest and most ingenious steps being as young as Newton was when he conceived his fundamental ideas. The method by which the wanted generalization of Newtonian mechanics has been achieved consists, as is well known, in interpreting the kinematical and dynamical variables of the classical equations of motion as linear operators and in relating Planck's constant solely with the non-commutability of the multiplication of any two conjugated variables. An immediate consequence of this formalism is that no unambiguous interpretation of both variables of such a conjugated pair can appear in the description of physical phenomena, but that there will be a reciprocal relationship between the latitudes within which they can be fixed, finding its quantitative expression in Heisenberg's principle of indeterminacy. The whole scheme known as quantum mechanics rivals Newton's and Maxwell's theories in consistency and beautiful simplicity and has permitted us to bring order in an amount of new experimental evidence comparable with that encompassed by the theories of classical physics.

The radical departure from accustomed demands as regards the explanation of physical phenomena and, in particular, the wide-going renunciation of a causal description and even of a visualization in space and time have naturally given rise to doubts whether quantum mechanics satisfies all the requirements which a complete description has to fulfil. It must be realized, however, that in atomic physics we are confronted with an essentially new situation in which, as regards proper quantum effects, it is in principle excluded to separate sharply between intrinsic properties of the objects and their interaction with the measuring instruments which are indispensable for the very definition of the phenomena. Of course, there is no question of any arbitrary renunciation of Newton's famous postulate of equality of action and reaction, but only of the infeasibility of taking the reaction of the objects on the measuring instruments into account if these shall serve their purpose.

It is, indeed, compelling to recognize that by an experiment we simply understand an event about which we are able in an unambiguous way to state the conditions necessary for the reproduction of the phenomena. In the account of these conditions, there can, therefore, be no question of departing from the Newtonian way of description and, in particular, it may be stressed that by the clocks which together with the scales are used to define the space-time frame, we simply understand some piece of machinery as regards the working of which classical mechanics can be entirely relied upon and where, consequently, all quantum effects have to be disregarded.

Under these circumstances, questions of terminology become of paramount importance, just as was the case in Newton's pioneer work. All confusion arises, in fact, from the use of such utterances as 'disturbance of phenomena by their observation', a phrase equally irreconcilable with any unambiguous meaning of the very words 'observation' and 'phenomenon'. Indeed, in a

59

[223]

rational account of evidence, the latter word must simply be taken to involve a complete description of the experimental arrangement as well as the observed results, and it is also the establishment of the statistical rules governing such results which is the only aim of quantum mechanics.

The well-known paradoxes concerning apparently incompatible properties of the objects observed under different experimental conditions find in this way their elucidation in the mutual exclusion of the experimental arrangements concerned, and, in particular, all arbitrariness in the interpretation of quantum mechanics is reduced to our freedom in handling the instruments of observation, to the variety and ingenuity of which Newton has himself set such a great example. Although the phenomena in quantum physics can no longer be combined in the customary manner, they can be said to be complementary in that sense that only together do they exhaust the evidence regarding the objects, which is unambiguously definable. Truly, the view-point of complementarity may be said to present a rational generalization of the very ideal of causality.

The reason for entering in such length on this point is that we are here concerned with a general lesson of epistemology which, like that of Newtonian mechanics, has a bearing far beyond the limited domain of physical science. Especially in the study of psychical experience we are confronted with an observational problem which exhibits a deep analogy with that in atomic physics. Thus, in introspection there will always be a question of separation between the theme of conscious analysis and the background from which it is judged, reminding us of the necessity of distinguishing in atomic research between the objects under investigation and the measuring instruments. More concretely speaking, the use of words like 'thoughts' and 'emotions' exhibits striking analogy to the complementary application of kinematical and dynamical variables in quantum mechanics. In particular, the latitude for volition may, in our terminology, be said to be simply afforded by the very circumstance that situations where we can speak of freedom of will are mutually exclusive to situations where any kind of causal analysis of psychical experience can be reasonably attempted.

We are here most acutely confronted with the problem of our own position in existence, which occupied Newton so deeply in his later years. Above all, it may in this connexion be stressed that speaking of the attitude of a scientist in terms such as rationalism and mysticism is essentially ambiguous. In fact, in the ceaseless striving for harmony between content and form we shall always have to do with a fluent border of the regions where some degree of order is established. In this respect, indeed, it is most suggestive that the analysis and synthesis in atomic theory, which in certain respects deals with the simplest of all human experience, has so seriously reminded us of the old wisdom that in the great drama of life we are at the same time actors and spectators.

I am afraid that by such remarks I have perhaps come too far away from our central subject and, in concluding, I feel that I have to apologize for not having

60

[224]

been able, like the foregoing speakers, at this great occasion to bring before your consideration new materials from the inexhaustible source provided by the writings Newton left to posterity and the real exploration of which demands a knowledge of the state of science and philosophy at his time far deeper than I possess. I hope, however, that I have conveyed to you an impression of the living inspiration which Newton's work still exerts on all endeavours aiming at the progress of science in its widest sense. At this occasion, where the international character of such endeavours has found such impressive and welcome a manifestation, it is also in my heart to say how much it has meant to me to have the early and intimate contacts which it was my good fortune to entertain with the great school of English physicists, which has so brilliantly upheld the traditions for which Newton stands as the unsurpassed ideal.

X. NIELS BOHR

Slovenski Poročevalec, Ljubljana, 29 November 1958

Greeting on the National Day of the Republic of Yugoslavia

TEXT AND MANUSCRIPT

See Introduction to Part II, p. [105].

[227]

Niels Bohr

Danski fizik, pionir atomske znanosti — Nobelov nagrajenec

V krogu danskih fizikov smo že vse doslej spremljali z velikim občudovanjem navdušena in smiselna prizadevanja Federativne ljudske republike Jugoslavije, da bi tudi ona sodelovala v sodobnem razvoju ene izmed znanosti, v kateri ima sicer vaša dežela že tako bogate tradicije.

Fiziki po vsem svetu resnično zelo spoštujejo temeljno delo Rudjerja Boškovića in Jožefa Stefana, kot tudi velikanski prispevek k uporabi naše znanosti v korist skupnosti, povezan z velikima imenoma Nikole Tesle in Mihajla Idvorskega Pupina.

Na Danskem smo v zadnjih letih doživeli veliko veselja zaradi plodnega sodelovanja med jugoslovanskimi in danskimi ustanovami za atomsko fiziko, ki delajo za napredek na področju, ki ne vsebuje samo

PRVA NAGRADA ZA MIROLJUBNO UPORABO ATOMSKE ENERGIJE

izredno velikih obetov za poglobljeno razumevanje naravnih zakonov, temveč za ostvaritev človeške blaginje.

V tej zvezi so se ustvarile številne prijateljske vezi med jugoslovanskimi in danskimi znanstveniki. Ob tej priložnosti bi rad izrazil tudi svoje najtoplejše želje za nadaljnje in še bolj učinkovito sodelovanje, ki bi omogočilo vzajemno pomoč pri našem delu ter krepitev prijateljskih čustev med narodi naših dveh dežel.

MANUSCRIPT

[Niels Bohr]^1

Within the circle of Danish physicists we have with great admiration followed the enthusiastic and purposeful endeavours in the Federal Peoples' Republic of Yugoslavia to participate in the modern development of a branch of science, for which your country possesses such great traditions.

Indeed, physicists all over the world venerate the fundamental work of Rudjer Bošković and Jozef Stefan^2 as well as the ingenious contributions to the application of our science for the benefit of the community, connected with the great names of Nikola Tesla and Mihajlo Idworsky Pupin.^{3}

In Denmark, we have in later years derived much pleasure from the fruitful collaboration between Yugoslavian and Danish atomic physics institutions working for the progress in a field which not only contains such great promises for deepening the understanding of the laws of nature, but also for the promotion of human welfare.

---

^{1} This heading was added by the newspaper. So was the text in the square and the figure caption, which read, respectively: "Danish physicist, pioneer of atomic science – Nobel laureate"; "The first award for peaceful use of atomic energy". The newspaper presumably refers to the award described in *The Presentation of the first Atoms for Peace Award to Niels Henrik David Bohr, October 24, 1957*, National Academy of Sciences, Washington, D.C. 1957. Reproduced in part in Vol. 11, pp. [637]–[644].
^{2} (1835–1893).
^{3} (1858–1935).

In this connection, many bonds of friendship have been created between Yugoslavian and Danish scientists, and I wish to express the warmest wishes for the continuous and ever more effective collaboration to mutual help in our work and the strengthening of the friendly feelings between the peoples of our countries.

Copenhagen, November 12, 1958.[4]

Niels Bohr.

[4] This is the date of the manuscript, which was not reproduced in the newspaper article.

XI. R.J. BOŠKOVIĆ

"Actes du symposium international R.J. Bošković, 1958",
Académie Serbe des sciences,
Académie Yougoslave des sciences et des arts,
Académie Slovène des sciences et des arts,
Belgrade, Zagreb, Ljubljana 1959, pp. 27–28

See Introduction to Part II, p. [105].

NIELS BOHR

R. J. BOŠKOVIĆ

Extrait des Actes du Symposium
International R. J. Bošković
1 9 5 8

1 9 5 9
BEOGRAD, ZAGREB, LJUBLJANA

[232]

R. J. BOŠKOVIĆ

Niels Bohr, Kobenhavn

Ruder Bošković, whose life-work is receiving greater and greater attention in the scientific world of today, was one of the most prominent figures among the 18th century natural philosophers who enthusiastically elaborated the fundamental conceptions of Newtonian mechanics. Indeed, he did not only make important contributions to mathematics and astronomy, but strove with remarkable imagination and logical power to develop a systematic account of the properties of matter on the basis of interactions of mass points through central forces. In this respect, Bošković's ideas exerted a deep influence on the work of the next following generation of physicists, resulting in the general mechanistic views which inspired Laplace and, perhaps less directly, even Faraday and Maxwell.

It is true that in our days the approach to such problems has undergone essential changes. Above all, it has been recognized that the consistent description of atomic processes demands a feature of indivisibility, symbolized by the quantum of action and which goes far beyond the old, much debated doctrine of a limited divisibility of matter. This development has revealed an unsuspected limitation of the scope of mechanical pictures and even of the deterministic description of physical phenomena. However, it has been possible, through a most efficient collaboration between physicists from many countries, gradually to develop a rational generalization of the classical theories of mechanics and electrodynamics, which has proved capable of accounting for an ever increasing wealth of experimental data concerning the properties of matter.

When, against this background, one reflects on the development of natural philosophy through the ages, one appreciates the wisdom of the cautious attitude towards atomic problems, which reigned until the last century. I think not only of the belief that, owing to the coarseness of our tools and sense organs, it would never be possible to obtain direct evidence of phenomena on the atomic scale, but also of the often expressed skepticism as to the adequacy of pictorial models in a domain so far removed from ordinary experience. Although the marvellous development of experimental technique has permitted us to record effects of single atomic objects, we are here in a novel situation which has necessitated a radical revision of the fundaments for the unambiguous use of the elementary

[233]

conceptions, like space and time, and cause and effect, embodied in the language adapted to our orientation in practical life.

The elucidation of the situation with which we are confronted in atomic physics has been obtained by raising anew the old problem of what answers we can receive to questions put to nature in the form of experiments. Of course, no physicist from earliest times has ever thought that he could augment physical knowledge in any other way than by accounting for recordings obtained under well-defined experimental conditions. While, in this respect, there is no change of attitude since Bošković's time, we have in our days, as is well known, received a new lesson regarding our position as to analysis and synthesis of such knowledge.

Our esteem for the purposefulness of Bošković's great scientific work, and the inspiration behind it, increases the more as we realize the extent to which it served to pave the way for later developments. In friendly and fruitful international cooperation physicists are working today, in Yugoslavia as in all other countries, for the progress of our knowledge of the atomic constitution of matter and for the application of this knowledge, which holds out promises surpassing even those of the technology based on classical physics. In the pursuit of such novel developments, it is essential that we not only keep an open mind for unforeseen discoveries, but that we are conscious of standing on the foundations laid by the pioneers of our science.

The 200th anniversary of the publication of Bošković's famous Theoria Philosophiae Naturalis could hardly be commemorated in a more fitting manner than by an international congress in the country of his birth, convened on the occasion of the opening of the museum in Dubrovnik with its historical treasures. In pointing to the future, it is also a most fortunate omen that the great occasion could be combined with the inauguration of the modern research institute in Zagreb, which bears Ruder Bošković's name and where Meštrović's impressive statue will daily remind students of the traditions on which they are building and inspire them to fruitful contributions to common human knowledge.

XII. [TRIBUTE TO TESLA]

"Centenary of the Birth of Nikola Tesla 1856–1956",
Nikola Tesla Museum, Belgrade 1959, pp. 46–47

See Introduction to Part II, p. [106].

NIELS BOHR

At this international conference in memory of the centenary of the birth of Nicola Tesla, the Danish Academies of Science and Technology and the Universities in Denmark want to join academic institutions and industrial organizations from all over the world in expressing our warmest felicitations.

Nicola Tesla's most ingenious invention of polyphase transmission of electric power and his researches on the marvellous phenomena of high freqency electric oscillations have been fundamental for that development which in our days has created quite new conditions for industry and radiocommunication and so deeply influences our whole civilization.

Niels Bohr

With deepest admiration we think of how Nikola Tesla could accomplish such great achievements and exert such influence in countries then more advanced in scientific and industrial developments than the country of his birth and youth, from where he brought his searching and independent spirit.

We rejoice in the energetic and enthusiastic efforts which now are taking place in Yugoslavia in science and technology and want with our felicitations at this solemn occasion to express our sincere hope for fruitful future collaboration with the people of Yugoslavia, so rich in heroic and cultural traditions.

XIII. [TRIBUTE TO BERING]

"Vitus Bering 1741–1941",
H. Hagerup, Copenhagen 1942, pp. 49–53

TEXT AND TRANSLATION

See Introduction to Part II, p. [106].

[237]

VITUS BERING

1741 · 1941

H. HAGERUP · KØBENHAVN
1942

VI er i Dag samlede for at mindes en af de største Bedrifter, nogen dansk Mand har øvet, og vel den, hvis Ry er naaet videst om. Hvert Menneske paa Jorden, der blot har mindste geografiske Kundskab, har hørt om Bering-Strædet, hvis Opdagelse blev en af de store Milepæle i vor Klodes Kortlægning. Den Opdagerfærd, som Vitus Bering under de vanskeligste Forhold og med Indsats af sit eget Liv gennemførte, har for stedse sikret ham en Plads blandt Menneskeslægtens Foregangsmænd og vil til sene Tider vidne om de gamle danske Traditioner til Lands og til Søs, som var Baggrunden for den Tillid, der i det fremmede blev ham til Del, og til hvilken han viste sig saa værdig.

At holde henfarne Slægters Bedrifter i Ære er ikke blot vor Pligt imod dem, men ogsaa en Hovedkilde til Styrkelse for os selv. Uadskillelig fra Beundringen og Taknemmeligheden er jo Opmuntringen til at følge efter i samme Spor og Fortrøstningen om, at vi netop i vore Traditioner kan finde Kraften dertil. Det skulde da ogsaa blive en af de Mænd, der i vor Tid har kastet Glans over vort Land ved Forskerfærd i samme Aand, der paa en Maade, der fandt Genklang hos hele vort Folk, først slog til Lyd for, at vi her i Vitus Berings Fædreland rejste ham et Minde, der kunde fortælle Verden om hans Landsmænds Værdsættelse af hans Daad.

49

Knud Rasmussen, hvis uforglemmelige Bedrifter og begejstrende Personlighed har gjort ham til et Ideal for dansk Ungdom, holdt jo for lidt mere end 10 Aar siden ved en af de aarlige Fester i Ræbild Bakker for danske, bosatte i fjerne Lande, den ildnende Tale, der gav Anledning til, at der til at varetage Bestræbelserne for at løse denne Opgave paa rette Maade nedsattes en Komité med Repræsentanter for vide Kredse af vort Samfund. For denne Komité overtoges Protektoratet af Hs. Kgl. Højhed Kronprins Frederik, der stadig har fulgt Sagen med den varmeste Interesse og meget beklager at være forhindret i at komme til Stede ved denne Mindefest.

Oprindelig var det jo Tanken, at Mindet skulde rejses ikke alene ved Bidrag herhjemme fra, men ogsaa fra de mange danske, der som Vitus Bering selv havde fundet Virkefelt og opnaaet Anseelse i det Fremmede. Som Forholdene i Verden udviklede sig, blev dette dog uigennemførligt, og i den Prøvelsens Tid, som vi nu maa gennemleve, adskilt fra vore Landsmænd i Udlandet, vil Gennemførelsen af Sagen med egne beskednere Midler staa som et Symbol for vor faste Vilje til at hævde vore Traditioner og samtidig til danske Verden over være Bud om de ubrydelige Mindernes Baand, der knytter deres Skæbne til vor.

Før denne for alle danske saa vigtige Opgave har kunnet løses paa smuk og værdig Maade, har Vejen været lang, og især blev Komitéens Arbejde vanskeliggjort derved, at dens selvskrevne Formand, Knud Rasmussen, til hele Landets Sorg skulde gaa saa tidligt bort. Interessen holdtes dog stadig vedlige, ikke mindst ved de utrættelige Bestræbelser, der udvistes af Politimester Krause her i Horsens og af Statsbibliotekar Steensgaard fra Aarhus. Da Tiden nærmede sig for 200-Aarsdagen for Vitus Berings Død, blev

50

paa Statsminister Staunings Initiativ Komitéen, hvoraf mange Medlemmer efterhaanden var afgaaet ved Døden, reorganiseret, og man samlede sig hurtigt om mere bestemte Planer for at hædre Vitus Berings Minde.

For det første kunde Medlemmerne fra Horsens oplyse, at man indenfor Bestyrelsen af det af Overretssagfører Danjelsen og Hustru til Forskønnelse af Horsens By oprettede store Legat var sindet at yde en meget betydelig Sum til et Mindesmærke for Vitus Bering her i hans Fødeby. For det andet besluttede man gennem en Indsamling indenfor en snævrere Kreds tillige at oprette et Vitus Bering Mindefond, der skulde overdrages Det Kongelige Danske Geografiske Selskab, til Bekostning af en Vitus Bering Medaille med tilhørende Pengepris, hvis Uddeling for store Fortjenester indenfor den geografiske Forskning vil give en stadig tilbagevendende Lejlighed til at minde Verden om Vitus Berings Tilknytning til Danmark.

Gennemførelsen af begge disse Planer er nu sikret. Ved et særligt Møde paa Horsens Raadhus i Formiddag har Byraadet gennem Borgmester Juliussen, der er Formand for Bestyrelsen af det Danjelsenske Legat, modtaget Meddelelse om, at Legatet har besluttet at stille indtil 300.000 Kr. til Raadighed for Opførelsen af et storslaaet Mindesmærke paa Grundlag af Planer udkastede af Billedhugger Mogens Bøggild og Arkitekt Kaare Klint. Dette Mindesmærke, som paa egenartet kunstnerisk Maade fremstiller Vitus Berings Fart over det ukendte Hav, bliver en Pryd for Horsens og vil i Fremtiden fortælle om, hvorledes Vitus Berings Daad holdtes i Ære i hans gamle By og om det Borgersind, senere Slægter her har lagt for Dagen.

Endvidere kan jeg meddele, at der ved Bidrag fra forskellige Institutioner og Privatpersoner hidtil er indsamlet

51

35.000 Kr. til Mindefondet, til hvilket yderligere Bidrag naturligvis vil være særdeles kærkomne. Paa Komitéens Vegne vil jeg her udtale en Tak til alle Bidragyderne, der har gjort det muligt at sikre Mindefondet en Grundkapital, og uden at gøre nærmere Rede for de enkelte Bidrag vil

jeg gerne nævne, at et af de største af disse stammer her fra Horsens, hvor Fabrikant Aggerbeck straks efter den oprindelige Komités Oprettelse hensatte et Beløb, som han, efter at tilstrækkelige Midler til Mindesmærket nu er fremskaffet paa anden Maade, velvilligt har overdraget til Mindefondet.

Til Vitus Bering Medaillen er der efter Komitéens Opfordring af Medailleur ved den Kongelige Mønt, Harald

52

Salomon, udarbejdet en Model, der paa den ene Side viser
et Kort over Beringstrædet og paa den anden Side en Laur-
bærkrans, der skal omgive Navnet paa den, **Medaillen
vil blive tilkendt.** Gipsmodellerne af Medaillens Sider i
forstørret Maalestok er udstillet her i Salen, og Komitéen

ønsker at overdrage Horsens en Gengivelse af disse Mo-
deller i Bronce til Opbevaring sammen med andre Minder
om Byens berømte Søn.

Jeg vil gerne slutte med at udtale Haabet om, at Mindesmærket og Mindefondet tilsammen paa lykkelig Maade
maa løse den Opgave at ære Mindet om vor store Lands-
mand og tjene til at styrke Sammenholdet mellem alle
Danske inden og uden for Landets Grænser.

53

TRANSLATION

We are gathered together today to remember one of the greatest exploits any Danish man has accomplished and probably the one that is most widely known. Every person on Earth who has only the slightest knowledge of geography has heard of the Bering Strait, whose discovery became one of the great milestones in the mapping of our planet. The voyage of discovery that Vitus Bering made under the most difficult conditions and with his own life at stake, has ensured him a place forever among the pioneers of the human race and will in distant times give witness to the old Danish traditions on land and at sea, which were the background for the trust which was shown to him abroad and for which he proved himself so worthy.[1]

To honour the deeds of departed generations is not only our duty towards them but also a main source of invigoration for ourselves. Inseparable from the admiration and the gratitude is of course the encouragement to follow in the same path and the reassurance that precisely in our traditions we can find the strength to do so. It was also to be one of the men who in our times have lent lustre to our country with expeditions in the same spirit who, in a way that won acclaim with all the people, was the first to suggest that we here in Vitus Bering's native country should erect a monument for him, which could tell the world about the appreciation of his achievement by his countrymen.

Knud Rasmussen, whose unforgettable exploits and inspiring personality have made him an ideal for Danish youth, gave, a little more than ten years ago, at one of the annual celebrations at Rebild Bakker for Danes living in distant countries,[2] the animating speech that gave rise to the setting up of a committee, with representatives from wide circles in our society, to look after the efforts for solving this task in the right way. The patronage of this committee was assumed by His Royal Highness Crown Prince Frederik, who has constantly taken the warmest interest in the matter and greatly apologizes for being unable to be present at this memorial celebration.

Originally, the intention was that the monument should be erected not only with contributions from here at home but also from the many Danes who, like Vitus Bering himself, have found a sphere of activity and achieved sta-

---

[1] This sentence was to serve as an inscription for the monument for Bering. However, the monument was never erected.

[2] The Fourth of July, the Independence Day of the United States, has been celebrated in Rebild Bakker (Rebild Hills) in northern Jutland every year since 1912, with the exception of the years during the World Wars.

tus abroad. As conditions in the world developed, this became impossible,[3] however, and in the trying time through which we must now live, cut off from our countrymen abroad, the realization of the matter with our own more modest means will stand as a symbol of our strong determination to uphold our traditions and will at the same time bear witness to Danes world-wide of the unbreakable ties of the memories linking their fate to ours.

It has been a long road before this task, so important for all Danes, could be solved in a beautiful and fitting way, and, in particular, the committee's work was made difficult by the early death, to the sorrow of the whole country, of Knud Rasmussen, the eminently qualified chairman. However, interest was maintained, not least by the untiring efforts displayed by Chief Constable Krause here in Horsens[4] and by State Librarian Steensgaard from Aarhus.[5] As the day of the 200-year anniversary of Vitus Bering's death approached, the committee, many of whose members had died in the meanwhile, was reorganized on the initiative of Prime Minister Stauning,[6] and soon more definite plans to honour the memory of Vitus Bering were agreed upon.

Firstly, the members from Horsens could relate that the board of directors of the large charitable trust set up by Barrister Danjelsen[7] and his wife[8] for the embellishment of the town of Horsens intended to donate a very considerable sum to a monument for Vitus Bering here in the town of his birth. Secondly, it was decided, through subscription within a small circle, also to establish a Vitus Bering Memorial Foundation,[9] which should be transferred to the Royal Danish Geographical Society,[10] for the funding of a Vitus Bering Medal with associated prize money. The awarding of this medal for great services within geographical research will provide a constantly recurring opportunity to remind the world of Vitus Bering's connection to Denmark.

The execution of both these plans is now ensured. At a special meeting at Horsens Town Hall this morning the Town Council has received the information, through Mayor Juliussen,[11] the chairman of the board of the Danjelsen

---

[3] The celebrations took place in December 1941, in the midst of World War II and the German occupation of Denmark.

[4] Richard C. Krause (1878–1945).

[5] Erling Steensgaard (1876–1966).

[6] The Social Democrat Thorvald Stauning (1873–1942) was Prime Minister of Denmark from 1924 to 1926 and from 1929 until his death.

[7] Christian Daniel Danjelsen (1842–1921).

[8] Karen Danjelsen, née Steenberg.

[9] Established with means collected by the Vitus Bering Committee, of which Bohr was Chairman.

[10] Established in 1876.

[11] Jens Christian Juliussen (1884–1968) was Mayor of Horsens from 1939 to 1949.

Trust, that the Trust has decided to make up to 300,000 kroner available for the construction of a magnificent monument on the basis of plans drawn up by the sculptor Mogens Bøggild and the architect Kaare Klint. This monument, which in a singularly artistic fashion demonstrates Vitus Bering's journey across the unknown sea, will adorn Horsens and will in the future relate how Vitus Bering's achievement was honoured in his old town and about the spirit of citizenship later generations have displayed here.

I can further report that until now, through contributions from various institutions and private persons, the sum of 35,000 kroner has been collected for the memorial foundation, to which further contributions will of course be extremely welcome. On behalf of the committee I here express thanks to all the contributors who have made it possible to secure the memorial foundation a basis capital, and without specifying the individual contributions I would like to mention that the largest of these comes from Horsens, where factory owner Aggerbeck,[12] immediately after the setting up of the original committee, provided a sum, which he, after sufficient funds for the monument have now been acquired in another way, has generously transferred to the memorial foundation.

On the committee's behest Harald Salomon,[13] medallist at the Royal Mint, has made a model for the Vitus Bering Medal, showing a map of the Bering Strait on the one side and, on the reverse, a laurel wreath which will encircle the name of the person to whom the medal will be awarded. The plaster casts of the two faces of the medal on a larger scale are on display here in the hall, and the committee wishes to hand over to Horsens a bronze reproduction of these models to be kept together with other mementos of the town's famous son.

I would like to close by expressing the hope that the monument and the memorial foundation may together accomplish in a happy way the task of honouring the memory of our great countryman and serve to strengthen the concord between all Danes inside and outside the country's borders.

[12] Einar Harry Aggerbeck (1889–1977).
[13] Salomon (1900–1990) worked as medallist at the Royal Mint from 1928 to 1968.

XIV. FOREWORD

FORORD
"Johan Nicolai Madvig: Et mindeskrift",
Royal Danish Academy of Sciences and Letters
and the Carlsberg Foundation, Copenhagen 1955, p. vii

With J. Pedersen

TEXT AND TRANSLATION

See Introduction to Part II, p. [107].

FORORD

Da Johan Nicolai Madvig døde d. 12. December 1886, 82 år gammel, havde han allerede længe været dansk videnskabs store gamle mand. Overalt i den videnskabelige verden havde han vundet autoritet ved sin indsats i den klassiske filologi som sprogmand, som textkritiker, som historiker og som samfundsskildrer. Som universitetslærer havde han præget en hel generation af skolens lærere, som undervisningsinspektør og minister havde han haft en dybtgående indflydelse på skolens ordning, som politiker havde han virket for reformer efter datidens nye idealer uden skarpe brud med fortiden. Overalt nød han den største tillid og han havde fået statens højeste udmærkelser. Kernen i alt hvad han udrettede var og blev imidlertid hans videnskabelige værk, og i nær tilknytning til det stod hans virksomhed i Det Kongelige Danske Videnskabernes Selskab. Efter at han havde forladt universitetsgerningen og opgivet andre hverv, bevarede han til sin død stillingen som præsident i Videnskabernes Selskab og som formand i Carlsbergfondets direktion. Det er derfor naturligt at disse to nær forbundne institutioner har forenet sig om at skabe og overgive til eftertiden et billede af Madvigs person og hans mangesidige virksomhed.

Det var hensigten, at denne publikation skulde fremkomme i forbindelse med 150-året efter Madvigs fødsel, idet hans værk nu er kommet så meget på afstand at vi kan se det i historiens lys, uden at det er blevet så fjernt at vi helt har mistet følingen med det. Imidlertid har forskellige omstændigheder medført nogen udsættelse af publikationen. Til den foreliggende 1. del, der omfatter skildringer af Madvigs levned og hans politiske virksomhed, af hans forhold til skolen og til Videnskabernes Selskab og Carlsbergfondet, vil senere komme en 2. del, der behandler hans forhold til Universitetet og hans virksomhed som filolog.

Maj 1955.

    Niels Bohr   *Johs. Pedersen*

[248]

TRANSLATION

FOREWORD

When Johan Nicolai Madvig died on 12 December 1886, 82 years of age, he had already been the grand old man of Danish scholarship for a long time. He had won authority everywhere in the scholarly world with his efforts in classical philology, as linguist, as text critic, as historian and as portrayer of society. As university teacher he had influenced a whole generation of school teachers, as Chief Inspector of Schools and Minister he had had a profound influence on the school system, as politician he had worked for reforms according to the new ideals of his time without sharp breaks with the past. He enjoyed the greatest trust everywhere and he had received the highest honours of the State. The core of everything he achieved, however, was and remained his scholarly work, and closely related to this was his activity in the Royal Danish Academy of Sciences and Letters. After he had retired from his university career and given up other posts, he kept until his death his post as President of the Academy of Sciences and Letters and as Chairman of the Board of the Carlsberg Foundation. It is therefore natural that these two closely connected institutions have joined to create and present to the future a portrait of Madvig as a person and of his many-faceted activities.

The original objective was that this publication should appear in connection with the 150th anniversary of Madvig's birth, as his work is now at such a distance that we can see it in the light of history, without it having become so remote that we have completely lost touch with it. However, due to various circumstances the publication has been subject to some delay. In addition to the present first part, which comprises accounts of Madvig's life and his political activity, of his relation to school education and to the Academy of Sciences and Letters and the Carlsberg Foundation, a second part will appear later, which deals with his relation to the University and his activities as philologist.[1]

May 1955.

Niels Bohr   *Johannes Pedersen*

---

[1] This second volume was published in 1963.

3. PHYSICS TEACHERS AND COLLEAGUES

XV. SIR J.J. THOMSON'S SEVENTIETH BIRTHDAY

Nature **118** (1926) 879

See Introduction to Part II, p. [108].

[251]

Sir J. J. Thomson's Seventieth Birthday.

MESSAGES OF CONGRATULATION FROM ABROAD.

Prof. NIELS BOHR, For.Mem.R.S.,

University Institute of Theoretical Physics, Copenhagen.

It is with great pleasure that I accept the invitation of the editor of NATURE to take part in the universal celebration of the seventieth birthday of Sir J. J. Thomson, to whom every one interested in the problem of the constitution of the atom is so greatly indebted. Not to speak of the leading part he has taken in the discovery of the electron as a common constituent of all atoms, we owe to him an abundance of ideas which have proved fruitful in the attempts to develop a detailed theory of atomic constitution based on this fundamental discovery. At a time when even the existence of atoms was regarded by many prominent scientists with scepticism, Thomson had the courage to venture on an exploration of the inner world of the atom. Guided by his wonderful imagination and leaning on the new discoveries of the cathode rays, Röntgen rays and radioactivity, he opened up an unknown land to science. By following electrical particles and ether-waves on their way through atoms he obtained the first estimate of the number of electrons contained in an atom and of the forces by which they were bound, laying in this way the foundation of that elaborate structure which has been built up in recent years through the joint efforts of a large number of workers. We find in his famous attempt to account for the remarkable periodicity of the physical and chemical properties exhibited by the elements when arranged in the order of their atomic weights, the germ of the ideas characteristic of the modern interpretation of the periodic table. Indeed, it is difficult for scientists of the younger generation, who are working on the new land to which Thomson has opened the gates, fully to realise the magnitude of the task with which the pioneers were confronted.

XVI. PROFESSOR SIR ERNEST RUTHERFORD AND HIS SIGNIFICANCE FOR THE RECENT DEVELOPMENT OF PHYSICS

PROFESSOR SIR ERNEST RUTHERFORD
OG HANS BETYDNING FOR
FYSIKENS NYERE UDVIKLING
Politiken, 18 September 1920

TEXT AND TRANSLATION

See Introduction to Part II, p. [109].

Professor Sir Ernest Rutherford og hans Betydning for Fysikens nyere Udvikling.

Den verdensberømte engelske Fysiker, Professor Sir Ernest Rutherford begynder i Dag paa vort Universitet en kort Række Forelæsninger efter Indbydelse af Danmarks naturvidenskabelige Samfund. Nedenfor giver hans danske Kollega Professor Niels Bohr en Skildring af Rutherford og hans Betydning for Fysikens nyere Udvikling.

Den Mand, hvis Navn staar over disse Linjer, og som i disse Dage gæster Kjøbenhavn for efter Indbydelse af Danmarks naturvidenskabelige Samfund at holde en Række Forelæsninger paa Universitetet, har bag sig et videnskabeligt Livsværk, som selv paa Baggrund af den vidunderlige Udvikling, som Naturvidenskaben har undergaaet i de sidste Decennier, tegner sig paa den mest straalende Maade.

For blot 20—30 Aar tilbage var det en almindelig udbredt Anskuelse, at den fysiske Videnskab maatte søge sine Fremtidsmuligheder igennem Forfinelse af Maalingerne af de kendte fysiske Fænomener og af den matematiske Behandling af Maalingsresultaterne, idet man mente sig berettiget til den Antagelse, at man kunde overse Omraadet af iagttagelige fysiske Fænomener og de for dem gældende Naturlove. Vel var man klar over, at de iagttagne Fænomener i og for sig ikke var af primær Natur, og at de fremkom som Sammenspil af Virkninger af det uhyre Antal Smaadele, hvoraf de iagttagelige Legemer maatte tænkes at bestaa; men man følte sig overbevist om, at det paa Grund af disse Smaadeles Lidenhed stadig vilde ligge uden for den af Menneskene opnaaelige Grænse at iagttage Virkningerne af de enkelte Smaadele hver for sig. Vidnesbyrd om dette Synspunkt finder man i Lærebøger i Fysik og Kemi fra den omhandlede Tid, hvor det stadig fremhæves, at man til Trods for Atomhypotesens tilsyneladende Nødvendighed for Beskrivelsen af Naturfænomener, dog altid maatte holde for Øje, at man havde med en saakaldt Arbejdshypotese at gøre, selv om det af alle maatte indrømmes, at det her drejede sig om en saadan af enestaaende Frugtbarhed. Dette Forhold har imidlertid fuldstændig forandret sig i den forløbne Tid, idet vi nu om Stunder ikke alene har uomtvistelige Beviser paa de enkelte Atomers Eksistens, ja behersker Metoder til med stor Nøjagtighed at bestemme Antallet af Atomer i de iagttagelige Legemer, men endda besidder et indgaaende Kendskab til de Bygestene, hvoraf de forskellige Grundstoffers Atomer er sammensat.

Denne Udvikling er fremkommet ved Sammenvirkning af mange forskellige Omstændigheder. Dels er man igennem teoretiske Studier ført til nye Synspunkter, der har aabenbaret en uanet Sammenhæng mellem tilsyneladende vidt forskellige Grupper af Fænomener, der, idet de maa tilskrives Virkninger af de samme Smaadele, sammenholdt aabner Mulighed for at drage bestemte Slutninger om de enkelte Smaadeles Virkninger. Som et Resultat, naaet paa denne Maade, kan nævnes, at det har været muligt ved Sammenligning mellem Resultater, hentet fra den mekaniske Varmeteori og den elektromagnetiske Varmestraalingsteori med stor Nøjagtighed at bestemme Antallet af Atomer i et Gram af et givet Grundstof; for et Gram Brint er dette Tal cirka seks Hundred Tusind Milliard Milliarder. Desuden har man lært et Antal nye Fænomener at kende, der, om end betinget af Tilstedeværelsen af de sædvanlige, kendte Stoffer, i Modsætning til de tidligere kendte Fænomener giver os Vidnesbyrd af mere direkte Karakter om de enkelte Atomers Virkninger. Dette gælder navnlig de mange smukke Fænomener, som man har opdaget ved det nærmere Studium af elektriske Udladninger gennem Luftarter. Jeg skal blot nævne, at Sir J. J. Thomson, hvis banebrydende Arbejder vi først og fremmest skylder vort Kendskab til og Forstaaelse af disse Fænomener, paa Basis af et Studium af disse ikke alene har kunnet uddanne en forholdsvis direkte Metode til Bestemmelse af Atomernes Antal, der har givet Resultater, der stemmer nøje overens med dem, der udledes ad den nys omtalte, mere indirekte teoretiske Vej, men ogsaa har kunnet fastslaa Tilstedeværelsen i alle Grundstoffers Atomer af smaa, ensartede Partikler, de saakaldte *Elektroner*, der besidder en negativ elektrisk Ladning og en Masse, som er meget lille i Forhold til hele Atomets Masse. Dette sidste Resultat har haft den største Betydning, idet man saaledes for første Gang kunde iagttage og undersøge i alt Fald visse af Atomernes Byggestene.

Disse Resultater skulde imidlertid hurtigt blive fuldstændiggjorte og overfløjne ved Opdagelsen og Undersøgelsen af et Antal Grundstoffer med helt nye Egenskaber, nemlig de saakaldte *radioaktive* Stoffer. Mange Læsere vil kunne erindre, hvor stor Interesse Opdagelsen af Radium, der navnlig skyldtes Mme. Curie's indsigtsfulde og energiske Arbejde, vakte omkring Aarhundredskiftet, og den Opsigt, de vidunderlige og tilsyneladende ubegribelige Virkninger af dette Stof fremkaldte. I Øjeblikket staar disse Virkninger dog ikke mere i et saadant mystisk Skær som den Gang, idet der gennem det enestaaende geniale Arbejde, som Sir *Ernest Rutherford* i Løbet af den siden da forløbne Tid har udført, er kastet det klarest tænkelige Lys over de fleste af de Fænomener, der knytter sig til de radioaktive Stoffer. Hans Studier af disse Fænomener har ledet til Resultater, der ikke alene ikke strider imod de Slutninger, man havde draget af tidligere Erfaringer, men netop belyser og tillader en dybere Forstaaelse af disse, samtidig med at de har aabnet helt nye og uanede Veje til at trænge ind i Spørgsmaalet om Atomernes Bygning.

Rutherfords første store Indsats paa detteOmraade bestod deri, at han erkendte, at den tilsyneladende lunefulde Optræden og Foranderlighed af de iagttagne Fænomener fuldkommen forsvandt, dersom man antog, at man i de radioaktive Stoffer havde at gøre med Grundstoffer, hvis Atomer uden ydre Indgreb undergik eksplosionsagtige Forandringer, ved hvilke elektrisk ladede Dele af Atomet, de saakaldte A(alfa)- og B(beta)-Partikler, udslyngedes, samtidig med at der af det tilbageblevne dannedes et Atom af et nyt Grundstof. Medens det nye Stof i mange Tilfælde vil være af en lige saa stabil Natur som de sædvanlige Grundstoffer, vil det ofte selv igen være radio-aktivt og dets Atomer undergaa en ny eksplosionsagtig Forvandling o. s. v. For denne dristige Hypotese, der med et Slag bragte Orden og Sammenhæng i det tilsyneladende Virvar, lykkedes det Rutherford og hans Medarbejdere at faa afgørende Bevis ved at isolere et stort Antal saadanne Forvandlingsprodukter, hvis kemiske Undersøgelse viste, at de besad Egenskaber, der ganske svarede til de kendte Grundstoffer, fra hvilke de kun skilte sig ved, at deres Atomer havde en begrænset „Levetid", der før eller senere afbrydes af en Eksplosion.

I de følgende Aar undersøgte Rutherford igennem en Række storslaaet anlagte og glimrende udførte Forsøg de Fænomener, som ledsager Atomernes Eksplosion. Som et særlig smukt Resultat af disse Undersøgelser kan nævnes, at det viste sig, at de ved Eksplo-

sionen udsendte Partikler besad saa stor Bevægelsesenergi, at Virkningen af en enkelt udslynget Partikel kunde iagttages. Herved kom man i Besiddelse af en direkte Metode til at tælle Atomerne, hvis Resultater viste sig at stemme fuldkommen overens med dem, der var fundet ved Hjælp af de før omtalte Metoder. Det Hovedresultat, Rutherford opnaaede, var dog de Slutninger, han formaaede at drage om Atomernes Byggestene. Idet de enkelte Partiklers Baner kunde følges paa deres Gang gennem Luftarter og andre Stoffer, viste det sig nemlig, at man kunde faa Oplysning om Bygningen af de Atomer, de mødte. Det fandtes saaledes, at et Atom i Almindelighed var meget let gennemtrængelig for Partiklerne, idet disse oftest uden mærkelig Retningsforandring kunde gennemtrænge mangeTusinde Atomer, Men en enkelt Gang fandtes Partiklen ved Sammenstødet med et Atom at lide storAfbøjning, hvilket, som Rutherford viste, maatte tydes derved, at Atomet foruden de negativ ladede Elektroner besad en Kerne, som var ladet med positiv Elektricitet, og som indeholdt næsten hele Atomets Masse. Da Kernen er uhyre lille i Forhold til hele Atomet, er dette bygget paa ganske lignende Maade som et Solsystem; Atomet bestaar af et Antal lette Elektroner, der bevæger sig omkring den tunge Atomkerne, som Planeterne omkring Solen. Ved denne Rutherford's Opdagelse af *Atomkernen*, der blev bestyrket paa mangfoldige Maader, maa man antage at have lært samtlige Atomets Byggestene at kende, og man besidder derfor nu et Grundlag for at trænge ind i Spørgsmaalet om Forklaringen af Stoffernes Egenskaber ved Hjælp af nøjere udarbejdede Forestillinger om Atombygningen.

Medens det ikke vil være muligt her at komme nærmere ind paa den teoretiske Behandling af det sidste Spørgsmaal og de forjættende Udsigter, der aabner sig for at skabe en fælles Basis for Forklaringen af saa vel Kemiens som Fysikens Fænomener, skal jeg endnu blot omtale det nye vigtige Resultat, som Rutherford har naaet ved Undersøgelser i den allersidste Tid. Ved overordentlig fint udførte og vanskelige Eksperimenter har han kunnet iagttage, hvorledes ved et Sammenstød mellem et Atom og en Alfa-Partikel, hvor Partiklen rammer lige ind i Atomkernen — hvad naturligvis sker uhyre sjældent — denne sidste under Omstændigheder kan sønderdeles i Brudstykker, der hver for sig vil danne Kerner for Atomer af andre Grundstoffer. Idet man gennem dette Resultat har fundet Midler til vilkaarlig at fremkalde en Proces, der ganske svarer til den, der spontant foregaar i den ved de radio-aktive Stoffers Forvandlinger iagttagne Eksplosioner af Atomkernen, kan Naturvidenskaben for saa vidt siges at være indtraadt i en ny Epoke, som Bestandigheden af Grundstoffernes Atomer overfor ydre Angreb hidtil har været anset for den faste Grund for fysiske og kemiske Betragtninger. Disse Linjers Hovedformaal har dog netop været at forsøge at give Læseren Indtryk af, hvor naturligt og direkte Rutherfords nye Resultat slutter sig til hans tidligere ovenfor nævnte Resultater, hvorigennem de Grundforestillinger, hvormed de kemiske og fysiske Videnskaber arbejder, er udviklet til en vidunderlig og uanet Klarhed.

Det ovenstaaende er naturligvis kun en kort Skitse af ganske faa Hovedresultater af Sir Ernest Rutherfords Arbejder, der ved Omfang og Idérigdom er uden Sidestykke i Nutidens eksperimentalfysiske Forskning. Naar det har været muligt for en enkelt Mand at gøre en saa enestaaende Indsats i Videnskaben, skyldes det naturligvis først og fremmest hans overlegne videnskabelige Geni, men samtidig tillige en Række almenmenneskelige Egenskaber, der har sat ham i Stand til at samle et stort Antal Elever omkring sig og vejlede og begejstre dem. Det ligger uden for min Evne blot at forsøge at beskrive disse Egenskaber nærmere, men dette er jo ogsaa saa meget mindre nødvendigt, som den danske videnskabeligt interesserede Offentlighed i disse Dage vil faa Lejlighed til selv at lære hans karakterfulde Personlighed at kende. Som en af de mange, der har haft den Lykke at arbejde i hans Laboratorium under hans Ledelse, vil jeg blot gerne give Udtryk for den Taknemmelighed og Beundring, hvormed Sir Ernest Rutherford omfattes af alle hans Elever.

N. Bohr.

TRANSLATION

Professor Sir Ernest Rutherford and his Significance for the Recent Development of Physics.

Professor Sir Ernest Rutherford, the world-famous British physicist, starts today at our University a short series of lectures at the invitation of the Science Society of Denmark. Below, his Danish colleague Professor Niels Bohr gives a description of Rutherford and his significance for the recent development of physics.

The man whose name stands above these lines and who is visiting Copenhagen at the present time on the invitation of the Science Society of Denmark in order to give a series of lectures at the University, has behind him a life in science which, even on the background of the wonderful development that science has gone through in the last decades, appears in the most brilliant fashion. Only 20–30 years ago it was generally assumed that the future of physical science was to be sought through the refinement of the measurements of the known physical phenomena and of the mathematical treatment of the measurement results, as one felt entitled to assume that one could command the area of observable physical phenomena and the laws of nature appertaining to them. It was clear, of course, that the observed phenomena on the whole were not of a primary nature, and that they arose as a combination of effects of the enormous number of small parts that observable bodies must be thought to consist of; but one felt convinced that, because of the minuteness of these small parts, it would remain beyond the limits of humans to observe the effects of each small part. Evidence for this viewpoint is to be found in physics and chemistry textbooks from the time in question, where it is constantly emphasized that despite the seeming necessity of the atomic hypothesis for the description of natural phenomena, one should nevertheless always keep in mind that one had to do with a so-called working hypothesis, even though this must be admitted by all to be one of singular fruitfulness. However, this state of affairs has changed completely in the meantime, as we nowadays not only have irrefutable evidence for the existence of single atoms – indeed, we master methods for determining the number of atoms in observed bodies with great accuracy – but even have detailed knowledge about the building blocks which make up the atoms of the various elements.

This development has come about through the interplay between many different circumstances. Partly, one is led through theoretical studies to new viewpoints that have revealed an unsuspected relationship between seemingly completely different groups of phenomena which, as they must be ascribed to effects of the same small parts, taken together open the possibility of drawing definite conclusions about the effects of the individual small parts. As a result reached in this way can be mentioned that it has been possible, by comparison of results taken from the mechanical theory of heat and the electromagnetic theory of heat radiation, to determine with great accuracy the number of atoms in one gramme of a known element; for one gramme of hydrogen, this number is about six hundred thousand billion billions. In addition, a number of new phenomena have become known, which, although contingent on the presence of the usual known substances, contrary to previously known phenomena provide evidence of a more direct character about the effects of single atoms. This holds in particular for the many beautiful phenomena which have been discovered by the close study of electric discharges in gases. I will only mention that Sir J.J. Thomson, to whose pioneering contributions we first and foremost owe our knowledge and understanding of these phenomena, has on the basis of a study of these not only been able to develop a relatively direct method for determining the number of atoms, which has produced results agreeing exactly with those derived in the more indirect theoretical way just discussed, but has also been able to determine, in the atoms of all elements, the presence of small identical particles, the so-called *electrons*, which have a negative electrical charge and a mass which is very small in relation to the mass of the whole atom. This last result has had the greatest significance, as it has been possible for the first time to observe and investigate at least some of the building blocks of the atoms.

These results should, however, quickly be rounded off and surpassed by the discovery and investigation of a number of elements with entirely new properties, namely the so-called *radioactive* elements. Many readers will remember how great an interest the discovery of radium, which was due in particular to Mme Curie's perspicacious and energetic work, aroused at the turn of the century, and the sensation created by the wonderful and seemingly incomprehensible effects of this substance. At present, though, these effects are no longer surrounded by such a mysterious aura as at that time, since through the unique brilliant work performed by Sir *Ernest Rutherford* during the time elapsed since then, the clearest imaginable light has been thrown on most of the phenomena which are linked to radioactive substances. His studies of these phenomena have led to results which not only do not conflict with the conclusions drawn from earlier experience, but precisely illuminate and allow a deeper

[259]

understanding of them, while they have opened quite new and unimagined ways to penetrate more deeply into the question of the constitution of atoms.

Rutherford's first great success in this field consisted of his recognition that the seemingly capricious behaviour and variability of the observed phenomena disappeared completely if it was assumed that, in dealing with radioactive substances, one had to do with elements whose atoms, without external influence, underwent explosion-like changes, by which electrically charged parts of the atom, the so-called A (alpha) and B (beta) particles, were ejected, while an atom of a new element was formed from the remaining part. While in many cases the new element will be of just as stable a nature as the ordinary elements, it will often itself also be radioactive and its atoms will suffer a new explosion-like transmutation and so on. For this daring hypothesis, which at a single blow brought order and cohesion into the seeming chaos, Rutherford and his collaborators succeeded in obtaining decisive proof by isolating a great number of such transmutation products, the chemical investigation of which showed that they possessed properties corresponding exactly to the known elements, from which they only differed in that their atoms had a limited "lifetime" which sooner or later is terminated by an explosion.

In the following years, through a series of magnificently arranged and brilliantly conducted experiments, Rutherford explored the phenomena which accompany the explosion of the atoms. As an especially beautiful result of these investigations can be mentioned that it turned out that the particles emitted on the explosion had such a large kinetic energy that the action of a single emitted particle could be observed. Thus a direct method for counting atoms was obtained, whose results turned out to agree completely with those found by the methods previously mentioned. The major result achieved by Rutherford, however, was the conclusions he was able to draw about the building blocks of atoms. As the paths of the individual particles could be followed on their way through gases and other substances, it namely turned out that information could be obtained about the constitution of the atoms they met. It was thus discovered that an atom was usually very easily penetrable for the particles, as these could penetrate many thousand atoms without any noticeable change of direction. But now and again on collision with an atom the particle was found to suffer a large deflection which, as Rutherford showed, must be interpreted to mean that the atom, apart from the negatively charged electrons, possessed a nucleus which was charged with positive electricity, and which contained nearly all the mass of the atom. As the nucleus is extremely small in relation to the whole atom, the latter is constructed in a way quite similar to a solar system: the atom consists of a number of light electrons which move around the heavy atomic nucleus as the planets do around the sun. With this discovery by Rutherford of the *atomic nucleus*, which was corroborated in many ways, it must be assumed that all the building blocks of the atom have become known, and therefore there is now a basis for mastering the question of the explanation of the properties of substances by means of more exactly developed ideas about the constitution of atoms.

While it will not be possible here to go into detail as regards the theoretical treatment of the latter question and the marvellous prospects opening up for the creation of a common basis for the explanation of the phenomena of chemistry as well as physics, I will still just mention the new important result that Rutherford has reached by very recent investigations. With exceptionally well-conducted and difficult experiments he has been able to observe how, on a collision between an atom and an alpha particle in which the particle hits the atomic nucleus directly – which of course happens extremely rarely – the latter can on occasion be shattered into pieces which each separately will form nuclei belonging to atoms of other elements. Since with this result means have been found to bring forth arbitrarily a process corresponding exactly to that occurring spontaneously in the explosions of the atomic nucleus observed on the transmutations of radioactive substances, science can more or less be said to have entered a new epoch, since the stability of the atoms of elements with respect to external influence has hitherto been regarded as the solid foundation for physical and chemical considerations. The main purpose of these lines, however, has precisely been to try to give the reader an impression of how naturally and directly Rutherford's new result falls in with his earlier results mentioned above, through which the basic ideas used in the chemical and physical sciences have been developed into a marvellous and unsuspected clarity.

Of course, the above remarks are only a brief sketch of some few main results of Sir Ernest Rutherford's work, which in their extent and richness of ideas are unparalleled in contemporary research in experimental physics. That it has been possible for one man alone to make such a unique achievement in science is naturally due first and foremost to his superior scientific genius, but at the same time also due to many human qualities which have enabled him to gather a great number of students around him and to guide and inspire them. It is beyond my abilities even to try to describe these qualities in detail, but this is of course not at all necessary, as the Danish public interested in science will in these days have the opportunity of personally becoming acquainted with his distinct personality. As one of the many who have had the joy of working in his laboratory under his leadership, I would just like to express the gratitude and admiration whereby Sir Ernest Rutherford is regarded by all his pupils.

N. Bohr.

XVII. SIR ERNEST RUTHERFORD, O.M., P.R.S.

Nature (Suppl.) **118** (18 Dec 1926) 51–52

See Introduction to Part II, p. [109].

Sir Ernest Rutherford, O.M., P.R.S.

By Prof. Niels Bohr, For.Mem.R.S., University, Copenhagen.

FOLLOWING the kind invitation of the editor to write a few words in appreciation of the work and influence of the present director of the Cavendish Laboratory, I presume that the readers of NATURE will not need any detailed exposition of his achievements. As, however, I am one of those who have had the good fortune to come into close personal and scientific contact with Sir Ernest Rutherford, it is a great pleasure to me to try to describe briefly how we, who are proud to count ourselves among his pupils, regard him.

My own acquaintance dates from the period when Rutherford, after years of ardent and successful collaboration with Sir J. J. Thomson in the Cavendish Laboratory, had left Cambridge, and —after his stay at McGill, where his work on radio-active substances had established his fame—in Manchester had founded a school for investigations in radioactivity. This centre attracted young scientists from all parts of the world. In the spring of 1912, on my first visit to Manchester, the whole laboratory was stirred by one of the great discoveries which in so full a measure have been the fruits of Rutherford's endeavours. Rutherford himself and his pupils were eagerly occupied with tracing out the consequences of his new view of the nuclear structure of the atom. It would give only a poor impression of our trust in his judgment for me to say that nobody in his laboratory felt the slightest doubt about the correctness and fundamental importance of this view, although naturally it was much contested at that time. I remember being told by Hevesy soon after my arrival the story circulating in the laboratory of how Rutherford, shortly before his discovery, in a conversation with Moseley expressed the opinion that after all the troublesome investigations of the preceding years—during which he had such faithful assistance from Geiger — one would have had quite a good notion of the behaviour of an α-ray, were it not for the return of a minute number of these rays from a material surface exposed to an α-ray bombardment. This effect, though to all appearances insignificant, was disturbing to Rutherford, as he felt it difficult to reconcile it with the general ideas of atomic structure

Photo]   [*J. Russell and Sons.*
FIG. 6.—SIR ERNEST RUTHERFORD, O.M., P.R.S., Director 1919– .

then favoured by physicists. Indeed, it was not the first, nor has it been the last time, that Rutherford's critical judgment and intuitive power have called forth a revolution in science by inducing him to throw himself with his unique energy into the study of a phenomenon, the importance of which would probably escape other investigators on account of the smallness and apparently spurious character of the effect. This confidence in his judgment, and our admiration for his powerful personality, was the basis of the inspiration felt by all in his laboratory, and made us all try our best to deserve the kind and untiring interest he took in the work of every one. However modest the result might be, an approving word from him was the greatest encouragement for which any of us could wish.

When the War broke out, the little community in his laboratory was dispersed. Having, however, then taken up a lecturing post in Manchester, I had the opportunity in the succeeding years of witnessing the undaunted spirit and never-failing cheerfulness of Rutherford even in the most difficult times. Although the study of the more practical physical problems arising in connexion with the defence of his country took up practically all his time and energy in those years, he could still towards the end of the War find leisure to prepare for, and finally accomplish, perhaps his greatest scientific achievement, the transmutation of an element through the disintegration of the atomic nucleus by impact with α-rays; an achievement which may be said indeed to open up a new epoch in physical and chemical science.

Just at this time Rutherford was, on Thomson's retirement, offered the directorship of the Cavendish Laboratory as his unrivalled successor. I remember on a visit to Manchester during the Armistice hearing Rutherford speak with great pleasure and emotion about the prospect of his going to Cambridge, but expressing at the same time a fear that the many duties connected with this central position in the world of British physics would not leave him those opportunities for scientific research which he had understood so well how to utilise in Manchester. As everybody knows, the sequel has shown that this fear was unfounded. The powers of Rutherford have never manifested themselves more strikingly than in his leadership of the Cavendish Laboratory, the glorious traditions of which he has upheld in every way. Surrounded by a crowd of enthusiastic young men working under his guidance and inspiration, and followed by great expectations of scientists all over the world, he is in the middle of a vigorous campaign to deprive the atoms of their secrets by all the means which stand at the disposal of modern science.

XVIII. [TWO SPEECHES FOR RUTHERFORD, 1932]

A.S. Eve, "Rutherford: Being the Life and Letters
of the Rt Hon. Lord Rutherford, O.M.",
Cambridge University Press, Cambridge 1939, pp. 361–363

See Introduction to Part II, p. [109].

Rutherford gave two lectures at Copenhagen on 14 and 15 Sept. 1932, and at the close it fell to his old pupil Niels Bohr to move a vote of thanks. He described Rutherford's successive triumphs as being

crowned by the fundamental discovery of the atomic nucleus which has caused so unexpected and immense development of physical and chemical science. He knew always how to use some new phenomena, which to others might appear insignificant, as a key to further developments. He was able to prove that the atomic nucleus may be broken up by alpha particles, by which a new epoch in science was begun, as it was shown, for the first time, that an element can be changed by external agencies. In his lectures he has given us a most vivid impression of the vigorous and fruitful attack on the problem of the inner structure of the nucleus, which he and his collaborators are making. When we think of the smallness of the nucleus, we remember that its size compared with the atom by Lord Rutherford was illustrated by a pin head compared with a well-sized lecture room, it is amazing how great an amount of knowledge has already been collected in this quite new domain of human experience, a domain which is perhaps the most important and most promising of all....

It is always a new and wonderful experience to hear him lecture, not only on account of the new and important progress which he every time, as in these lectures, has to tell about, but we always get a renewed enthusiasm for the beauty of the subject and a fresh courage to the best of our modest powers to contribute to the great common aim.

In the evening Bohr again warmed to his subject at the Banquet of Danmarks Naturwidenskabelige Samfund given in honour of Rutherford:

This afternoon I had the honourable but hopeless task in a few words to picture the unique position of Lord Rutherford in the world of science, a position which rests for all times on his many great discoveries, of which several may be said to be epoch making, if the word 'epoch' could be reasonably used in relation with the short time intervals he had allowed to pass between them. I shall not expand on this subject to-night, but Lord Rutherford's discoveries do not exhaust the picture of him, in which his pupils and his friends rejoice. When Lord Rutherford visited Copenhagen twelve years ago, I had opportunity in this society to reproduce a number of stories which circulate among the pupils in his laboratory, in the inspiring atmosphere of which I had the good fortune to work for a time. I shall not repeat these stories to-day, the more so because Lord Rutherford last time reproached me for telling stories against him. Still the jest of these stories is not a malicious one, but an emphasis of his purity of zeal and directness of methods in dealing with difficulties in his work and the deficiencies of his collaborators. If a single word could be used to describe so vigorous and manysided a personality, it would certainly be 'simplicity'. Indeed, all aspects of his life are characterised by a simplicity of a similar kind to that which he has claimed of nature, which he is able to discover, where others before were not able to see it. It is his simplicity and self-forgetfulness which allow him to use all his moods in the best way for the progress of his work. In his usual bright spirit he sees always new and hopeful openings in any kind of difficulties, and his courage and cheerfulness extend themselves to everyone who works with him; at the same time he knows how to use his temper, which he does not try to hide, to a sound criticism of his own efforts and those of his assistants. To the pupils in his laboratory it is often as if the sun suddenly began to shine when he arrives in the morning; but sometimes it is as if the sky was darkened by a thunder cloud....In such moments everyone is of course afraid of him, but not more so than we dare to confess. All know that, if he has time and leisure, as he surprisingly often has, he is just as willing to attend to the youngest student and if possible to learn from him, as he is to listen to any recognised scientific authority. Not least is this entire disregard of social standing, the background not only for the unlimited admiration of his pupils but for their complete devotion. These same characteristics which make him the truest of friends have not only been essential for his work and for the successful administration of the great institution he is leading, but have also proved most helpful in treating matters at large relating to scientific education and research in his country, from the government of which he has received the external recognition which corresponds to the unique position he has long ago acquired in the minds and hearts of all his fellow-scientists. We wish him all happiness which life can bring; that in many years to come he may retain his power and his youth for the progress of science and the benefit of mankind and to the pleasure of his friends.

XIX. [OBITUARY FOR RUTHERFORD]

Nature **140** (1937) 752–753

See Introduction to Part II, p. [111].

WITH the passing away of Lord Rutherford*, the life of one of the greatest men who ever worked in science has come to an end. For us to make comparisons would be far from Rutherford's spirit, but we may say of him, as has been said of Galileo, that he left science in quite a different state from that in which he found it. His achievements are indeed so great that, at a gathering of physicists like the one here assembled in honour of Galvani, where recent progress in our science is discussed, they provide the background of almost every word that is spoken. His untiring enthusiasm and unerring zeal led him on from discovery to discovery, and among these the great landmarks of his work, which will for ever bear his name, appear as naturally connected as the links in a chain.

Those of us who had the good fortune to come into contact with Rutherford will always treasure the memory of his noble and generous character. In his life all honours imaginable for a man of science came to him, but yet he remained quite simple in all his ways. When I first had the privilege of working under his personal inspiration, he was already a physicist of the greatest renown, but nevertheless he was then, and always remained, open to listen to what a young man might have on his mind. This, together with the kind interest he took in the welfare of his pupils, was indeed the reason for the spirit of affection he created around him wherever he worked.

Rutherford passed away at the height of his activity, which is the fate his best friends would have wished for him, but just on account of this he will be missed more, perhaps, than any scientific worker has ever been missed before. Still, together with the feeling of irreparable loss, the thought of him will always be to us an invaluable source of encouragement and fortitude.   NIELS BOHR.

*A short tribute given at the Galvani celebrations in Bologna on October 20.

XX. [OBITUARY FOR RUTHERFORD]

Nature (Suppl.) **140** (1937) 1048–1049

See Introduction to Part II, p. [111].

Prof. Niels Bohr, For.Mem.R.S.

University of Copenhagen

I AM thankful for the invitation of the Editor of NATURE to write a few words about my relations with Lord Rutherford that have been so decisive for my work and have filled so large a place in my life. Indeed, neither in the short article about Rutherford's relationship to his pupils, which I had the pleasure of contributing to the Cavendish Laboratory Supplement to NATURE of December 19, 1926, nor in the short tribute to Rutherford's memory, which I had the sad duty of giving at the Galvani Congress on the announcement of his untimely death and which appeared in NATURE of October 30, 1937, did I find opportunity to give a proper expression of my personal indebtedness to him, who was to me everything that an inspiring leader and a fatherly friend could be.

From the moment I was admitted into the group of students from very different parts of the world working under Rutherford's guidance in his laboratory in Manchester, he has to me appeared as the very incarnation of the spirit of research. Respect and admiration are words too poor to describe the way his pupils regarded the man whose discoveries were the basis of the whole development in which they were enthusiastically striving to partake. What we felt was rather a boundless trust in the soundness of his judgment, which, animated with his cheerfulness and good will, was the fertile soil from which even the smallest germ in our minds drew its force to grow and flourish. His simplicity and disregard of all external appearance perhaps never disclosed themselves more spontaneously than in discussions with his students, who were through his straightforwardness even tempted in youthful eagerness to forget with whom they were talking until, by some small remark, the point of which they often first fully understood after they had left him, they were reminded of the power and penetration of his insight.

The stimulus Rutherford gave his pupils was, however, in no way limited to times of daily intercourse. Thus when, returned to Denmark, I pursued the line of work which I had taken up in Manchester, it was to me a most encouraging feeling to know that I could always count on his warm interest and invaluable advice. Indeed, looking through our correspondence from those days, I can hardly realize how in the midst of all his work he could find time and patience to answer in the kindest and most understanding way any letter with which a young man dared to augment his troubles. Especially close our relations became during my stay for the first years of the Great War as lecturer in Manchester and when, in times full of anxieties, he kept up the spirits of the small group left in the laboratory and, in the short moments of leisure from the great practical duties entrusted upon him, steadily went on preparing the road to new discoveries which should soon lead to such great results.

In later years, it was each time to me the greatest source of renewed encouragement to visit him in his home in Cambridge, where, in spite of never-ceasing work and an ever heavier burden of duties, he shared so quiet and simple a life with the companion who, always in contact with what was deepest in his character, from early days stood by him in every joy and sorrow. With age the vigour of his spirit did not abate, but found outlet in ever new ways, and his genial understanding and sympathy with all honest human endeavour gave to his advice in any scientific or practical matter a value treasured in wider and wider circles. To every one of us to whom he extended his staunch and faithful friendship an approving smile or a humorous admonition from him was enough to warm our hearts, and for the rest of our lives the thought of him will remain to inspire and guide us.

XXI. [TRIBUTE TO RUTHERFORD]

"Hommage à Lord Rutherford, sept huit et neuf novembre MCMXLVII",
La fédération mondiale des travailleurs scientifiques, Paris [1948], pp. 15–16

See Introduction to Part II, p. [111].

[275]

La parole est au Professeur Møller, Membre de l'Académie des Sciences Danoise, Directeur de l'Institut de Physique de Copenhague.

C. MØLLER

NIELS BOHR

Ladies and Gentlemen,

Professeur Bohr who unfortunately found himself unable to be present here to-day has asked me to read the following address :

" As one who had the great privilege of close personal relation through many years with the late Lord Rutherford, and benefited so much from his inspiring influence, I deeply regret to be unable to attend this Commemoration arranged by the " Wold Federation of Scientific Workers " at the tenth anniversary of his death. I am grateful, however, for the opportunity to send a message to the meeting and to try with a few words to give expression of our veneration for the great genius to whose pioneering life-work we so largely owe the foundation on which science builds to-day.

The whole history of science presents, indeed, few parallels to the achievements of Ernest Rutherford who, by a series of fundamental discoveries, opened up a whole new field of knowledge, which should give us such a deep insight into the atomic constitution of matter and, within the last years, has brought the mastery of immense forces of nature into the hands of man.

The germ to the new development grew out of the great discoveries made in various countries towards the end of the last century and, at this occasion, there is special reason to recall the great work of Henri Becquerel and of Marie and Pierre Curie who so eminently carried on the glorious traditions of the old centre of culture and learning chosen as meeting place for this assembly.

Grasping the importance of the phenomena of radio-activity, which entailed so great surprises, Rutherford threw his prodigious energy into the exploration of the new field of research and, with unique power of penetration and imagination, succeeded within a few years in disentangling the multi-various evidence and, especially, in clearing up the genetic relationships within the families of radio-active substances.

With untiring perseverance and incomparable mastery Rutherford followed up this great advance by developing the study of the penetrating radiations emitted in radioactive transformations into a powerful tool for the exploration of the atom itself. The climax of these wonderful researches which by their ingenuity created the utmost admiration and, at the same time, by their extreme simplicity carried general conviction, was the epoch-making discovery of the atomic nucleus and of the possibility of producing nuclear transmutations under controlled circumstances.

The marvellous story of the development thus initiated is known to everybody and has for all times secured Rutherford a place among the greatest scientists mankind has fostered. Our picture of Rutherford as a leader in science of his time embraces, however, also the memory of a true human character. The inspiration Rutherford exerted on all who came into contact with him was founded on our admiration for his unique powers and our affection for his straightforward personality and, above all, on his rare gift of conveying his own enthusiasm to everyone of the collaborators whom he gathered around him from all parts of the world.

International scientific co-operation was to Rutherford not only a way to progress in Science, but also a means of promoting that mutual understanding between nations which is vital for the harmonious development of human culture. It was a great loss to science that he should leave us in the midst of his fruitful activities, but he has left us an inheritance from which we derive encouragement and inspiration which is so greatly needed to-day.

15

Indeed, the advance of science does not only hold out the brightest prospects for the improvement of human welfare, but also may bring with it ominous menaces to world security, unless mankind can adjust itself to the exigencies of the new situation. In this matter of such great bearing on the future of civilization, goodwill from the side of all nations will be demanded, but we hope that the world-wide bonds created between scientists in their common search for truth may assist in bringing about that the challenge to our civilization is met in a spirit worthy of the ideals for which the memory of Lord Rutherford stands ".

Copenhagen, November 1947.

Niels Bohr.

XXII. THE GENERAL SIGNIFICANCE OF
THE DISCOVERY OF THE ATOMIC NUCLEUS

"Rutherford at Manchester" (ed. J.B. Birks),
Heywood, London 1962, pp. 43–44

See Introduction to Part II, p. [112].

PARTICIPANTS AT COMMEMORATIVE SESSION

(*left to right*) Sir James Chadwick, E. N. da C. Andrade, Sir Charles Darwin, N. Bohr, Sir Ernest Marsden

THE GENERAL SIGNIFICANCE OF THE DISCOVERY OF THE ATOMIC NUCLEUS

By NIELS BOHR

IT has been a pleasant and moving experience to listen to the talks of some of Rutherford's closest collaborators from the Manchester time, each of whom in his own way has described Rutherford's genius as a scientist and his great human personality. As one who had the good fortune, a few months after Rutherford's discovery of the nucleus of the atom, to join the group working here in Manchester under his inspiration, and during many subsequent years enjoyed his warm friendship, I am grateful to have been asked to add a few words on this occasion.

Having had the opportunity in a Rutherford Memorial Lecture* to dwell on some of my most treasured remembrances, I shall here only stress the point which formed the main theme for the Lecture. In completing to such an unsuspected degree our conception of the constitution of the atom, Rutherford's discovery from the very beginning presented physicists with the challenge to establish a comprehensive explanation of the experiences, accumulated through the centuries, about the physical and chemical properties of matter.

As we all know, the utilization for this purpose of the nuclear model of the atom was to rest essentially on the novel feature of natural philosophy introduced by the recognition of the universal quantum of action to which Planck was led in the first year of this century. Looking back on the development we might ask ourselves why the modern methods of quantum mechanics and electrodynamics, which now appear indispensable in dealing with atomic problems, were not already established at the time of Rutherford's discovery.

* Reprinted in this volume.

43

It is true that already a few years after Planck's pioneer work, the foundation for subsequent progress was laid by Einstein's ingenious analysis of the apparent paradoxes involved in the exchange of momentum and energy in elementary radiative processes, which had led him to the concept of the photon. The situation remained, however, so puzzling and strange that even a preliminary orientation as regards the mechanism responsible for the remarkable stability of atomic systems, on which all properties of matter ultimately depend, would hardly have been possible without such definite knowledge about the structure of the atom as that which Rutherford's great work gave us.

Indeed, the discovery of the atomic nucleus offered a decisive impetus at all stages of the ensuing development, by which it became gradually possible to achieve the incorporation of the quantum in a consistent generalization of the classical approach. As is well known, the realization of this great task demanded the co-operation of a whole generation of physicists from many parts of the world.

It has therefore been a most proper decision to connect the celebration of the fiftieth anniversary of Rutherford's epoch-making discovery with an international congress to discuss the present state of nuclear science. Great expectations are attached to this conference, and already from the first sessions we have got a strong impression of the spirit in which new knowledge and viewpoints have been exchanged between the participants.

We of the older generation have been vividly reminded of the atmosphere in Rutherford's own laboratory in the Manchester days, when we received such deep inspiration from his openmindedness as regards physical inquiry. Equally those of us who have benefited from Rutherford's unique gifts as a leader of scientific co-operation have had much occasion to remember his beliefs and hopes regarding the opportunities which science offers for promoting understanding and confidence between individuals from different nations.

44

XXIII. HENRY GWYN JEFFREYS MOSELEY

Phil. Mag. **31** (1916) 173–176

See Introduction to Part II, p. [112].

XXV. *Intelligence and Miscellaneous Articles.*

Henry Gwyn Jeffreys Moseley.

THE brilliant young physicist who was killed in action on August 10th at the Dardanelles was the only son of my close friend and fellow student the late Professor Henry Nottidge Moseley, F.R.S., of Oxford. He was educated at Eton, where he entered as a scholar, and at Trinity College, Oxford, where he gained a Millard scholarship. He obtained a First Class in Mathematical Moderations and Honours in Natural Science.

Harry Moseley was a most loveable boy and early showed great enthusiasm for science and marked originality. On leaving Oxford in 1910, he was appointed by Professor Rutherford of Manchester as lecturer and demonstrator in the Physics department of the University. After two years he resigned his lectureship and was awarded the John Harling Fellowship, which enabled him to devote his energies entirely to research. In 1913 he returned to Oxford to live with his mother, and continued his experiments in the laboratory of Professor Townsend. He went in the summer of 1914 to the meeting of the British Association in Australia; but on the outbreak of war made as speedily as possible for England, and resigning all thought of continuing the researches in which he was so successfully engaged, applied for and obtained a Commission in the Royal Engineers. He was, I am assured, offered work suited to his scientific capacities at home, but deliberately chose to share with others of his age the dangers of active service. He was made signalling officer to the 38th Brigade of the First Army, and left for the Dardanelles on June 13th, 1915. There, in the beginning of August, he was instantaneously killed by a bullet through the head as he was in the act of telephoning an order to his division. He was only 27 years of age.

Sir Ernest Rutherford writes that Moseley was "one of those rare examples of a man who was a born investigator." This quality he inherited from his father; and it is noteworthy that his grandfather Canon Moseley, F.R.S., was a distinguished and original mathematical physicist, whilst his grandfather on his mother's side, Mr. Gwyn Jeffreys, was also a Fellow of the Royal Society and for many years a leader in the study of Oceanography and Marine Zoology. Professor Rutherford tells us that Harry Moseley's "undoubted originality and marked capacity as an investigator were very soon ungrudgingly recognized by his co-workers in the laboratory, while his cheerfulness and willingness to help in all possible ways endeared him to all his colleagues." Mr. C. G. Darwin, a grandson of Charles Darwin—who worked with him at Manchester and produced with him a joint paper in which they mapped out accurately, for the first time, the spectrum of the characteristic X-radiation from an X-ray tube with a platinum anticathode,—writes that he was without exception or exaggeration the most brilliant man whom he had ever come across. Others who came to know him well in Manchester write of his charm of manner and personality, of his kindliness and unselfishness and care for the interests of others. As a boy (when I knew him best) he was a keen and observant naturalist and knew every bird and bird's nest in the neighbourhood of his home. In this and in the collection of flint implements he was enthusiastically aided by his sister. His last letters home from the East were full of observations on the plant-life, the birds, the beasts, and the flint implements *of all ages* which he found in a day's ramble on the hills where he was encamped. A most happy life of experimental research, with natural history and his garden as relaxations, was assured to one so greatly gifted and beloved, for he had private means and an ideal home. Whilst abroad on active service, in view of possibilities now alas! realized, Harry Moseley expressed the wish (which will be eventually carried out) to bequeath any property at his disposal to the Royal Society of London " for the furtherance of scientific research."

I am enabled by the kindness of Dr. Bohr of the University of Manchester, who has been aided by Dr. Makower, to add to this short personal sketch a notice of Moseley's scientific work which will, I am sure, be highly valued by the readers of the Phil. Mag.

E. RAY LANKESTER, Dec. 24, 1915.

MOSELEY came to Manchester in the spring of 1910 as lecturer and demonstrator in the Physical Laboratories of the University. At that time a great number of scientists from all parts of the world were working in the laboratories under the direction and inspiration of Sir Ernest Rutherford. Moseley at once caught the spirit of the laboratories and applied himself with characteristic energy and enthusiasm to the difficult and important problem of determining the number of β-particles emitted by a radioactive atom on disintegration. This problem had been investigated only for the active deposit of radium, and even in this case no high degree of accuracy could be claimed for the result of previous investigators. Moseley succeeded in improving the method and in obtaining much more accurate results. He also extended the investigation so as to include most of the radioactive products emitting β-rays [*]. His results have since gained additional importance on account of their bearing on theories of the origin of β- and γ-rays. The knowledge and experience which he gained in these experiments Moseley subsequently used to obtain high potentials in vacuo by means of the charge acquired by a radioactive substance during the emission of β-rays [†]. He was thus able to obtain higher steady potentials than had previously been reached in this or any other way. Whilst engaged in these difficult researches Moseley still found time to devote attention to other problems. In collaboration with Fajans he developed a most interesting method of determining the life of very rapidly decaying radioactive products [‡]; with Makower he discovered that radium B emitted γ-rays which were so easily absorbed that they had not been previously detected [§]; and with Robinson he measured the total ionization produced by the β and γ radiation from radium B and radium C [‖].

In 1912 Moseley resigned his lectureship, and having obtained the John Harling Fellowship at Manchester University, he was able to devote all his time to scientific investigation. In the same year a new field of Physical research was created by Laue's discovery of the interference of X-rays in crystals; and interest was soon intensified by the brilliant work of W. H. and W. L. Bragg on the constitution of X-rays and the structure of crystals. Very soon after Laue's discovery Moseley, working with Darwin, started a thorough investigation of the properties of X-rays by means of the new method. Their results were published in June 1913 [¶]; and although many of the results were discovered and published earlier by the Braggs, this paper contained a great number of most interesting experimental details and theoretical considerations, and constituted an important step in the rapidly increasing knowledge about the nature of X-rays. Immediately after the completion of this work Moseley undertook a systematic investigation of the characteristic X-radiation from as many different elements as possible. This investigation involved great experimental difficulties, partly on account of the fact that the chemical nature of many

---

[*] Moseley, Proc. Roy. Soc. A. lxxxvii. p. 230 (1912).
[†] Moseley, Proc. Roy. Soc. A. lxxxviii. p. 471 (1913).
[‡] Moseley and Fajans, Phil. Mag. xxii. p. 629 (1911).
[§] Moseley and Makower, Phil. Mag. xxiii. p. 312 (1912).
[‖] Moseley and Robinson, Phil. Mag. xxviii. p. 327 (1914).
[¶] Moseley and Darwin, Phil. Mag. xxvi. p. 210 (1913).

elements makes them rather unsuitable as anticathodes in X-ray tubes, and partly on account of the extreme absorbability of the radiation from many of the elements. Nevertheless, by his wonderful energy and by an ingeniously simple experimental arrangement for photographing the X-ray spectra, Moseley in less than half a year obtained measurements of the wave-lengths of the most intense lines in the high-frequency spectra of the greater part of the known elements and discovered the fundamental laws, which will always bear his name*. This work, begun in Manchester, was completed in Oxford in the early spring of 1914. As is well known, Moseley found that the frequencies of the principal lines in the high-frequency spectra are simple functions of the whole number which represents the position of the elements in the periodic table of Mendelejeff. The extreme importance of this result is that it reveals a relation between properties of different elements far simpler than any which could be expected from properties previously investigated, all of which, including the ordinary visible spectra, vary in an intricate manner from element to element. Moseley's discovery therefore gives a most important clue to the question of the internal structure of the atom which has received so much attention in recent years. While it is hardly the place here to enter into this problem in any detail, the general importance of Moseley's results is perhaps best illustrated by the fact that they enabled him to predict with certainty the number of possible elements hitherto unknown and their position in the periodic series. In this way he was, for instance, able to fix the number of possible elements in the group of the rare earths; and just before he went to Australia he was occupied in collaboration with Prof. Urbain on an investigation of the high-frequency spectra of the elements of this group, which no doubt will throw very much light upon this field of investigation which hitherto has given chemists so much trouble. A full account of this investigation has not yet been published, but Moseley gave a paper on the general question before the Meeting of the British Association in Sydney.

Every reader of Moseley's papers will be strongly impressed by his penetrating theoretical understanding and his great experimental skill which, together with his unique capacity for work, have secured him a place among the foremost workers in science of his time, although he was not able to devote more than four short years to scientific investigations.

* Moseley, Phil. Mag. xxvi. p. 1024 (1913), and xxvii. p. 703 (1914).

XXIV. PROFESSOR MARTIN KNUDSEN

Berlingske Aftenavis, 14 February 1941

TEXT AND TRANSLATION

See Introduction to Part II, p. [113].

Professor
Martin Knudsen

PROFESSOR *Martin Knudsen*, der i Morgen fylder 70 Aar, har et enestaaende Livsværk i Videnskabens Tjeneste bag sig. Ikke alene har han udført grundlæggende Arbejder indenfor et vigtigt Omraade af den nyere Fysik og gjort en meget betydningsfuld Indsats ved Udarbejdelsen af fysiske Metoder til Løsning af mange af de samfundsnyttige Problemer, som den moderne Havforskning har taget op; men tillige har han gennem et usædvanligt langt Spand af Aar øvet en overmaade virksom og resultatrig Gerning indenfor de talrige danske og internationale videnskabelige Organisationer, hvor han har været og stadig er betroet ledende Hverv.

Baggrunden for denne vidtforgrenede Virksomhed og for den store Tillid, hvortil Professor Knudsen har gjort sig saa fortjent, er foruden hans sjældne Arbejdslyst og Arbejdsevne særlige Træk i hans Begavelse og Karakter, der præger hans Behandling af alle videnskabelige og praktiske Spørgsmaal. Altid genfinder vi hos ham en ejendommelig Blanding af nøgternt Omdømme og skabende Fantasi, der Gang paa Gang har ledet ham til at finde sikre Udgangspunkter for nye Veje til frugtbart Fremskridt. Denne Indstilling kommer maaske allerklarest frem i Professor Knudsens rent videnskabelige Afhandlinger, hvor han ved egen Kraft og næsten helt alene har gjort saa store Landvindinger for vor fysiske Viden og Indsigt.

Professor Martin Knudsen.

Da Martin Knudsen omkring Aarhundredskiftet begyndte indgaaende at studere Luftarters Egenskaber ved lave Tryk, var den mekaniske Varmeteori allerede almindelig anerkendt, og man havde især opnaaet tilfredsstillende Forklaring af mange af de ofte saa simple Love, der behersker Luftarternes Tilstandsform. Efter disse Forestillinger bevæger de enkelte Molekyler af en Luftart sig i retlinede Baner, der kun ændres, naar Molekylerne enten støder mod hverandre eller rammer Væggene af den Beholder, hvori Luftarten er indesluttet. Ved de Lufttryk, hvormed man sædvanligt har at gøre, er imidlertid den saakaldte frie Vejlængde mellem to Sammenstød overordentlig lille i Forhold til Beholderens Dimensioner, og saadanne Egenskaber som Luftarternes indre Gnidning og Varmeledning afhænger derfor praktisk talt alene af Molekylernes indbyrdes Sammenstød. Naar Trykket derimod er overmaade lavt, omkring en Milliontedel Atmosfæretryk, bliver Sammenstødene saa faa, at Molekylerne næsten altid vil bevæge sig uhindret fra den ene Væg af Beholderen til den anden, og Luftarternes Egenskaber bliver derfor ganske anderledes. Ved saadanne uhyre lave Tryk bereder dog alle Maalinger de største Vanskeligheder, men med en overlegen Eksperimenterkunst formaaede Martin Knudsen at overvinde dem og derved skabe Mulighed for at undersøge en Række dybtliggende Problemer, der hidtil havde unddraget sig al Udforskning.

Ved disse Undersøgelser, der i lige høj Grad bærer Præg af den yderste Omhu og Udholdenhed som af den mest levende Forestillingsevne, lykkedes det ham ikke alene at opspore og efterprøve mange hidtil upaaagtede Konsekvenser af den mekaniske Varmeteori, men tillige at vise, hvorledes vigtige Oplysninger af indtil da ganske utilgængelig Art vedrørende Sammenstød mellem Luftmolekyler og Overflader af faste Legemer og Vædsker kunde opnaas ved Eksperimenter med stærkt fortyndede Luftarter. I en lang Række Arbejder efter Retningslinier, der her kun er antydet, tog Martin Knudsen stadig nye Problemer op og løste dem alle med største Mesterskab. Disse klassiske Arbejder, hvorved den kinetiske Luftteori naaede en uanet Uddybning og Afrunding, vakte ved deres Fremkomst den største Opsigt og Beundring blandt Fysikerne i alle Lande.

Ganske de samme Egenskaber, der kendetegner saavel Professor Knudsens sjældent selvstændige Forskergerning som hans Medvirken ved Tilrettelæggelsen af den store tekniske og organisatoriske Virksomhed ved de internationale Havundersøgelser, møder vi gennem hele hans omfattende Arbejde indenfor Københavns Universitet og andre danske videnskabelige Institutioner. Endnu før sin Ansættelse som Universitetslærer fik han Lejlighed til at lægge sine store Organisationsevner for Dagen, da han i Aaret 1900 valgtes til Formand for „Selskabet for Naturlærens Udbredelse". Allerede de første Aar lykkedes det ham at genoplive de fra Selskabets Stifter, H. C. Ørsted, stammende store Traditioner, og i Løbet af de næsten 40 Aar, hvor han virkede som Formand, blev paa hans Initiativ Selskabets Virksomhed stadig udvidet til at omfatte nye og vigtige kulturelle Opgaver som Oprettelsen af H. C. Ørsteds 100 Aars Fond til Støtte for dansk Naturvidenskab, og Stiftelsen af H. C. Ørsted-Medaillen til Opmuntring og Anerkendelse af danske Naturvidenskabsmænd. Samme Begejstring og Virketrang viste Professor Martin Knudsen ogsaa ved sin Deltagelse i Stiftelsen og Ledelsen af Danmarks naturvidenskabelige Samfund, der har været et nyt og betydningsfuldt Bindeled mellem videnskabelige og tekniske Bestræbelser her i Landet.

Vidnesbyrdene om den Tillid, som Professor Knudsen nyder indenfor videnskabelige Kredse, og som er grundet lige fuldt paa Anerkendelsen af hans Forskergerning som paa hans støtte og retlinede Personlighed, er mangfoldige. Ikke alene har han i de sidste 24 Aar været Sekretær og derigennem det faste Samlingspunkt i Det Kgl. Danske Videnskabernes Selskab, men tillige har han paa lignende Maade virket i forskellige internationale Institutioner, hvoraf flere er skabt væsentligt paa hans Initiativ. Foruden i Organisationen af de internationale Havundersøgelser, hvor Danmark ikke mindst paa Grund af hans Indsats indtager saa fremskudt en Stilling, deltog han saaledes igennem mange Aar i Ledelsen af Solvay-Stiftelsen i Bruxelles, der har bidraget saa meget til at styrke og fremme det internationale Samarbejde paa mange af Videnskabens Omraader.

Dersom de øjeblikkelige Forhold ikke i saa væsentlig Grad havde vanskeliggjort og afbrudt Forbindelsen mellem de forskellige Lande, vilde der utvivlsomt i disse Dage være strømmet Lykønskninger og Vidnesbyrd om Beundring og Anerkendelse til Professor Martin Knudsen fra Videnskabsmænd og Institutioner Verden over. Til Gengæld vil vi herhjemme i Tider som disse med særlig Taknemmelighed tænke paa, hvor meget en Mand som han har bidraget til vort Lands Anseelse, og enhver af os, der har haft Lejlighed til nært Samarbejde med ham, føler Trang til sammen med vor varmeste Lykønskning at sige ham en hjertelig Tak.

NIELS BOHR

TRANSLATION

Professor
Martin Knudsen

Professor Martin Knudsen, who will be 70 years old tomorrow, has a unique lifework in the service of science to his credit. Not only has he carried out fundamental work within an important field of recent physics and made a very important contribution by developing physical methods for solving many of the problems of public utility that modern oceanic research has taken up; but in addition he has throughout an unusually long span of years carried out an extremely active effort, rich in results, within the numerous Danish and international scientific organizations where he has been and still is entrusted leading posts.

The background for this widely branching activity and for the great trust, which Professor Knudsen so deservedly enjoys, is, apart from his rare zeal and capacity for work, special traits in his talents and character that mark his handling of all scientific and practical questions. We always recognize in him a remarkable mixture of sober judgement and creative imagination which time and time again has led him to find solid starting points for new ways to fruitful progress. This attitude appears perhaps most clearly in Professor Knudsen's purely scientific papers, where with his own effort and almost entirely alone he has made such great conquests for our physical knowledge and insight.

When, at around the turn of the century, Martin Knudsen started to study in detail the properties of gases at low pressure, the mechanical theory of heat was already generally accepted, and, in particular, a satisfactory explanation of many of the often quite simple laws valid for the gaseous state had been achieved. According to these ideas the individual molecules of a gas move in straight paths only altered when the molecules either collide with each other or hit the walls of the container in which the gas is held. At the air pressures usually involved, however, the so-called mean free path between two collisions is extremely small compared to the dimensions of the container, and properties such as the internal friction and thermal conduction of the gases depend practically speaking only upon the mutual collisions of the molecules. On the other hand, when the pressure is extremely low, around one millionth atmospheric pressure, the collisions are so few that the molecules will almost always move undisturbed from one wall of the container to the other, and the properties of the gases therefore become quite different. At such extremely low pressures, however, all measurements involve the greatest difficulties, but using his brilliant experimental artistry, Martin Knudsen managed to overcome these and thus create the possibility for investigating a series of fundamental problems which previously had escaped all exploration.

With these investigations, which are marked to an equally high degree by the most extreme care and persistence as by the most vivid imagination, he succeeded not only in tracking down and testing many hitherto unnoticed consequences of the theory of mechanical thermodynamics, but in addition to show how important information, of an until then quite inaccessible kind, regarding collisions between molecules of air and the surface of solid bodies and liquids, could be obtained by experiments with extremely rarefied gases. In a long series of investigations following the principles only hinted at here, Martin Knudsen took up still new problems and solved them all with the greatest mastery. These classic papers, whereby the kinetic theory of gases reached an unsuspected elaboration and consolidation, caused on their appearance the greatest attention and admiration among physicists in all countries.[1]

Quite the same qualities which characterize Professor Knudsen's singularly independent research work, as well as his participation in the preparation of the great technical and organizational activity in connection with the international ocean studies, we meet throughout the whole of his extensive work at the University of Copenhagen and other Danish scientific institutions. Even before his employment as a university teacher he had the opportunity to demonstrate his organizational talents when in the year 1900 he was elected as President of the "Society for the Dissemination of Natural Science".[2] Already in the first years he was successful in reviving the great traditions stemming from H.C. Ørsted, the Society's founder, and in the course of the nearly 40 years when he served as President, the activity of the Society was constantly extended on his initiative to encompass new and important cultural tasks, such as the setting up of the H.C. Ørsted Centennial Foundation for the support of Danish science and the establishment of the H.C. Ørsted Medal for the encouragement and recognition of Danish scientists. Professor Martin Knudsen also showed the same enthusiasm and diligence in his participation in the establishment

---

[1] A bibliography of Knudsen's scientific publications can be found in Fys. Tidsskr. **47** (1949) 159–164.
[2] Knudsen remained in this post until Bohr took over in 1939. See Vol. 11, p. [353].

and leadership of the Science Society of Denmark,[3] which has provided a new and important link between the scientific and the technical endeavours in this country.

The testimonials of the trust Professor Knudsen enjoys within scientific circles, which is based in equal measure on the recognition of his research work and on his steadfast and upright personality, are manifold. Not only has he for the last 24 years been Secretary and thereby the steady rallying point in the Royal Danish Academy of Sciences and Letters, but he has also been active in a similar fashion for various international institutions, of which several are created essentially on his initiative. Besides participating in the organization of international oceanic research, where Denmark not least on account of his efforts holds such a prominent position, he thus took part for many years in the direction of the Solvay Foundation in Brussels,[4] which has contributed so much to strengthening and advancing international cooperation in many of the fields of science.

If the current conditions had not to such a substantial degree obstructed and cut off communication between the various countries,[5] congratulations and testimonies of admiration and recognition would undoubtedly have poured in to Professor Martin Knudsen at this time from scientists and institutions all over the world. On the other hand, in times like these we, here at home, will think with special gratitude of how much a man such as he has contributed to our country's standing, and each of us, who has had the opportunity of close cooperation with him, feels the urge to give him, together with our warmest congratulations, our heartfelt thanks.

NIELS BOHR

---

[3] Knudsen and the industrialist Gustav Adolph Hagemann (1842–1916) established *Danmarks naturvidenskabelige Samfund* in 1911, after which Knudsen served as its Vice-President until 1937.
[4] The International Solvay Institute for Physics was established by Ernest Solvay in 1912, in the wake of the first successful Solvay Conference the year before. One important task of the Institute was to organize the subsequent Solvay Conferences. Knudsen became Secretary of the Institute in 1912.
[5] Bohr wrote this article during the German occupation of Denmark in World War II.

XXV. MEETING ON 19 OCTOBER 1945

MØDET DEN 19. OKTOBER 1945
Overs. Dan. Vidensk. Selsk. Virks. Juni 1945 – Maj 1946, pp. 31–32

M. Knudsen's retirement as Secretary of the Royal Danish Academy

TEXT AND TRANSLATION

See Introduction to Part II, p. [113].

Sekretæren meddelte, at han paa Grund af Alder ønskede sig fritaget for sit Hverv som *Selskabets Sekretær*, og bad om Tilladelse til at sætte paa Mødesedlen til næste Møde: Første Behandling af Valg af Sekretær.

Præsidenten udtalte i denne Anledning følgende:

»Alle Medlemmerne vil med Beklagelse have hørt om Professor Martin Knudsens Beslutning om paa Grund af sin høje Alder at trække sig tilbage fra sin Gerning som Selskabets Sekretær. Vi forstaar jo, at han ønsker at blive fritaget for nogle af de Byrder, han i saa mange Aar har baaret, og vi kan kun imødekomme hans Henstilling om, at Valget af en ny Sekretær maa blive sat paa Dagsordenen for næste Møde. Det ligger imidlertid os alle dybt paa Hjerte at give Udtryk for den Taknemmelighed, som Selskabet skylder Professor Knudsen for hans trofaste Tjeneste igennem næsten en Menneskealder ved Varetagelsen af hans for vor Virksomhed saa betydningsfulde Hverv. Mange Ord vil ikke behøves i Dag, da de fleste af os i frisk Erindring bevarer de Vidnesbyrd om, hvor højt hans Gerning paaskønnes, der fandt saa smukke Udtryk ved den Fest, som Selskabet afholdt, efter at Professor Knudsen i 25 Aar havde virket som vor Sekretær. Ikke alene fremdroges, hvor meget han ved sit eget Initiativ og sin Lydhørhed for alt, hvad der ligger Medlemmerne paa Sinde, har betydet for den rige Udvikling af Selskabets Virksomhed, der har fundet Sted i de Aar, hvor han har været dets Sekretær, men det, som fra alle Sider maaske stærkest betonedes, var den inderlige Interesse, ja vi tør vel sige Kærlighed, hvormed Professor Knudsen omfatter alt, hvad der angaar vort gamle Selskab og dets mangfoldige Opgaver. Til vor Tak ønsker vi i Dag at føje de varmeste Ønsker for endnu mange lykkelige Arbejdsaar til Fortsættelse af det store Livsværk, hvormed han har bidraget saa overordentlig til Fremgang for Videnskaben og til dens Anvendelse i Selskabets Tjeneste.«

TRANSLATION

The Secretary announced that due to age he wished to be released from his post as *Academy Secretary*, and requested permission to put on the agenda for the next meeting: First discussion of election of Secretary.

The President stated the following in this connection:

"All the members will have heard with regret of Professor Martin Knudsen's decision to retire from his work as Academy Secretary because of his great age. We understand, of course, that he wishes to be relieved of some of the burdens he has borne for so many years and we can only acquiesce to his request that the election of a new Secretary should be put on the agenda for the next meeting. However, it is a heartfelt wish of us all to express the gratitude that the Academy owes Professor Knudsen for his faithful service throughout nearly a lifetime in the execution of his task so important for our activity. Many words will not be needed today, as most of us have fresh in mind the evidence of how greatly his work is appreciated which was so beautifully expressed at the festivity held by the Academy when Professor Knudsen had acted as Secretary for 25 years.[1] Not only was it pointed out how much, by his own initiative and his responsiveness to everything of importance to the members, he has meant for the rich development in the activity of the Academy that has taken place during the years in which he has been Secretary, but what was perhaps most strongly emphasized from all sides was the sincere interest, we might even say devotion, with which Professor Knudsen embraces everything concerning our old Academy and its manifold tasks. To our thanks today we wish to add the warmest wishes for many more happy working years for the continuation of the great lifework with which he has contributed so profusely to the advance of science and to its use in the service of the Academy."

---

[1] *Meeting on 8 May 1942*, Overs. Dan. Vidensk. Selsk. Virks. Juni 1941 – Maj 1942, p. 44.

XXVI. PROFESSOR MARTIN KNUDSEN DIED YESTERDAY

PROF. MARTIN KNUDSEN DØD I GAAR
Politiken, 28 May 1949

TEXT AND TRANSLATION

See Introduction to Part II, p. [114].

Prof. Martin Knudsen
død i gaar

Niels Bohr skriver mindeord om sin mangeaarige
ven og kollega

Professor gennem en menneskealder ved Københavns Universitet, fysikeren, dr. phil. Martin Knudsen fik for nogen tid siden et hjertetilfælde og sov i gaar morges — 78 aar gammel — stille ind i døden.

V ED *Martin Knudsens* død har dansk videnskab mistet en af sine forgrundsskikkelser, der ikke alene har givet skelsættende bidrag til forskningen, men som tillige paa mangfoldige maader har virket til gavn for vort samfund.

Især naaede Martin Knudsen ved sine overalt i verden beundrede un-

Prof. Martin Knudsen.

dersøgelser over luftarter ved yderst lave tryk resultater af overordentlig betydning. Medens luftmolekylerne ved de tryk, som man almindeligvis har med at gøre, kun er i stand til at bevæge sig ganske korte strækninger uden at støde sammen, vil de

ved tryk paa omkring en milliontedel atmosfære tilbagelægge saa lange afstande uden indbyrdes sammenstød, at de i regelen vil bevæge sig uhindret fra væg til væg i de beholdere, hvori de er indesluttet. Under saadanne forhold faar luftarterne, som Martin Knudsen paaviste, i mange henseender helt andre egenskaber end dem, vi er vant til, og ved en række undersøgelser, planlagt med dyb indsigt og mesterligt gennemført, lykkedes det ham at klarlægge de love, der gælder for luftmolekylernes tilbagekastning og opfangning, naar de træffer faste legemer eller vædskeoverflader. Disse resultater, der fik den største betydning for udviklingen inden for den fysiske videnskab, fandt ogsaa vigtige anvendelser paa mange praktiske omraader, for hvilke Martin Knudsens sindrige apparatkonstruktioner i stedse stigende grad er blevet taget i brug.

De samme sjældne evner for tilrettelæggelse og løsning af forskningsopgaver lagde Martin Knudsen for dagen ved sin deltagelse i den for mange samfundsformaal saa vigtige udforskning af havene. At Danmark blev hovedsæde for de internationale havundersøgelser, skyldtes saaledes i første linje hans indsats, og anerkendelsen af hans virke paa dette omraade fandt ligeledes udtryk ved, at han i mange aar valgtes som **præsident for den internationale Association for Fysisk Oceanografi. Ogsaa paa mange andre maader var Martin Knudsen virksom ved organisation af internationalt samarbejde, og blandt andet valgtes han allerede ved stiftelsen af Institut International**

de Physique Solvay til medlem af bestyrelsen for dette institut, der har haft saa stor betydning for udviklingen af samarbejdet mellem fysikerne i de forskellige lande.

Ved siden af hans forskerarbejde og undervisningsvirksomhed blev der ogsaa herhjemme lagt stor beslag paa hans organisatoriske og administrative evner. Saaledes fik i de næsten 40 aar, Martin Knudsen ledede Selskabet for Naturlærens Udbredelse, dette af H. C. Ørsted grundede gamle selskab en ny opblomstring, og ved stiftelsen af H. C. Ørstedfondet og oprettelsen af H. C. Ørstedmuseet viste Martin Knudsen en levende interesse for at hædre den store forgænger og bevare minderne om ham. Tillige var Martin Knudsen i over 25 aar vicepræsident i Danmarks Naturvidenskabelige Samfund, hvis formaal er at fremme forbindelsen mellem videnskab og ingeniørvirksomhed herhjemme. Især ofrede dog Martin Knudsen sin tid og sine kræfter paa Det Kgl. Danske Videnskabernes Selskab, for hvilket han i næsten 30 aar var sekretær og utrætteligt med altid frisk initiativ arbejdede for Selskabets anliggender.

Overalt hvor Martin Knudsen virkede, lagde han en sjælden arbejdslyst og arbejdsevne for dagen, der sammen med hans ligefremme væsen og venlige sind skaffede ham beundrere og venner i vide krese, hvor mindet om hans store gerning i videnskabens og samfundets tjeneste altid vil holdes i ære og længe bevares i taknemmelighed.

Niels Bohr.

TRANSLATION

Professor Martin Knudsen
died yesterday
Niels Bohr writes commemorative words about his
long-standing friend and colleague

> The physicist Martin Knudsen, D.Phil., professor at the University of Copenhagen for a generation, had a heart attack some time ago and died peacefully in his sleep yesterday morning – 78 years of age.

On the death of *Martin Knudsen*, science in Denmark has lost one of its prominent figures, who not only has made epoch-making contributions to research but has also in manifold ways worked for the benefit of our society.

In particular with his investigations about gases at extremely low pressure, admired all over the world, Martin Knudsen reached results of extraordinary importance. While air molecules at normal pressures are only able to move about for quite short distances without colliding with each other, at pressures of about one millionth of an atmosphere they will travel so large distances without mutual collisions that as a rule they will move unhindered from wall to wall in the vessels they are held. Under such conditions, as Martin Knudsen demonstrated, the gases have in many respects properties totally different from those we are used to, and by a series of experiments, planned with deep insight and masterly performed, he succeeded in revealing the laws that hold for the reflection and capture of air molecules when they hit solid bodies or surfaces of liquids. These results, which had the greatest significance for the development within physical science, also found important applications in many practical fields, for which Martin Knudsen's ingenious apparatus constructions have been used to a steadily increasing degree.

The same remarkable talents for planning and solving research problems were shown by Martin Knudsen in his participation in the exploration of the oceans so important for many general purposes. That Denmark became headquarters for international oceanic research is due first and foremost to his efforts, and recognition of his work in this area was similarly demonstrated by the fact that for many years he was the elected President of the international

Association of Physical Oceanography.[1] In many other ways, too, Martin Knudsen was active in the organization of international cooperation, and among other things, he was already at the founding of the Institut International de Physique Solvay elected member of the board of directors for this institute, which has been of such great importance for the development of the cooperation between physicists in the various countries. Besides his research work and educational activity, great calls were also made here at home on his organizational and administrative talents. Thus in the nearly 40 years Martin Knudsen led the Society for the Dissemination of Natural Science, this old society, founded by H.C. Ørsted, enjoyed a renaissance, and with the establishment of the H.C. Ørsted Foundation and the creation of the H.C. Ørsted Museum, Martin Knudsen showed a vivid interest in honouring the great predecessor and to preserve the memories of him. Additionally, for more than 25 years Martin Knudsen was Vice-President of the Science Society of Denmark, whose objective is to advance the connection between science and engineering here at home. In particular, though, Martin Knudsen gave his time and energy to the Royal Danish Academy of Sciences and Letters, for which he was Secretary for nearly 30 years and worked tirelessly for the Academy's affairs with ever fresh initiative.

Wherever Martin Knudsen was active he showed a remarkable appetite and ability for work, which together with his forthright character and friendly attitude brought him admirers and friends in wide circles, where the memory of his great work in the service of science and society will always be honoured and long held in gratitude.

Niels Bohr.

---

[1] This institution was set up in 1919 as a section of the International Union of Geodesy and Geophysics under the auspices of the newly established International Research Council, the predecessor of the current International Council of Scientific Unions created in 1931. Knudsen served as President from 1930 to 1936. In 1967, the Association was renamed the International Association of Physical Sciences of the Ocean, under which name it continues today.

XXVII. MARTIN KNUDSEN 15.2.1871–27.5.1949

Fys. Tidsskr. **47** (1949) 145–147

TEXT AND TRANSLATION

See Introduction to Part II, p. [114].

Martin Knudsen

15. 2. 1871—27. 5. 1949

Den 27. maj 1949 døde professor *Martin Knudsen* i sit hjem i Gentofte, 78 år gammel.

Onsdag d. 5. oktbr. 1949 afholdt Selskabet for Naturlærens Udbredelse og Fysisk Forening en mindeaften for professor Knudsen med taler af professor Niels Bohr og professor R. E. H. Rasmussen.

Professor *Niels Bohr* udtalte:

Fysisk Forening og Selskabet for Naturlærens Udbredelse har ønsket ved dette fælles møde at mindes Martin Knudsen, ved hvis bortgang dansk videnskab har mistet en af sine mest fremragende dyrkere, og til hvem alle fysikere her hjemme står i dyb taknemmeligheds gæld.

Vi ønsker at udtrykke vor tak for, at Martin Knudsens hustru og børn og nogle af hans slægt og nærmeste venner har villet modtage vor indbydelse og er kommet til stede her i aften, og vi byder dem hjertelig velkommen. Vi er også taknemmelige for, at Fysisk Forenings formand, professor R. E. H. Rasmussen, der i mange år var en så trofast medarbejder ved Martin Knudsens undersøgelser og siden på selvstændig og frugtbar måde har arbejdet videre på de områder, hvor Martin Knudsen brød nye baner, i aften vil give os en beretning om hans store videnskabelige livsværk.

Jeg skal derfor ikke her komme nærmere ind på de sindrige og med så stor eksperimenterkunst udførte undersøgelser, hvormed Martin Knudsen på vide områder nåede resultater, der overalt i den videnskabelige verden omfattedes med største beundring. Som en af dem, der gennem mange år har haft lejlighed til at følge Martin Knudsens forskergerning fra bedrift til bedrift og samarbejde med ham på den opgave, der lå ham så meget på hjerte, at fremme videnskaben her i landet og udbrede kendskabet til dens fremskridt, vil jeg imidlertid gerne dvæle ved nogle af de mange minder om Martin Knudsens videnskabelige bebejstring og hans arbejdsglæde og virketrang, som såvel i Selskabet for Naturlærens Udbredelse som i Fysisk Forening altid vil bevares i taknemmelig erindring.

I de smukke ord, som Fysisk Forenings formand udtalte ved Foreningens møde umiddelbart efter Martin Knudsens bortgang, fortalte han om, hvorledes Fysisk Forening voksede ud af en mindre studiekreds, i hvilken Martin Knudsen var den bærende kraft. Det var også ved de første sammenkomster i den nystiftede forening, at han berettede om sine undersøgelser over luftarternes egenskaber, der førte ham ind på den vej, hvor han skulle nå så store resultater. For Fysisk Forening, for hvilken Martin Knudsen i nogle år var formand, bevarede han altid den varmeste interesse.

Hvad Selskabet for Naturlærens Udbredelse angår, er det jo vor stolthed, at dets virksomhed går helt tilbage til H. C. Ørsteds dage. Den beretning, som Selskabet udgav ved 100-årsdagen for elektromagnetismens opdagelse, og hvori der gjordes rede for Selskabets historie, giver et levende billede af H. C. Ørsteds bestræbelser for at udbrede kendskabet til naturvidenskaben i vide kredse og virke for dens anvendelse i samfundets tjeneste. Ikke mindst giver beretningen et dybt indtryk af, hvor megen tid og arbejde Ørsted anvendte for at udbrede sådant kendskab overalt i landet til unge mennesker, som dengang ikke havde samme muligheder som nu til at søge tekniske skoler og lignende uddannelse.

Siden H. C. Ørsteds tid skylder Selskabet ingen så meget som Martin Knudsen, der i næsten 40 år virkede som formand i direktionen. Han indtrådte i dette hverv efter sin højt beundrede lærer, Christian Christiansen, som også på mange andre af os har udøvet så inspirerende en indflydelse. Straks efter at Martin Knudsen overtog formandsskabet, bragte han fornyelse i Selskabets hele virksomhed, og møderne, der hidtil havde været afholdt på søndage om eftermiddagen, blev for at muliggøre deltagelse for en større tilhørerkreds forlagt til hverdagsaftener, således som det siden har været skik. På få år lykkedes det også at tredoble Selskabets medlemstal, der nåede op på omtrent den størrelse, det nu har.

I den samme ånd, som karakteriserede H. C. Ørsteds egne bestræbelser, blev i de følgende år under Martin Knudsens ledelse mange nye opgaver taget op. Således foranstaltedes aftenkurser i eksperimentalfysik for skoleelever, og denne virksomhed, for hvilken der føltes et stort behov, opgav Selskabet først, da eksperimentalfysisk undervisning var blevet obligatorisk i danske skoler. Også på mange andre måder kom Martin Knudsens vågne initiativ til

udtryk. Blandt andet overtog Selskabet udgivelsen af Fysisk Tidsskrift der først var startet af en snævrere kreds men som Martin
Knudsen fra begyndelsen af havde omfattet med største interesse.
Der indførtes tillige en årsberetning om Selskabets virksomhed med
gengivelse af de i årets løb holdte foredrag.

En sag der lå Martin Knudsen særligt på sinde var at bevare
og styrke den tradition som H. C. Ørsted havde skabt herhjemme.
Til opmuntring af dansk videnskabelig indsats på de områder hvor
Ørsted havde virket stiftede Selskabet således på Martin Knudsens
forslag H. C. Ørsted-medaljen. Efter at denne anerkendelse først
var blevet de to ældre danske fremragende videnskabsmænd kemi-
keren S. P. L. Sørensen og fysikeren Christian Christiansen til del,
fulgte direktionen det som en selvfølge, uden at formanden inddroges
i overvejelserne, at Ørsted-medaljen tildeltes Martin Knudsen selv.

100-årsdagen for opdagelsen af elektromagnetismen var naturligvis
en særlig anledning til at mindes H. C. Ørsted. Sammen med Danmarks tekniske Højskole foranstaltede Selskabet en uforglemmelig
mindefest, og samtidigt arrangeredes i København et møde af skandinaviske naturforskere. Endvidere stiftedes på Martin Knudsens
initiativ H. C. Ørsteds 100-års fond med det formål at virke for de
interesser, som havde ligget Ørsted på sinde. Tillige indrettedes
et H. C. Ørsted museum til bevarelse af de instrumenter fra
Ørsteds tid, som det er det danske samfunds stolthed at besidde,
og det lykkedes også Martin Knudsen ved forståelse og imøde-
kommenhed fra Ørsteds slægt til musæet at erhverve møbler og
malerier, der havde været i H. C. Ørsteds eje.

Med samme utrættelige interesse og initiativ virkede Martin Knudsen som Selskabets formand i de følgende tyve år, indtil han på
grund af svigtende helbred udtalte ønsket om at træde tilbage. Få
år efter, på hans 70-års fødselsdag, fandt Selskabet lejlighed til at
hædre Martin Knudsen ved udgivelsen af et festskrift, som frem-
trådte som en udvidet årgang af den foredragsberetning, hvis
tilblivelse skyldtes hans eget initiativ, og direktionen havde tillige
den glæde at kunne meddele Martin Knudsen, at man enstemmigt
havde besluttet at bede ham være Selskabets æresmedlem.

I danske fysikerkredse vil Martin Knudsens rige forskergerning
og minderne om hans virksomhed iblandt os altid bevares i dyb
beundring og taknemmelighed.

TRANSLATION

Martin Knudsen
15.2.1871–27.5.1949

On 27 May 1949 Professor *Martin Knudsen* died in his home in Gentofte,[1] 78 years old.

On Wednesday 5 October 1949 the Society for the Dissemination of Natural Science and the Physical Society arranged a memorial evening for Professor Knudsen with speeches by Professor Niels Bohr and Professor R.E.H. Rasmussen.

Professor *Niels Bohr* stated:

The Physical Society and the Society for the Dissemination of Natural Science wish at this joint meeting to honour the memory of Martin Knudsen, on whose death Danish science has lost one of its most eminent practitioners, and to whom all physicists here at home owe a great debt of gratitude.

We wish to express our thanks to Martin Knudsen's wife and children and some of his relations and closest friends for having accepted our invitation and being here this evening, and we welcome them heartily. We are also grateful that Professor R.E.H. Rasmussen,[2] President of the Physical Society, who for many years was such a faithful collaborator in Martin Knudsen's investigations and who later in an independent and fruitful way has continued the work in the fields in which Martin Knudsen broke new ground, will give us an exposition this evening of his great scientific lifework.[3]

I will therefore not describe in detail here the ingenious investigations, carried out with such great experimental art, whereby Martin Knudsen in broad fields achieved results which were regarded with the greatest admiration everywhere in the scientific world. As one of those who throughout many years have had the opportunity to follow Martin Knudsen's research work from achievement to achievement, and to cooperate with him in the task, which was so close to his heart, of advancing science in this country and disseminating

---

[1] Township just outside Copenhagen.

[2] (1901–1972).

[3] Rasmussen's obituary for Knudsen follows the present contribution by Bohr on pp. 148–158 in the original publication.

the knowledge of its progress, I would, however, like to dwell on some of the many memories of Martin Knudsen's scientific enthusiasm and his joy in work and his diligence, which will be held in grateful memory in the Society for the Dissemination of Natural Science as well as in the Physical Society.

In the beautiful words that the President of the Physical Society spoke at the Society's meeting immediately following Martin Knudsen's death, he related how the Physical Society grew out of a small study circle in which Martin Knudsen was the mainstay. It was also at the first gatherings of the newly established Society that he told of his investigations into the properties of gases which led him on to the path where he should achieve such great results. Martin Knudsen always retained the warmest interest in the Physical Society, for which he was President for some years.

As regards the Society for the Dissemination of Natural Science, we pride ourselves, of course, that this activity goes right back to the days of H.C. Ørsted. The account that the Society published on the occasion of the centennial of the discovery of electromagnetism, and in which the history of the Society was reviewed, gives a vivid picture of H.C. Ørsted's endeavours to disseminate the knowledge of science in wide circles and to work for its application in the service of society. Not least does the account give a profound impression of how much time and effort Ørsted used to disseminate such knowledge throughout the country to young people, who at that time did not have the same opportunities as now to attend technical schools and similar education.[4]

Since H.C. Ørsted's time the Society owes nobody so much as Martin Knudsen, who acted as chairman of the board for nearly 40 years. He took up this task following Christian Christiansen, his highly esteemed teacher, who has also been such a great inspiration for many others among us. Immediately upon assuming the chairmanship, Martin Knudsen brought renewal to all activity of the Society, and the meetings which had hitherto been held on Sunday afternoons were moved to weekday evenings to allow attendance by a larger audience, as has been the custom since. Within a few years it proved possible to triple the membership of the Society, which reached about the size it is now.

In the same spirit as that which characterized H.C. Ørsted's own endeavours, many new tasks were taken up in the following years under Martin Knudsen's direction. Thus, evening classes in experimental physics for school children were organized, and this activity, for which there was felt a great need, was

---

[4] M.C. Harding, *Selskabet for Naturlærens Udbredelse: H.C. Ørsteds Virksomhed i Selskabet og dettes Historie gennem hundrede Aar*, Gjellerup, Copenhagen 1924.

only given up by the Society when teaching in experimental physics had become obligatory in Danish schools. Martin Knudsen's vigilant initiative was also expressed in many other ways. Among other things, the Society assumed responsibility for the publication of Fysisk Tidsskrift, which was originally started by a small circle, but in which Martin Knudsen had taken the greatest interest from the beginning.[5] In addition, an annual proceedings of the Society's activities was introduced with a printed version of the talks held in the course of the year.

One cause especially close to Martin Knudsen's heart was to preserve and strengthen the tradition created by H.C. Ørsted here at home. Thus, at the suggestion of Martin Knudsen, the Society established the H.C. Ørsted Medal in order to encourage Danish scientific effort in the fields in which H.C. Ørsted had been active. After this distinction had first been bestowed on two eminent senior Danish scientists, the chemist S.P.L. Sørensen and the physicist Christian Christiansen, the board took it for granted, without involving the chairman in the deliberations, that the Ørsted Medal be awarded to Martin Knudsen himself.[6]

The centennial for the discovery of electromagnetism was naturally a special occasion for honouring the memory of H.C. Ørsted. Together with the Technical University of Denmark, the Society arranged an unforgettable memorial celebration, and at the same time a meeting of Scandinavian scientists was arranged in Copenhagen.[7] Furthermore, the H.C. Ørsted Centennial Foundation was established on Martin Knudsen's initiative, with the objective of working for the interests that had been close to Ørsted's heart. In addition, an H.C. Ørsted Museum was organized for the preservation of the instruments from Ørsted's time, which it is the pride of Danish society to possess, and Martin Knudsen was also successful, with the understanding and generosity of Ørsted's family, in acquiring for the museum furniture and paintings that had been in H.C. Ørsted's possession.[8]

---

[5] The first issue of Fysisk Tidsskrift was published in 1902, with Kirstine Meyer as editor and with some financial support from the Society. The journal was incorporated into the Society in 1905.

[6] See Vol. 11, p. [353].

[7] At this "First Meeting of Nordic Physicists" Bohr gave the lecture, N. Bohr, *On the Result of Collisions between Atomic Systems and Free Electrical Particles* in *Det nordiske H.C. Ørsted Møde i København 1920*, H.C. Ørsted Komiteen, Copenhagen 1921, pp. 120–121. Reproduced in Vol. 8, pp. [195]–[198] (Danish original) and [199]–[200] (English translation).

[8] See N. Bohr, *Ninth Presentation of the H.C. Ørsted Medal*, Fys. Tidsskr. **51** (1953) 65–67, 80. Reproduced in Vol. 11, pp. [477]–[481] (Danish original) and [482]–[485] (English translation). The original H.C. Ørsted Museum has since become part of the Technical Museum of Denmark in Elsinore, north of Copenhagen.

Martin Knudsen worked as the Society's Chairman for the next twenty years, with the same untiring interest and initiative, until he expressed the wish to retire because of failing health. A few years later, on his 70th birthday, the Society found the opportunity of honouring Martin Knudsen with the publication of a Festschrift, which appeared as an expanded volume of the proceedings of lectures whose origin was due to his own initiative,[9] and the board also had the pleasure of being able to inform Martin Knudsen that it had been unanimously decided to ask him to become an honorary member of the Society.

In Danish physics circles, Martin Knudsen's rich research and the memories of his activity among us will always be held with profound admiration and gratitude.

---

[9] *Selskabet for Naturlærens Udbredelse, 1939–1940*, J. Jørgensen, Copenhagen 1941.

XXVIII. [OBITUARY FOR M. KNUDSEN]

Overs. Dan. Vidensk. Selsk. Virks. Juni 1949 – Maj 1950, pp. 61–65

TEXT AND TRANSLATION

See Introduction to Part II, p. [114].

NORDISK KUNST– OG LYSTRYK

Martin Knudsen

[314]

2.

Af **Niels Bohr.**

Det rige videnskabelige livsværk, hvormed Martin Knudsen
brød nye baner paa saa forskellige omraader som udviklingen af
metoder til at udforske de fysiske forhold i havenes vældige vand-
masser og klarlæggelsen af de love, der behersker opførselen
af de enkelte usynlige molekyler under deres hastige bevægelse
i luftformige stoffer, vakte største anerkendelse og beundring
blandt fagfæller verden over og er efter Martin Knudsens død
blevet mindet fra mange forskellige sider.

Her i Videnskabernes Selskab, hvor vi netop har lyttet til
en beretning — fremført med sagkundskab af hans efterfølger
som professor i fysik ved Københavns Universitet — om Martin
Knudsens forskningsgerning paa fysikens omraade, mindes vi
alle ogsaa hans frugtbare og mangesidige virksomhed i det danske
samfunds tjeneste og især i vort selskab, hvor han med vaagent
initiativ og utrættelig pligttroskab i næsten 30 aar varetog sekre-
tærens betydningsfulde hverv.

Martin Knudsen, der selv konstruerede og delvis forfærdigede
mange af de sindrige apparater, han benyttede ved sine undersøgelser, var paa flere maader i sin forskning mere uafhængig
af andres hjælp, end det almindeligvis er tilfældet. Men samtidig
gav netop hans aabne blik for enkeltheder og hans store evne
til at sammenfatte disse udfra simple og klare synspunkter ham

særlige betingelser for at organisere det samarbejde, der er væsentligt for videnskabens støtte fremgang og uundværligt for løsningen
af samfundets mangfoldige opgaver.

En af dem, der tidligt fik øje paa de evner, der boede i Martin
Knudsen, og maaske den, der fik den største betydning for hans
livsbane, var professor Christian Christiansen, hvis vidtfavnende
interesser og dybe indblik saavel i videnskabens væsen som i
menneskelivets almindelige kaar var saa stor en inspiration for
mange af os, der havde den lykke at komme i nærmere forhold
til ham. Martin Knudsen har selv i et interview i anledning af
sin 70 aars fødselsdag, hvori han paa sin jævne og samtidig
stemningsfulde maade kommer ind paa erindringer fra sin barndom og tidlige ungdom, beskrevet, hvor meget det betød for ham,
at han som ung student af Christiansen blev indbudt til at deltage
i undersøgelser af de ejendommelige straalefænomener, der ledsager elektriske udladninger i fortyndede luftarter og hvis opdagelse i slutningen af forrige aarhundrede skulle blive indledningen til en helt ny epoke paa fysikens omraade.

Før Martin Knudsen skulle komme til at udføre de berømte
undersøgelser, som bidrog saa meget til forstaaelse af luftarternes
egenskaber, ledtes han imidlertid ved deltagelsen i Ingolf-ekspe-
ditionen 1895 og 1896 ind i arbejdet paa et andet forskningsomraade, hvor han snart skulle bringes ind i et verdensomspændende
samarbejde. Ved behandlingen af de opgaver, som betroedes ham
paa ekspeditionen, fandt han lejlighed til at lægge evner til at
planlægge og udføre nøjagtige og systematiske undersøgelser for
dagen paa en maade, der tiltrak sig stor opmærksomhed, og hans
metoder til maaling af vandmassernes temperatur og saltholdighed
blev hurtigt taget i brug overalt. Naar Danmark er kommet til at
indtage en saa fremtrædende plads i de fysiske undersøgelser af
havene, skyldes det, at man i Martin Knudsen fandt ikke alene
en fremragende eksperimentator, men tillige en organisator,
hvem man trygt kunne betro varetagelsen af store fælles opgaver.

I sit hele virke fandt Martin Knudsen en fast støtte i sin
norskfødte hustru, der allerede før giftermaalet var knyttet til
hans arbejde ved havundersøgelserne og senere saavel i det
gæstfrie hjem som i forbindelse med de mange repræsentations-
pligter, der kom til at hvile paa dem begge, ved sin varme
menneskelige forstaaelse skabte en baggrund, der i saa høj grad

bidrog til at muliggøre den forening mellem intensiv forskergerning og omfattende samfundsvirksomhed, der karakteriserede
Martin Knudsens liv.

For de opgaver, som frembød sig her hjemme i forbindelse med
den højere fysikundervisning, viste Martin Knudsen altid stor
interesse, og han virkede især med energi for allerede paa undervisningens tidligste trin at bibringe eleverne kendskab til eksperimentelle metoder. I anerkendelse af det frugtbare initiativ, han
her fra første færd udviste, blev han efter Christiansens forslag
i en alder af knapt 30 aar valgt til formand for det gamle, af
H. C. Ørsted stiftede Selskab for Naturlærens Udbredelse, og den
tillid, der ved dette valg i saa ung en alder blev Martin Knudsen
til del, viste sig ogsaa at være fuldtud berettiget.

De næsten 40 aar, han ledede Selskabet for Naturlærens
Udbredelse, blev en ny blomstringsperiode for dets virksomhed,
under hvilken mange nye opgaver blev taget op i overensstemmelse
med den aand, der havde besjælet selskabets stifter, og Martin
Knudsen fandt ikke mindst mangfoldig lejlighed til at værne om
H. C. Ørsteds minde. Saaledes indstiftedes paa hans forslag
H. C. Ørsted-medaillen til anerkendelse af og opmuntring for
dansk videnskabelig indsats paa de omraader, hvor Ørsted havde
virket. I forbindelse med 100-aarsdagen for opdagelsen af elektromagnetismen indrettedes endvidere et H. C. Ørsted-museum
til bevarelse af instrumenter, som Ørsted havde anvendt ved
sine undersøgelser, tillige med personlige ejendele, som fra hans
tid forefandtes inden for Ørsteds familie.

Imidlertid skulle Martin Knudsen komme i forbindelse med
stedse større forskerkredse, efterhaanden som hans ry i den
videnskabelige verden voksede, især som følge af hans grundlæggende undersøgelser over luftarternes forhold ved lave tryk,
hvorved han viste sig at være saa stor en mester i den fysiske
eksperimentalkunst. De første beretninger om disse undersøgelser,
der nu overalt betragtes som klassiske, fremkom i de Mathematisk-
fysiske Meddelelser i 1909, samme aar som Martin Knudsen blev
indvalgt i Videnskabernes Selskab, og den store opmærksomhed,
som de vakte, gav sig blandt andet udtryk i, at han fik sæde i
den første komité til organisationen af de internationale møder i
Solvay Instituttet for Fysik, der oprettedes i 1911, og ved hvilket
Martin Knudsen i over 20 aar virkede som komiteens sekretær.

64

Sin virksomme interesse for mellemfolkeligt samarbejde paa videnskabens omraade lagde Martin Knudsen ogsaa for dagen som generalsekretær fra 1915 til 1928 for den danske bestyrelse af de skandinaviske naturforskermøder, der gaar mere end 100 aar tilbage, og i hvilke H. C. Ørsted fra begyndelsen havde taget saa aktiv en del. Ligeledes medvirkede Martin Knudsen ved stiftelsen af Danmarks naturvidenskabelige Samfund, hvis formaal det er saavel at skabe nøjere forbindelse mellem videnskabelige institutioner og tekniske virksomheder her i landet som at styrke baandene med tilsvarende udenlandske kredse. I mere end 25 aar virkede Martin Knudsen som vicepræsident i Danmarks naturvidenskabelige Samfund og paa hans 70-aars fødselsdag i 1941 fandt Samfundet lejlighed til at hylde ham ved en fest, hvorunder den beundring og taknemmelighed, man følte over for ham, kom til orde fra alle sider.

Inden for Videnskabernes Selskab mødtes Martin Knudsen fra første færd med den største tillid, og i stadig stigende grad blev der lagt beslag paa hans kundskaber og arbejdskraft. Efter at have været revisor fra 1911 indvalgtes han i 1914 i kassekommissionen, hvor han fra 1916 virkede som formand, indtil han ved H. G. Zeuthens tilbagetræden i 1917 valgtes til dennes efterfølger som Selskabets sekretær. De særlige forudsætninger, som Martin Knudsen medbragte til dette hverv, skulle blive af største betydning for Selskabets virksomhed. Ikke alene lykkedes det ham paa heldigste maade at videreføre de mange forbedringer i Selskabets organisation, som Zeuthen havde indført under sin 39-aarige, saa fortjenstfulde sekretærgerning, men tillige at løse mange nye opgaver, som ikke mindst genopbygningen af det internationale videnskabelige samarbejde efter afslutningen af den første verdenskrig frembød.

Jeg skal her blot minde om den energi, hvormed Martin Knudsen virkede i det af Selskabet nedsatte udvalg for Danmarks deltagelse i Union Académique Internationale, til hvilket Académie des Inscriptions et Belles-Lettres i Paris tog initiativet i 1919, og i det kort forinden paa initiativ fra Royal Society i London oprettede International Research Council, samt om hans virksomhed inden for flere af de danske nationalkomiteer for de internationale videnskabelige fagunioner. For de talrige andre maader, hvorpaa Martin Knudsen initiativrigt arbejdede for Selskabets opgaver, er der udførligt gjort rede i de Samlinger til Selskabets Historie, der udgaves til 200-aars jubilæet i 1942, og hvis tilblivelse især skyldes det nære samarbejde mellem Martin Knudsen og fuldmægtigen i Selskabets sekretariat, Asger Lomholt, der har tilegnet Samlingernes nyligt udkomne andet bind mindet om Martin Knudsen.

Som personlig erindring vil jeg ogsaa gerne anføre nogle sætninger fra en tale, det tilfaldt mig at holde ved den festaften, hvorved Selskabet i maj 1942 hyldede Martin Knudsen, da han i 25 aar havde virket som dets sekretær:

»For de fleste af os, der er her til stede, er tanken om sekretærposten i Selskabet uadskillelig fra Martin Knudsens person, og jeg selv har ofte maattet tænke paa, at han oprindelig valgtes til sekretær netop den aften, hvor jeg som nyvalgt medlem for første gang havde lejlighed til at være til stede ved et af Selskabets møder. I alle disse aar har Martin Knudsen fulgt alt, hvad der angik Selskabet med den dybeste interesse for ikke at sige den inderligste kærlighed. Hans sjældne administrative evner — der sammen med hans grundlæggende videnskabelige arbejde var baggrunden for, at sekretærhvervet blev ham betroet — har været Selskabet til største gavn og har medført, at den overordentlige udvikling af Selskabets virksomhed ikke mindst med hensyn til de internationale forbindelser, der i hans sekretærtid har fundet sted, har været saa vel tilrettelagt, at de fleste medlemmer næppe nogensinde har mærket, hvor store byrder der var lagt paa hans skuldre. Alle føler vi taknemmelighed for den naturlige elskværdighed, hvormed han er kommet hver af os i møde, og for den selvforglemmende maade, hvorpaa han i alle sager, hvor forskellige opfattelser er kommet til orde, altid har bestræbt sig for at bidrage til det for Selskabets virksomhed og dets indre sammenhold gavnligste resultat.«

Lige til Martin Knudsen i 1945 paa grund af svigtende helbred ønskede at nedlægge sekretærhvervet, virkede han paa samme utrættelige og pligttro maade for Selskabets mange opgaver, og her i Videnskabernes Selskab, ligesom i langt videre kredse, vil mindet om hans store og frugtbare gerning i videnskabens og samfundets tjeneste altid holdes i ære og bevares med dyb taknemmelighed.

TRANSLATION

2
By **Niels Bohr**

The rich scientific lifework whereby Martin Knudsen broke new paths in so diverse fields as the development of methods to investigate the physical conditions in the enormous water masses of the oceans and the clarification of the laws controlling the behaviour of the individual invisible molecules during their rapid movement in gaseous substances, aroused great appreciation and admiration among his peers worldwide and after Martin Knudsen's death has been commemorated in many different quarters.

Here in the Academy of Sciences and Letters, where we have just been listening to an account – delivered with authority by his successor as professor in physics at the University of Copenhagen[1] – of Martin Knudsen's research work in the field of physics, we all also remember his fruitful and many-faceted activity in the service of Danish society, and especially in our Academy where with keen initiative and tireless dedication he attended to the important task of being the Secretary for nearly 30 years.

Martin Knudsen, who himself constructed and partially completed many of the ingenious apparatuses he used in his investigations, was in his research in many ways more independent of the help of others than is usually the case. But at the same time, precisely his keen grasp of particulars and his great capacity for ordering them on the basis of simple and clear viewpoints, gave him special qualifications for organizing the collaboration that is important for the constant progress of science and invaluable for the solution of the diverse tasks of society.

One of those who early saw the talents Martin Knudsen had in him, and perhaps the one who had the greatest significance for his career, was Professor Christian Christiansen, whose wide-ranging interests and profound insight into the nature of science as well as the general conditions of human life were such a great source of inspiration for many of us who had the good fortune of getting on close terms with him. In an interview on the occasion of his 70th birthday,

---

[1] Bohr's talk reproduced here is preceded on pp. 55–61 of the Academy's Proceedings by an obituary written by Bohr's and Knudsen's close colleague, the physicist Jacob Christian Jacobsen.

where in his plain, but yet evocative, fashion he touches upon memories from his childhood and early youth, Martin Knudsen has himself described how much it meant for him as a young student to be invited by Christiansen to take part in the investigations of the peculiar radiation phenomena which accompany electrical discharges in rarefied gases and whose discovery at the close of the previous century would be the introduction to an entirely new epoch in the field of physics.[2]

Before Martin Knudsen should come to carry out the famous investigations which contributed so much to the understanding of the properties of gases, he was, however, led, through participation in the Ingolf Expedition in 1895 and 1896,[3] into work in another field of research where he soon was to be part of a worldwide cooperation. When dealing with the tasks entrusted to him on the expedition, he found the opportunity of displaying talents for planning and carrying out precise and systematic investigations in a way that attracted great attention, and his methods for the measurement of the temperature and salinity of the water masses were rapidly adopted everywhere. The reason for Denmark having come to occupy such a prominent position as regards physical investigations of the oceans is that in Martin Knudsen one found not only an excellent experimenter, but also an organizer to whom could safely be entrusted the handling of large shared tasks.

In all his work Martin Knudsen found solid support in his Norwegian-born wife,[4] who already before their marriage was connected with his work in oceanic research, and who later – in the hospitable home, as well as in connection with the many obligations to entertain which would become their joint responsibility – by her warm human understanding created a background which to such a great extent contributed to making possible the combination of intensive research work and extensive activity for society that characterized Martin Knudsen's life.

Martin Knudsen always showed great interest in the tasks offered here at home in connection with advanced education in physics, and, in particular, he worked energetically to impart knowledge of experimental methods to pupils at the earliest stage of education. In recognition of the fruitful initiative he thus displayed from the very beginning, he was, on Christiansen's recommendation, elected at an age of not quite 30 years as chairman of the old Society for the Dissemination of Natural Science established by H.C. Ørsted, and the trust

---

[2] The interview referred to by Bohr has not been identified.

[3] During the summer months of 1895 and 1896 the schooner Ingolf carried out a state-financed investigation of the seabed around Iceland north to Jan Mayen and outside western Greenland.

[4] Ellen, née Ursin (1880–1964).

shown to Martin Knudsen by this election at so young an age turned out to be fully justified.

The nearly 40 years he led the Society for the Dissemination of Natural Science became a new golden age for its activity, during which many new tasks were taken up in correspondence with the spirit that had inspired the Society's founder, and Martin Knudsen found not least many opportunities to cherish the memory of H.C. Ørsted. Thus the H.C. Ørsted Medal was instigated at his suggestion for the recognition and encouragement of Danish scientific effort in the fields where Ørsted had been active. In connection with the centennial for the discovery of electromagnetism, an H.C. Ørsted Museum was furthermore established for the preservation of instruments that Ørsted had used in his investigations, together with personal possessions which had been kept by the Ørsted family since his time.

In the meanwhile, Martin Knudsen was to come into contact with ever larger research circles, as his reputation in the scientific world grew, especially as a result of his fundamental investigations into the properties of gases at low pressure, whereby he proved to be such a great master of experimental technique in physics. The first accounts of these investigations, which now everywhere are regarded as classic, appeared in the Mathematical–Physical Communications[5] in 1909, the same year as Martin Knudsen was elected to the Academy of Sciences and Letters, and the great attention they caused manifested itself in, among other things, his obtaining a seat on the first committee for the organization of the international meetings at the Solvay Institute of Physics, which was established in 1911 and in which Martin Knudsen was active for more than 20 years as the committee Secretary.

Martin Knudsen also displayed his active interest for international cooperation in the field of science as General Secretary from 1915 until 1928 for the Danish board of the Scandinavian Meetings of Natural Scientists, which goes back more than 100 years and in which H.C. Ørsted had taken so active a part from the very beginning.[6] Similarly, Martin Knudsen took part in the founding of the Science Society of Denmark,[7] whose aim is to create closer connections between scientific institutions and technological enterprises in this country, as well as to strengthen the ties with comparable circles abroad. For more than 25 years Martin Knudsen acted as Vice-President of the Science Society of Denmark, and on his 70th birthday in 1941 the Society found the opportunity

---

[5] "Mathematisk–fysiske Meddelelser" of the Royal Danish Academy of Sciences and Letters.
[6] See Vol. 11, pp. [354]–[356].
[7] See p. [294], ref. 3.

to pay tribute to him with a celebration at which the admiration and gratitude felt for him was expressed from all quarters.

Within the Academy of Sciences and Letters Martin Knudsen was met with the greatest trust from the very beginning, and to a constantly increasing degree use was made of his proficiencies and capacity for work. After having been auditor from 1911, he was elected to the cash commission in 1914, where he acted as chairman from 1916 until, at H.G. Zeuthen's[8] retirement in 1917, he was elected his successor as Academy Secretary. The special qualifications Martin Knudsen brought to this task were to be of the greatest importance for the activity of the Academy. Not only did he succeed in the most fortunate way in continuing the many improvements in the Academy's organization that Zeuthen had introduced during his 39 years of meritorious work as Secretary, but also in solving many new problems, not least provided by the reestablishment of international scientific cooperation after the end of the First World War.

Here I will only recall the energy with which Martin Knudsen worked on the committee, set up by the Academy, for Danish participation in the Union Académique Internationale, for which the Académie des inscriptions et belles-lettres in Paris took the initiative in 1919, and in the International Research Council, established shortly before on the initiative of the Royal Society of London, as well as his activity on several of the Danish national committees for international scientific unions. The numerous other ways in which Martin Knudsen worked with great initiative for the Academy's tasks are described in detail in the "Samlinger til Selskabets Historie"[9] that was published for the bicentennial in 1942, and whose creation is due in particular to the close cooperation between Martin Knudsen and Asger Lomholt,[10] chief assistant in the Academy's secretariat, who has dedicated the recently published second volume of the "Samlinger" to the memory of Martin Knudsen.

As a personal memory I would also like to add some sentences from a speech it was allotted to me to give at the festive evening whereby the Academy in May 1942 paid tribute to Martin Knudsen, when he had worked as its Secretary for 25 years.[11]

"For most of us who are present here, the idea of the position as Secretary

---

[8] Hieronymus Georg Zeuthen (1839–1920), Danish mathematician.

[9] *Det Kongelige Danske Videnskabernes Selskab 1742–1942: Samlinger til Selskabets Historie*, five volumes, Munksgaard, Copenhagen 1942–1973. The title can roughly be translated "Collections of material documenting the history of the Academy".

[10] (1901–1990).

[11] The manuscript is in the BMSS. See *Inventory of Relevant Manuscripts in the Niels Bohr Archive*, below, p. [528], folder 28.

in the Academy is inseparable from Martin Knudsen's person, and I have often had to remind myself that he was originally elected Secretary on precisely the evening when as a newly elected member I had the opportunity of being present for the first time at one of the Academy meetings. In all these years, Martin Knudsen has followed everything concerning the Academy with the deepest interest, if not to say the most heartfelt devotion. His unique administrative abilities – which together with his fundamental scientific work were the background for the position of Secretary being entrusted to him – have been of the greatest benefit to the Academy and have entailed that the enormous development of the Academy's activity, not least as regards international relations, which has taken place in his time as Secretary, has been so well organized that most members have hardly ever noticed what great burdens have been laid on his shoulders. We all feel gratitude for the natural kindness with which he has met each of us and for the self-denying fashion in which he, in all matters where different opinions have been aired, always has striven to contribute to the result most beneficial for the Academy's activity and internal concord."

Right up until 1945, when, because of failing health, Martin Knudsen wished to resign from his position as Secretary, he worked in the same untiring and conscientious fashion for the tasks of the Academy, and here in the Academy of Sciences and Letters, as in much wider circles, the memory of his great and fruitful work in the service of science and society will always be held in esteem and remembered with deep gratitude.

XXIX. A PERSONALITY IN DANISH PHYSICS

EN PERSONLIGHED I DANSK FYSIK
Politiken, 7 September 1946

For H.M. Hansen

TEXT AND TRANSLATION

See Introduction to Part II, p. [114].

[325]

En Personlighed i dansk Fysik

Et Billede af Professor H. M. Hansens Gerning i Videnskabens og det danske Samfunds Tjeneste

Af Niels Bohr

PAA Professor *H. M. Hansens* 60 Aars Fødselsdag, hvor han kan se tilbage paa en lang og frugtbar Virksomhed i Videnskabens og vort Samfunds Tjeneste, vil han møde mange Vidnesbyrd om den Taknemmelighed, som ikke alene hans Kolleger og Elever skylder ham, men som deles af de vide Kredse, der i Aarenes Løb har søgt hans Raad og Støtte, og som stedse dybere har lært at værdsætte den saglige Indstilling, der hos ham er saa nøje forbundet med Hjertelighed og Hjælpsomhed.

Allerede i sin tidligste Studietid lagde han de Egenskaber for Dagen, som skulde danne Baggrunden for hele hans senere saa omfattende Virke og for den højt paaskønnede Stilling, som han i Dag indtager inden for den akademiske Verden herhjemme. Med sjældne Evner og aldrig svigtende Begejstring for den fysiske Videnskab forbandt han fra første Færd den varme Interesse for sine Medmennesker og deres Bestræbelser, der kom til at betyde saa meget for alle, med hvem han stod i Berøring.

Under hans Virksomhed som Assistent ved Den polytekniske Læreanstalts fysiske Laboratorium, hvor han aktivt deltog i den omfattende Reorganisation af Laboratoriet, der i de Aar fandt Sted under Professor K. Prytz' Ledelse, falder et Studieophold i Göttingen, hvor der i Begyndelsen af Aarhundredet udfoldedes et rigt videnskabeligt Liv, og som var et saa besøgt Samlingssted for Videnskabsmænd fra forskellige Lande, der her knyttede mange værdifulde personlige Forbindelser.

I Göttingen udførte han de magnetospektroskopiske Undersøgelser, for hvilke han efter Hjemkomsten til København gjorde Rede i sin Doktordisputats, der ved det righoldige Iagttagelsesmateriale og den omhyggelige Udarbejdelse vandt stor Anerkendelse og i høj Grad bidrog til at vække Interesse for saadanne Problemer blandt Fysikerne her i Landet. For mit eget Vedkommende blev H. M. Hansens Indsigt paa Atomspektrenes Omraade en Kilde til Inspiration, da jeg omtrent paa samme Tid kom tilbage fra et Studieophold i England, hvor Thomsons og Rutherfords Opdagelser angaaende Atomernes Byggestene havde indledt en ny Udvikling inden for den fysiske Videnskab og stillet Forskningen over for Problemet om at sammenfatte Erfaringer fra de mest forskellige Omraader af Fysiken og Kemien.

Dette Fælleskab i Interesser blev Indledningen til et langt Venskab og Samarbejde, og især skylder jeg H. M. Hansen dyb Taknemmelighed for den Beredvillighed, hvormed han stillede hele sin Erfaring til Raadighed ved Indrettelsen og Udviklingen af det Universitetsinstitut, der kort efter oprettedes for her i Landet at skabe Betingelser for Deltagelse i Atomforskningen. I Institutets første Aar, hvor H. M. Hansen var med i det daglige Arbejde, og hvor det var en saa vigtig Opgave i Undersøgelserne at inddrage en stedse voksende Kreds af yngre Medarbejdere,

Professor H. M. Hansen.

lærte jeg endvidere paa nærmeste Hold at kende og værdsætte den Indstilling, som ligger til Grund for hans senere frugtbare Bestræbelser for at forbedre Levevilkaarene for Studerende og unge Videnskabsmænd.

H. M. Hansens dybe Forstaaelse for Fysikens Betydning paa mangfoldige videnskabelige Grænseomraader og især for dens Anvendelse til Forbedring af Hjælpemidlerne for Lægekunsten førte ham i de følgende Aar ind paa nye Felter, hvorom han mere og mere koncentrerede sig efter sin Ansættelse som Professor i Biofysik ved Universitetet og Oprettelsen af det biofysiske Laboratorium i det ved Understøttelse fra Rockefellerfondet opførte nye store Institut paa Juliane Maries Vej. Sammen med sine dygtige og energiske Medarbejdere satte han her hurtigt en Række Undersøgelser i Gang, der spænder over Omraader lige fra den medicinske Straalebehandling til akustiske Problemer af Betydning blandt andet for Konstruktionen af bedre Apparater for Tunghøre.

Samtidig med at han har udfoldet ihærdige Bestræbelser for det biofysiske Laboratoriums stadige Udvikling, er der i Aarenes Løb lagt stedse større Beslag paa H. M. Hansens omfattende Indsigt og store organisatoriske Evner. Saaledes beror det først og fremmest paa hans Virksomhed inden for Hovedbestyrelsen af Landsforeningen til Kræftens Bekæmpelse og Direktionen af Finsen-Institutet, at Straalebehandlingen ved de danske Radiumstationer er bragt op til saa højt et Stade i teknisk-fysisk Henseende, ligesom hans Initiativ har været afgørende for det Foregangsarbejde paa Forbedring af Røntgenfotograferingen, der har været saa betydningsfuldt for Gennemførelsen af den landsomfattende Tuberkuloseundersøgelse. At han foruden sine mange andre Hverv tillige virker som Formand for Tunghøreforeningen, er saa betegnende et Udtryk for Omfanget af hans Følelser for sine Medmennesker.

Inden for Universitetet staar H. M. Hansen ikke alene som en højt skattet Lærer for de fysiske og medicinske Studerende, men hans varme Interesse for hele Universitetets Virksomhed og ikke mindst for de Unges Vel har givet sig Udtryk i mange betydningsfulde Tillidshverv. Saaledes har han i en Aarrække været Medlem af Konsistorium og Universitetets Legatbestyrelse, og netop i Aar beklæder han som bekendt Posten som Universitetets Prorektor, hvorved endnu større Kredse har faaet Lejlighed til at stifte Bekendtskab med hans ligefremme og hjertevarme Personlighed. Som en af de Sager, der har ligget ham mest paa Sinde, og ved hvis Gennemførelse han som Prorektor har været særlig virksom, kan nævnes det i det sidste Foraar paa bredt Grundlag paabegyndte Folkeoplysningsarbejde ved Universitetet, der som Led i Bestræbelserne for at skabe nøje Forbindelse mellem akademiske Kredse og den opvoksende Ungdom fra alle Samfundslag rummer saa store Løfter.

Selv om disse Linjer naturligvis kun kan give et meget ufuldstændigt Billede af Professor H. M. Hansens store Indsats for vort Samfund, haaber jeg dog, at de er egnet til at give et Indtryk af den Taknemmelighed, det skylder ham, og af Varmen af de Ønsker, hvori hans Kolleger og Medarbejdere i Dag forenes, om at han endnu i mange Aar med usvækket Kraft og Begejstring maa kunne fortsætte sin Virksomhed iblandt os.

NIELS BOHR.

TRANSLATION

A Personality in
Danish Physics

––––––

A Portrait of Professor H.M. Hansen's Work
in the Service of Science and Danish Society

––––––

By Niels Bohr

On Professor H.M. Hansen's 60th birthday, when he can look back on a long and fruitful activity in the service of science and our society, he will see many proofs of the gratitude not only his colleagues and pupils owe him, but which is also shared by wide circles who have asked his advice and support over the years and have learnt to value ever more deeply the professional approach which in him is so closely associated with cordiality and readiness to help.

Already in his earliest student days he showed the characteristics that would form the background for the whole of his later so extensive work and for the highly appreciated position he fills today within the academic sphere in this country. From the very beginning he combined his rare gifts and never-failing enthusiasm for physical science with the warm interest for his fellow beings and their endeavours, which came to mean so much to all with whom he came into contact.

While working as an assistant at the physical laboratory of the Technical University, where he took an active part in the extensive reorganization of the laboratory during those years under the leadership of Professor K. Prytz,[1] he went for a period to study in Göttingen, where there was a rich scientific activity at the beginning of this century, and which was a frequently visited meeting place for scientists from many countries, who there made valuable personal contacts.

---

[1] Peter Kristian Prytz (1851–1929).

In Göttingen he carried out the magneto–spectroscopical investigations which he described in his doctoral dissertation upon his return home. Due to its rich observational material and careful preparation, it was highly praised and contributed decisively to arousing interest in such problems among the physicists in this country. In my own case, H.M. Hansen's insight in the field of atomic spectra became a source of inspiration, as at about the same time I returned from a period of study in England, where Thomson's and Rutherford's discoveries regarding the constituents of atoms had started a new development within physical science and confronted research with the problem of combining experience from the most diverse fields within physics and chemistry.

This sharing of interests was the beginning of a long friendship and co-operation, and I owe H.M. Hansen deep gratitude especially for the readiness with which he placed all of his experience at my disposal during the setting up and development of the university institute which was established shortly thereafter in order to provide conditions for participation in atomic research in this country.[2] During the first years of the institute, when H.M. Hansen took part in the daily work and when it was such an important task to involve a steadily growing circle of younger collaborators in the research, I also got to know and appreciate at first hand the attitude which is the basis of his later fruitful endeavours to improve living conditions for students and young scientists.

H.M. Hansen's deep understanding of the importance of physics in numerous scientific borderline areas, and especially of its application for the improvement of auxiliary aids in medicine, led him in the following years into new fields on which he concentrated more and more after his appointment as Professor of Biophysics at the University and the establishment of the Biophysical Laboratory in the large new Institute on Juliane Maries Vej, built with support from the Rockefeller Foundation.[3] Together with his able and energetic collaborators he soon started here a series of investigations which cover fields ranging from medical radiation treatment to acoustical problems of importance for, among other things, the construction of better hearing aids for the hard of hearing.

At the same time as he has made unremitting efforts for the continued development of the Biophysical Laboratory, there has been a constantly growing demand over the years for H.M. Hansen's extensive insight and great organizational talents. Thus it is first and foremost due to his activity on the

---

[2] The University Institute for Theoretical Physics was established for Bohr in 1921.

[3] This institute was built to include all activities in physiology in the Natural Science and Medical Faculties of the University of Copenhagen. It was in full operation by 1930.

executive committee of the National Association for Combating Cancer[4] and on the board of the Finsen Institute,[5] that radiation treatment at the Danish radium stations has reached such a high standard as regards technology and physics, just as his initiative has been decisive for the pioneer work on the improvement of X-ray photography, which has been so important for the realization of the nation-wide tuberculosis screening.[6] The fact that in addition to his many other posts he is also the chairman of the Association of the Hard of Hearing[7] is a characteristic expression of the extent of his feelings for his fellow beings.

Within the University H.M. Hansen is not only a highly esteemed teacher for the students of medicine and physics, but his warm interest in the activities of the entire University, and not least in the well-being of young people, is apparent from his numerous important honorary posts. Thus he has been a member of the Academic Council and the Grant Committee of the University for a number of years, and, as is well known, this very year he holds the position of Prorector of the University, whereby even wider circles have the opportunity to become acquainted with his straightforward and warm-hearted personality. As one of the matters which have been most important to him, and for the realization of which he has been particularly active as Prorector, may be mentioned the work for public general education at the University begun on a broad basis last spring, which, as part of the endeavours to establish close contact between academic circles and the rising younger generation from all parts of society, holds such great promise.

Although these lines can of course only give a very incomplete picture of professor H.M. Hansen's great efforts for the benefit of our society, I do hope that they serve to give an impression of the gratitude society owes him and of the warmth of the wishes, in which his colleagues and collaborators unite today, that he may continue his activities among us with unfailing energy and enthusiasm for many years to come.

NIELS BOHR.

---

[4] On Bohr's involvement in this organization, see Vol. 11, pp. [373]–[375].

[5] The Finsen Institute was established in 1896 with the purpose of continuing the treatment with light instigated by Niels Finsen (1860–1904), the first Danish Nobel Prize laureate (1903). It was later expanded into a hospital for cancer treatment.

[6] The National Association for Combating Tuberculosis, established in 1901, carried out the screening from 1953 to 1954, criss-crossing Denmark with buses carrying X-ray equipment.

[7] Established in 1912.

XXX. A SHINING EXAMPLE FOR US ALL

ET LYSENDE FORBILLEDE FOR OS ALLE
Politiken, 14 June 1956, p. 13

For H.M. Hansen

TEXT AND TRANSLATION

See Introduction to Part II, p. [114].

[331]

Et lysende forbillede for os alle —

Niels Bohr skildrer sin ungdomsven, rektor, professor H. M. Hansen, der i gaar døde midt under sit store arbejde for dansk forskning

Et af de smukkeste billeder fra de sidste aar af professor H. M. Hansen med den gyldne rektorkæde.

H. M. HANSENs bortgang betyder et stort og smerteligt tab ikke alene for Københavns Universitet, som han med lykkelig haand har ledet i en længere aarrække, men for hele det danske samfund, som han har tjent saa trofast paa mangfoldige omraader. De af os, som stod ham personligt nær, har mistet en ven, der altid var rede, naar man søgte raad og hjælp.

Mine egne erindringer om H. M. Hansen gaar tilbage til vor fælles studietid, hvor han allerede viste de evner og den menneskelige indstilling, som skulle blive det bærende grundlag for hans rige livsgerning. Med taknemmelighed tænker jeg paa den inspiration og opmuntring som samtalerne med ham om arbejde og planer altid var for mig, og ikke mindst paa, hvad jeg skylder hans medarbejderskab ved opbygningen af virksomheden paa Institutet for teoretisk fysik.

I de følgende aar, hvor H. M. Hansen blev leder af det nyoprettede biofysiske laboratorium, førte hans omfattende kundskaber og interesser og hans medfølelse med lidende mennesker ham ind i en virksom og resultatrig deltagelse i bestræbelserne for at udnytte fysikkens fremskridt til hjælp for svage og syge. I Dansk Tunghøreforening virkede han fra dens oprettelse som formand, og i kampen mod kræften blev han en af de førende skikkelser. Det skyldes ikke mindst hans energi og hans fremsyn, at vi her i landet i dag besidder radiumstationer udstyret med de mest effektive hjælpemidler til sygdomsbehandlingen.

FRA sin første tilknytning til Københavns Universitet lagde H. M. Hansen den dybeste interesse for alle dets anliggender for dagen, og især laa de unge studerendes kaar ham paa hjerte. Efter i en aarrække at have været medlem af konsistorium og stipendieudvalget valgtes han med tilslutning fra alle sider i 1948 til rektor for vort gamle universitet. Hans særlige forhandlingsevner, der hvilede paa nøgtern saglighed og paa varm menneskelighed, kom her til rig udfoldelse og hjalp ham til at løse de mange opgaver, som tidens krav stillede. Ved denne utrættelige og frugtbare virksomhed, der vakte en stadig voksende beundring for H. M. Hansens personlighed, lykkedes det ham i stedse videre krese at skabe forstaaelse for videnskabeligt studiums og forsknings betydning for hele samfundet.

Det er næsten ufatteligt, at H. M. Hansen, samtidig med saa krævende en gerning, med lige kraft og fremgang kunne varetage de mange andre samfundshverv, som med aarene betroedes ham paa grund af den store tillid, alle nærede til ham. Foruden at være præsident for Dansk Røde Kors var han en bærende kraft i mangfoldige bestræbelser for at fremme mellemfolkeligt kulturelt samarbjde, og i dette arbejde stilledes han over for nye store opgaver, som han løste paa en maade, der videnom vandt anseelse for vort land.

DET gav anledning til stor bekymring, da H. M. Hansen for nogle aar siden blev angrebet af en alvorlig sygdom, men efter det første hospitalsophold formaaede han med sjælden udholdenhed og selvforglemmelse at fortsætte sit frugtbare virke. Selv da han for nogle maaneder siden atter maatte indlægges, ledede han fra sit sygeleje indtil de allersidste uger Universitetet og fulgte enhver sag med usvækket interesse.

Omend det var med tungt hjerte, man ved de seneste besøg paa hospitalet saa, at H. M. Hansens kræfter begyndte at svigte, forlod man ham hver gang med en stedse stærkere beundring for den sjælsstyrke, hvormed han bevarede haabet om helbredelse og om fremgang for alt det, han havde virket og kæmpet for. Dagen før sin død talte han baade om sine haab og om den taknemmelighed han følte for den forstaaelse og støtte, der var blevet hans bestræbelser til del.

For os andre var det klart, at det uundgaaelige nærmede sig, dog er tanken om, at H. M. Hansen ikke mere lever og virker iblandt os, vanskelig at fatte. Trods sorgen over hans bortgang og det dybe savn hans kolleger og venner føler, vil de rige og lyse minder, han har givet os, altid være en kilde til fornyet styrke. H. M. Hansen efterlader en tom plads i vort samfund, som det vil blive svært at udfylde, men han efterlader ogsaa et lysende forbillede til efterfølgelse for os alle.

Niels Bohr.

TRANSLATION

A Shining Example
for Us All –

Niels Bohr portrays the friend of his youth, Vice-Chancellor, Professor H.M. Hansen, who died yesterday in the middle of his great work for Danish research[1]

H.M. HANSEN's death means a great and painful loss not only for the University of Copenhagen, which he has led with a fortunate touch for many years, but for the whole Danish society, which he has served so faithfully in numerous fields. Those of us who were close to him personally have lost a friend who was always prepared when one sought advice and help.

My own memories of H.M. Hansen go back to our common student years when he already showed the talents and the human attitude that were to become the principal basis for his rich lifework. With gratitude I think of the inspiration and encouragement which the conversations with him about work and plans always were for me, and not least of what I owe his collaboration in building up the activities at the Institute for Theoretical Physics.

In the subsequent years, when H.M. Hansen became the leader of the newly established Biophysics Laboratory, his wide-ranging abilities and interests, and his sympathy with suffering people, led him into an active and productive participation in the endeavours to utilize the advances of physics for the benefit of the infirm and the sick. He acted as chairman of the Association of the Hard of Hearing from its establishment, and he became one of the leading figures in the fight against cancer. It is due not least to his energy and his foresight that we now have in this country radium stations equipped with the most efficient tools for the treatment of illness.

From his first affiliation with the University of Copenhagen, H.M. Hansen showed the deepest interest in all its affairs, and especially the fortunes of

[1] Translation of the figure caption on p. [332]: "One of the finest pictures from recent years of Professor H.M. Hansen with the gold chain of the Vice-Chancellor."

[333]

the young students lay close to his heart. After having been a member of the University Senate and the Grants Committee for several years, in 1948 he was elected Vice-Chancellor for our old university with support from all quarters. His special negotiating abilities, which rested upon matter-of-fact impartiality and upon warm humanity, were richly expressed here and helped him to solve the many tasks posed by the demands of the time. With this untiring and successful activity, which aroused a steadily growing admiration for H.M. Hansen's personality, he succeeded in ever widening circles to create understanding for the importance of scientific study and research for the whole society.

It is almost inconceivable that H.M. Hansen, concurrently with such a demanding occupation, with equal energy and success could attend to the many other public posts which were entrusted to him over the years because of the great confidence all had in him. Besides being President of Danish Red Cross, he was a mainstay in numerous endeavours to advance international cultural cooperation, and in this work he was faced with new great tasks which he solved in a way that far and wide won respect for our country.

It was cause for great anxiety when H.M. Hansen some years ago was attacked by serious illness, but after the first stay in hospital he succeeded with rare persistence and unselfishness to continue his successful work. Even when a few months ago he once again had to be admitted to hospital, he led the University from his sickbed until the very last weeks and followed every matter with unmitigated interest.

Even though it was with a heavy heart one saw, on the latest visits to the hospital, that H.M. Hansen's powers were beginning to fail, one left him every time with an ever stronger admiration for the strength of will whereby he kept up hopes of recovery and of progress for all that he had worked and fought for. The day before his death he spoke both of his hopes and of the gratitude he felt for the understanding and support that his endeavours had received.

For the rest of us it was clear that the inevitable was approaching, though the thought that H.M. Hansen is no longer living and working among us is difficult to grasp. Despite the grief at his death and the deep loss felt by his colleagues and friends, the rich and bright memories that he has given to us will always be a source of renewed strength. H.M. Hansen leaves an empty place in our community that will not be easily filled, but he also leaves a shining example for us all to follow.

Niels Bohr.

XXXI. [OBITUARY FOR H.M. HANSEN]

Fys. Tidsskr. **54** (1956) 97

TEXT AND TRANSLATION

See Introduction to Part II, p. [114].

H. M. Hansen.

[336]

Ved rektor, professor H. M. Hansens bortgang har ikke alene Københavns universitet, men mange andre danske kulturelle og humanitære institutioner lidt et stort tab. Savnet føles ikke mindst blandt hans fagfæller, der foruden på hans rige forskergerning og hans utrættelige virke for den fysiske videnskabs anvendelse i lægekunstens tjeneste tænker på, hvorledes han i årenes løb som redaktør af Fysisk Tidsskrift formåede at opretholde og styrke forbindelsen mellem den fysiske videnskabs dyrkere og de vide kredse, der varetager ungdommens indførelse i naturvidenskaben. I dette nummer af tidsskriftet, som bringer mindeord fra nogle af H. M. Hansens kolleger ved Københavns universitet, ønsker derfor både hans medarbejdere ved tidsskriftets redaktion og Selskabet for Naturlærens Udbredelse at være med til at give udtryk for den taknemmelighed, som hele vort samfund skylder ham.

NIELS BOHR.

---

[337]

TRANSLATION

On the death of Vice-Chancellor, Professor H.M. Hansen, not only the University of Copenhagen, but also many other Danish cultural and humanitarian institutions, have suffered a great loss. The loss is felt not least among his professional colleagues who remember his rich research work and his tireless activities for the application of physical science in the service of medicine as well as how, throughout the years as editor of Fysisk Tidsskrift, he was able to maintain and strengthen the connection between the practitioners of physical science and the wide circles responsible for introducing young people to science. In this issue of the journal, which carries obituaries by some of H.M. Hansen's colleagues at the University of Copenhagen, both his collaborators in the journal's editorial office and the Society for the Dissemination of Natural Science therefore wish to join in giving expression to the gratitude owed to him by our whole society.

NIELS BOHR.

XXXII. FRIEDRICH PASCHEN ON HIS SEVENTIETH BIRTHDAY

FRIEDRICH PASCHEN ZUM SIEBZIGSTEN GEBURTSTAG
Naturwiss. **23** (1935) 73

TEXT AND TRANSLATION

See Introduction to Part II, p. [115].

Sonderdruck aus Die Naturwissenschaften 1935. 23. Jahrg., Heft 5, S. 73
(Verlag von Julius Springer, Berlin W 9)
Printed in Germany

Friedrich Paschen zum siebzigsten Geburtstag.

FRIEDRICH PASCHENs siebzigster Geburtstag am 22. Januar 1935 gibt jedem Physiker Anlaß, in Bewunderung und Dankbarkeit auf seine für die Entwicklung unserer Wissenschaft so bedeutungsvolle Wirksamkeit zurückzublicken, die sich durch ein halbes Jahrhundert ununterbrochen entfaltete und immer noch mit unverminderter Kraft fortgesetzt wird.

In seinen Arbeiten zeigt PASCHEN sich nicht nur als der große Meister der Experimentierkunst, der immer in erfolgreichster Weise bestrebt war, die Versuchsmethoden zu vervollkommnen und unsere Erfahrungen auf neuen Gebieten zu bereichern, sondern das Merkmal seines Schaffens dürfte vor allem die glückliche Intuition sein, womit er stets solche Probleme experimentell verfolgt hat, deren Erforschung sich für die Ausbildung der allgemeinen theoretischen Vorstellungen von entscheidender Bedeutung erweisen sollte.

Dieses gilt in gleichem Maße für PASCHENs allererste Arbeit, in welcher er bei der Untersuchung des elektrischen Durchschlags von Gasen bei verschiedenen Drucken das Gesetz fand, das seinen Namen trägt und das so wichtig war für die Entwicklung der Theorie der Gasentladungen, wie für seine Untersuchungen auf dem für die Theorie des Atombaues fundamentalen Gebiet der Spektren, denen er fast seine ganze wissenschaftliche Tätigkeit widmete, und die ihm unter seinen Fachgenossen so großen Ruhm verschafft haben.

Seit seiner fruchtbaren Zusammenarbeit mit RUNGE hat PASCHEN wohl mehr als irgendein anderer Forscher zur Sicherstellung und Erweiterung der durch ihre Genauigkeit in der ganzen Physik ausgezeichneten empirischen Gesetzmäßigkeiten der Spektrallinien beigetragen. Dabei war besonders seine Entdeckung neuer Züge in der Struktur der Spektren fördernd für die allmähliche Entwicklung einer quantentheoretischen Systematik der möglichen Atomzustände, die das Entstehen einer rationellen Quantenmechanik vorbereitet hat. Auch kann der Wert der von PASCHEN in Zusammenarbeit mit BACK entdeckten Umwandlung der Zeeman-Effekte bei steigenden Feldstärken für die Aufklärung von tiefsten Problemen der Elektronentheorie nicht überschätzt werden.

Nach seiner ganzen, stets auf Fortschritte gerichteten Einstellung hat PASCHEN in seiner Arbeit immer wieder neue Bahnen eingeschlagen, und in den letzten Jahren finden wir ihn mit gewohnter Meisterschaft beschäftigt mit der Erforschung der feinsten Strukturen der Spektrallinien, die eine so wunderbare Quelle darbieten zur Auskunft über das Problem der Konstitution der Atomkerne, das zur Zeit im Zentrum des Interesses der Naturforscher steht.

Der Einfluß von PASCHENs Wirksamkeit beschränkt sich aber keineswegs auf seine eigenen wissenschaftlichen Leistungen, sondern es hat auf seinem Arbeitsgebiet wohl keiner in größerem Maße auf andere Physiker durch persönliche Belehrung befruchtend gewirkt, und es finden sich wenige physikalische Institute, die nicht in irgendeiner Weise die Hilfe der von PASCHEN geschaffenen Tradition täglich genießen.

Seine Fachgenossen in allen Ländern vereint heute der herzlichste Wunsch, daß FRIEDRICH PASCHEN noch viele glückliche und fruchtbare Arbeitsjahre im Dienste unserer Wissenschaft erleben möge. NIELS BOHR.

TRANSLATION

Friedrich Paschen on his seventieth birthday.

FRIEDRICH PASCHEN's seventieth birthday on 22 January 1935 offers an opportunity for every physicist to look back with admiration and gratitude on his activity which was so important for our science and which evolved without interruption through half a century and still continues with undiminished strength.

In his contributions PASCHEN not only appears as the great master of the art of experimenting who always successfully strove to perfect the experimental methods and to enrich our experience in new fields, but his work distinguishes itself above all by the happy intuition whereby he always pursued experimental problems whose investigation would prove to be of decisive importance for the development of the general theoretical ideas.

This is equally true of PASCHEN's very first contribution in which, by investigating the electric discharge in gases at various pressures, he discovered the law which carries his name and which was so important for the development of the theory of discharges in gases, as well as for his studies in the field of spectra, so fundamental for the theory of atomic constitution, to which he devoted almost his entire scientific activity and which earned him such high reputation among his colleagues.

After his fruitful collaboration with RUNGE,[1] PASCHEN surely contributed more than any other scientist to the confirmation and extension of the empirical regularities of spectral lines, which stand out in the entire field of physics for their precision. In this connection, especially his discovery of new features in the structure of spectra has advanced the gradual development of a quantum theoretical classification of the possible states of atoms, which prepared the way for the creation of a rational quantum mechanics. Moreover, the discovery by PASCHEN, in collaboration with BACK,[2] of the modification of the Zeeman effects with increasing field strengths is of an importance for the clarification of the most fundamental problems of electron theory that can hardly be overestimated.

---

[1] Carl Runge (1856–1927), German physicist.
[2] Ernst E.A. Back (1881–1959), German physicist.

In line with his general, always progress-oriented, approach PASCHEN again and again chose new paths in his work, and in the last few years we find him engaged, with his usual mastery, in the study of the hyperfine structure of spectral lines, which afford such a wonderful source of information about the problem of the constitution of the atomic nucleus, which at present stands at the centre of interest for natural scientists.

The impact of PASCHEN's activity is by no means limited to his own scientific achievements, but in his field of work surely no one has to a greater extent been an inspiration to other physicists through personal instruction, and there are few institutes of physics which do not enjoy daily, in one way or another, the help of the tradition created by PASCHEN.

Colleagues in his field in all countries join today in the most cordial wish that FRIEDRICH PASCHEN may experience many more happy and fruitful years of work in the service of our science. NIELS BOHR.

Niels Bohr, Friedrich Paschen and Francis William Aston at the 1927 Como conference.

XXXIII. SOMMERFELD AND THE THEORY OF THE ATOM

SOMMERFELD UND DIE ATOMTHEORIE
Naturwiss. **16** (1928) 1036

TEXT AND TRANSLATION

See Introduction to Part II, p. [115].

Sonderdruck aus Die Naturwissenschaften. 16. Jahrg., Heft 49, S. 1036

(Verlag von Julius Springer, Berlin W 9)

Sommerfeld und die Atomtheorie.

Von NIELS BOHR, Kopenhagen.

Durch die bahnbrechende Arbeit von SOMMER-FELD über die Feinstruktur der Wasserstofflinien wurde die Theorie des Atombaues nicht nur um ein besonders schönes und fruchtbares Ergebnis bereichert, sondern die Beteiligung eines Forschers von seiner Eigenart mußte der ganzen Arbeit auf diesem Gebiete notwendigerweise einen großen Aufschwung bringen. Seine weitgehende Beherrschung der Methoden der theoretischen Physik, die er schon früher auf die angrenzenden Gebiete der Mechanik und Elektrodynamik so erfolgreich angewandt hatte, kam hierbei besonders zur Geltung; vor allem aber sollte seine Gabe, seine Begeisterung auf den ihn umgebenden Kreis zahlreicher Schüler zu übertragen, die reichsten Früchte hervorbringen. Unter der Fülle der Ergebnisse der Theorie des Atombaues in den nächstfolgenden Jahren, die wir dem SOMMERFELDschen Kreise verdanken, dürfte es schwierig sein, einige besonders hervorzuheben. Der glücklichen Intuition des Führers entsprechend kann man als ihr gemeinsames Merkmal die Bestrebung erkennen, eine sinngemäße Klassifikation des Erfahrungsmaterials mit Hilfe ganzer Zahlen zu erreichen, welche für dessen Einordnung unter die Gesichtspunkte der Quantentheorie vor allem in Betracht kommt.

Während des Ringens mit der Entwicklung eines Gebietes wie dem der Quantentheorie, wo sogar die grundsätzlichen Begriffe sich erst allmählich klären konnten, war es wohl unvermeidlich, daß scheinbare Gegensätze zu Tage traten, indem die einzelnen Forscher besonders Gewicht auf verschiedene Seiten der Sache legten, um darin den Antrieb zu Weiterarbeit zu finden. So war es bei der damaligen Stufe der Entwicklung der Quantentheorie mehr eine Frage des Gefühls, wie stark man in deren Darstellung die Abweichungen von den klassischen Vorstellungen betonte, oder wie weit man bestrebt war, sie als eine natürliche Verallgemeinerung jener Vorstellungen erscheinen zu lassen. Tatsächlich waren diese beiden Seiten des Forschens untrennbar miteinander verknüpft, und es war ja eben die ständige Erweiterung der Systematik der Quantenzahlen, mit deren Hilfe SOMMERFELD unsere Einsicht in den Ursprung der Serienstruktur der Spektren und der Multipletaufspaltung der Serienlinien so entscheidend gefördert hat, daß die Mittel darbot, die Korrespondenz mit der klassischen Theorie immer weiter zu verfolgen. Erst die nachfolgende Entwicklungsstufe der Quantentheorie, die uns konsequente quantitative Methoden gegeben hat, konnte hier eine endgültige Klärung herbeiführen und die verschiedenen Seiten des Problems in ihrer vollen Harmonie erscheinen lassen.

Mit seinen Leistungen auf dem engeren Gebiete des Atombaus ist SOMMERFELDS unermüdliche und erfolgreiche Wirksamkeit in der Aufhellung der reizvollen Fragen, welche die Entdeckung der Elementarteilchen und die Entwicklung der Quantentheorie uns gestellt haben, noch keineswegs erschöpft. Mit jugendlicher Begeisterung hat er sich in letzter Zeit, unter Heranziehung der neuen Quantenstatistik, dem Problem der elektrischen Leitfähigkeit der Metalle zugewendet, das trotz der großen Hoffnungen, die sich einst an die klassischen statistischen Methoden geknüpft haben, so lange allen weiteren Angriffen widerstanden hatte. Bekanntlich ist es ihm hier gelungen, die alten Schranken zu durchbrechen und ein fruchtbares Arbeitsgebiet zu eröffnen, auf dem schon eine Schar von Nachfolgern tätig ist. Zu seinem sechzigsten Geburtstag werden alle Physiker ihm von Herzen wünschen, daß er zum Besten unserer Wissenschaft seine volle Kraft noch viele Jahre behalten möge.

TRANSLATION

Sommerfeld and the Theory of the Atom.

By NIELS BOHR, Copenhagen.

Through SOMMERFELD's epoch-making papers on the fine structure of the hydrogen lines, the theory of atomic constitution was not only enriched with an especially beautiful and fruitful result, but the participation of a scientist with his special gifts would necessarily advance greatly the entire work in this field. His extensive command of the methods of theoretical physics, which he had already earlier applied so successfully to the bordering fields of mechanics and electrodynamics, came into its own especially in this case; above all, however, his gift to inspire the wide circle of pupils around him with his enthusiasm has borne rich fruit. Among the wealth of results within the theory of atomic constitution in the following years, which is due to the circle around SOMMER-FELD, it is difficult to emphasize any one result in particular. In accordance with the happy intuition of the leader, we can recognize as its prevalent trait the endeavour to achieve a meaningful classification of the observational material by means of integral numbers, which finds its application above all in the arrangement of this material from the viewpoint of quantum mechanics.

During the struggle with the development of a field such as quantum theory, where even the basic concepts could be clarified only gradually, it was surely unavoidable that seeming contradictions appeared, as individual researchers placed special emphasis on different aspects of the problems, in order to find there an incentive for further work. Thus, at that stage of the development of quantum theory, it was mostly a question of personal preference how strongly the deviations from classical ideas were emphasized in the description, or to what extent one sought to make them appear as a natural generalization of these ideas. In fact, these two approaches to the search were inseparably related, and it was precisely the sustained extension of the classification of quantum numbers, by means of which SOMMERFELD so decisively advanced our understanding of the origin of the series structure of spectra and the multiplet splitting of the line series, which offered the means to pursue the correspondence with classical theory still further. Only the next stage in the development of quantum theory, which has given us consistent quantitative methods, could lead to a final clarification and allow the various aspects of the problem to appear in their full harmony.

With his achievements within the narrower field of atomic constitution, SOMMERFELD's untiring and successful effort to elucidate the exciting questions posed by the discovery of elementary particles and the development of quantum theory is still by no means exhausted. With youthful enthusiasm he recently turned, on the basis of the new quantum statistics, to the problem of the electric conductivity of metals which, in spite of the great hopes that were once attached to the methods of classical statistics, had resisted all further attacks for so long. As is well known, he has in this case succeeded in breaking through old barriers and in opening up a fruitful field in which a great number of his followers are already active. On the occasion of his 60th birthday all physicists will bring him the most cordial wishes that he may keep his full strength for many years to come for the benefit of our science.

Arnold Sommerfeld and Niels Bohr in Lund, Sweden, 1919.

XXXIV. ON THE DEATH OF HENDRIK ANTHONY KRAMERS

VED HENDRIK ANTON KRAMERS' DØD
Politiken, 27 April 1952

TEXT AND TRANSLATION

See Introduction to Part II, p. [116].

[347]

Ved Hendrik Anton Kramers' død

Niels Bohr skriver om den store hollandske fysiker, der var saa nær knyttet til Danmark

EFTERRETNINGEN om, at professor *H. A. Kramers* ved Leidens Universitet for faa dage siden er død, bragte ikke alene bud om et stort tab for den fysiske videnskab, men fremkaldte dyb sorg i vide krese, ikke mindst her i landet, hvor han i den aarrække, han opholdt sig og virkede i København, vandt saa mange nære venner.

I efteraaret 1916, hvor jeg kort forinden var vendt tilbage fra England for at overtage det nyoprettede professorat i teoretisk fysik ved Københavns Universitet, kom Kramers hertil som en ganske ung mand, der lige havde fuldført sit universitetsstudium i Leiden. Han var saaledes den første af den række udenlandske fysikere, der kom til at arbejde i København, og faa aar efter blev han den første assistent ved det i 1918 oprettede institut for teoretisk fysik, der, efter i en tid at have fundet husly i universitetets fysiske laboratorium paa Den polytekniske Læreanstalt, i 1920 flyttede ind i en egen bygning paa Blegdamsvej.

LIGE fra begyndelsen lagde Kramers de særlige evner for dagen, der hurtigt skulle give ham en fremtrædende stilling blandt vor tids fysikere. Hans gennem aarene saa talrige og ofte afgørende bidrag til atomfysikens udvikling vidner saavel om hans mesterskab i beherskelsen

Hendrik Anton Kramers fotograferet i København.

af de matematiske metoder som om en sjælden evne til at trænge ind til problemernes kerne. Under opholdet i København, der kom til at strække sig over ti aar, og hvorunder han efterhaanden som lektor ved universitetet ogsaa udfoldede en højt paaskønnet virksomhed ved de fysikstuderendes undervisning, arbejdede han særligt paa udviklingen af korrespondens-principet, hvis maal det er, trods alle særprægede afvigelser, at lade kvanteteorien fremtræde som en almindeliggørelse af den saakaldte klassiske fysik.

Efter et antal frugtbare undersøgelser over principets anvendelse paa spektrallinjernes finstruktur og røntgenstraalernes absorption, lykkedes det Kramers i 1923 at udvikle en almindelig kvanteteoretisk dispersionsteori, der skulle faa afgørende følger for den senere udvikling. Paa denne tid var der efterhaanden ved institutet samlet en gruppe yngre teoretiske fysikere fra forskellige lande, blandt hvilke navnlig Pauli fra Østrig og Heisenberg fra Tyskland udmærkede sig ved deres særlige, i mangt og meget om Kramers' evner mindende begavelse og indstilling. Ikke mindst blev samarbejdet mellem Kramers og Heisenberg paa den videre udvikling af dispersionsteorien et væsentligt udgangspunkt for udviklingen af den rationelle kvantemekanik, der som bekendt var Heisenbergs store bedrift.

UNDER sin senere virksomhed som professor i Utrecht og derefter i Leiden udførte Kramers mange resultatrige undersøgelser, der indbefattede de mest forskellige omraader af atomfysiken, som optiske, termodynamiske og magnetiske fænomener. Alligevel vedblev dispersionsproblemerne at fastholde hans dybe interesse, og den videre udvikling af hans tanker om disse problemer indebar kim til væsentlig afklaring inden for den, med udgangspunkt i Heisenbergs og Paulis og især Diracs grundlæggende arbejder, i de senere aar udbyggede kvanteelektrodynamik.

Det er næppe muligt at give nogen, der ikke er fortrolig med de enkelte skridt i atomfysikens hastige og mangesidige udvikling i den sidste menneskealder, et virkeligt indtryk af, hvad Kramers har betydet for denne udvikling, ikke blot gennem sine egne arbejder, men ogsaa gennem den inspirerende paavirkning, han udøvede i vide krese. Paaskønnelsen heraf fandt mangfoldige udtryk i videnskabelige æresbevisninger og den række tillidshverv, som blev betroet ham baade i hans eget land og inden for det internationale videnskabelige samarbejde. Saa sent som sidste sommer ledede han som præsident for den internationale fysiske union dens møder her i København, og ved den i forbindelse med dette møde arrangerede atomfysiske konferens paa Institut for teoretisk Fysik var han blandt de mange gamle medarbejdere stadig en af de mest virksomme, til hvem alle særlig lyttede.

ENHVER, med hvem Kramers kom i berøring, modtog et uforglemmeligt indtryk af hans aands rigdom og den varme menneskelighed, der prægede hele hans færd. Trods hans evne til koncentration om vanskelige videnskabelige problemer kendte hans interesse for alt, der havde med menneskelivet at gøre, ingen begrænsning. Hans viden om og indlevelse i de mest forskellige humanistiske problemer var helt sjælden for en saa aktiv naturforsker og fandt udtryk i en forstaaelse og medfølelse, der sammen med en aldrig svigtende hjælpsomhed dannede baggrunden for hans evne til at bringe medmennesker opmuntring og styrke. For mig selv er det mangeaarige samarbejde, der udviklede sig til et nært og fortroligt venskab, forbundet med dyrebare minder, som jeg altid vil omfatte med dybeste taknemmelighed.

Efter Kramers' hele indstilling var det naturligt, at ingen kunstnerisk stræben stod ham fjern; især følte han en dyb kærlighed til musiken og spillede selv med stort mesterskab paa flere instrumenter. Det var ogsaa inden for tonekunstnernes kres, han — kort efter at han var kommet til København — fandt sin hustru, gennem hvem de baand, der knyttede ham til Danmark, yderligere fæstnedes. I hjemmet i Leiden, hvor børnene hver paa sin vis delte forældrenes mangesidige interesser, fandt familiens danske venner altid den hjerteligste modtagelse, og lige til det sidste var professor Kramers' jævnlige besøg i København en kilde til glæde og inspiration for os alle. Tanken om, at vi ikke skal se ham mere, føles som en dyb smerte.

NIELS BOHR

TRANSLATION

On the death of Hendrik Anthony Kramers

Niels Bohr writes about the great Dutch physicist
who was so closely connected to Denmark[1]

The news that Professor *H.A. Kramers* at Leiden University died only a few days ago was not only a message of a great loss for physical science, but also caused deep sorrow in wide circles, not least in our country where, during the years he lived and worked in Copenhagen, he won so many close friends.

In the autumn of 1916, shortly after I had returned from England in order to take over the newly established professorship in theoretical physics at the University of Copenhagen, Kramers came here as a very young man who had just completed his university studies in Leiden. He was thus the first in the series of foreign physicists who came to work in Copenhagen, and only a few years later he became the first assistant at the Institute for Theoretical Physics, founded in 1918. After having found shelter for a while in the University's physical laboratory at the Technical University, the institute moved into its own building at Blegdamsvej in 1920.

From the very beginning, Kramers showed the special abilities which soon were to give him a prominent position among the physicists of our time. His numerous and often decisive contributions to the development of atomic physics through the years bear witness both to his mastery in the command of mathematical methods and to a remarkable ability to penetrate to the heart of the problems. During his stay in Copenhagen, which was going to last ten years, and as part of which he in time also displayed, as lecturer at the university, a highly appreciated activity in the teaching of physics students, he worked especially on the development of the correspondence principle, whose

---

[1] Translation of the figure caption on p. [348]: "Hendrik Anthony Kramers photographed in Copenhagen".

aim is to allow quantum theory, despite all strange deviations, to appear as a generalization of the so-called classical physics.

After a number of fruitful investigations into the application of this principle to the fine structure of spectral lines and the absorption of X-rays, Kramers succeeded in 1923 in developing a general quantum-theoretical dispersion theory which should have decisive consequences for the subsequent development. At this time a group of younger theoretical physicists from various countries was gathered at the institute, among whom notably Pauli[2] from Austria and Heisenberg from Germany distinguished themselves by their special talent and approach, which in many ways resembled Kramers's abilities. Not least the cooperation between Kramers and Heisenberg on the further development of dispersion theory became an important point of departure for the development of rational quantum mechanics, which, as is well known, was Heisenberg's great achievement.

During his later activity as professor in Utrecht and thereafter in Leiden, Kramers carried out many investigations which gave rich results and comprised the most diverse branches of atomic physics, such as optical, thermodynamic and magnetic phenomena. Nevertheless, the dispersion problems continued to hold his deep interest, and the further development of his ideas about these problems carried the germ of an important clarification within quantum electrodynamics, as this theory expanded in recent years after having originated from Heisenberg's and Pauli's, and particularly Dirac's, fundamental contributions.

It is hardly possible to give anybody not familiar with every step in the rapid and many-sided development of atomic physics during the last generation a true impression of what Kramers has meant for this development, not only through his own contributions, but also through the inspiring influence he exercised in wide circles. The appreciation of this found many expressions in scientific honours and the succession of honorary offices entrusted to him both in his own country and within international scientific cooperation. As late as last summer he conducted, as President of the International Union of Physics, its meeting here in Copenhagen, and at the conference on atomic physics arranged at the Institute for Theoretical Physics in connection with this meeting, he was among the many old collaborators still one of the most active, to whom all listened with special attention.

---

[2] Bohr's tribute to Pauli is reproduced on pp. [361] ff.

[351]

Anybody with whom Kramers came into contact received an unforgettable impression of the richness of his spirit and the warm human attitude which marked all his conduct. In spite of his ability to concentrate on difficult scientific problems, his interest in everything having to do with human life knew no limit. His knowledge of and empathy with the most diverse problems in the humanities were quite rare for such an active scientist and found expression in an understanding and sympathy which, together with a never-failing readiness to help, formed the basis of his ability to give fellow beings encouragement and strength. For me personally, the collaboration over many years that developed into a close and intimate friendship is associated with precious memories which I shall always regard with the deepest gratitude.

In accordance with Kramers's general attitude it was natural that no artistic endeavour was foreign to him; he felt especially a deep love for music and played several instruments with great mastery. It was also in musical circles that he – shortly after he had come to Copenhagen – met his wife, through whom the bonds which tied him to Denmark were strengthened further. In the home in Leiden, where the children each in their own way shared their parents' many-sided interests, the family's Danish friends were always received most cordially, and until the very end, Kramers's regular visits to Copenhagen were a source of joy and inspiration to us all. The thought that we shall not see him again is felt as a deep sorrow.

NIELS BOHR

XXXV. HENDRIK ANTHONY KRAMERS †

Ned. T. Natuurk. **18** (1952) 161–166

See Introduction to Part II, p. [116].

HENDRIK ANTHONY KRAMERS † (1952)

Versions published in English and Danish

English
A   Ned. T. Natuurk. **18** (1952) 161–166

Danish
B   Fys. Tidsskr. **52** (1954) 1–8

The translation *B* contains some minor improvements of language in relation to *A*.

Hendrik Anthony Kramers †

by N. Bohr *)

Hendrik Anthony Kramers's death is a great loss to physical science and a grief to the whole community of contemporary physicists. Not only did Kramers as few others contribute to the development of atomic physics in our generation, but, by his keen interest and deep insight in all aspects of this field of science, he exerted an inspiring influence in wide circles, and his noble and harmonious personality won the respect and friendship of colleagues all over the world. As one who, through a long period, has collaborated initimately with Kramers and, until his last days, stood in close personal relation to him, I am grateful, at this meeting of commemoration arranged by the Dutch Physical Society, to be given the opportunity of relating some of the recollections which are so vivid in my mind and which I treasure so highly.

Kramers came to Denmark in 1916, in the middle of the first world war, shortly after I had returned from England to take over a newly established chair in theoretical physics at the University of Copenhagen. He was then only 21 years of age, and I still remember his radiant youth so singularly combined with a maturity of mind, which by no means excluded eager expectations as to what life might hold out for him in the way of experience and achievements. In itself Kramers's journey to Copenhagen was a youthful adventure; indeed, he came without any notice to a place where nobody had heard of him before. Of course, the young man did not come from nowhere, but with the background of the great traditions of this University of Leiden, where Lorentz had worked and taught to the admiration of the world and where, after he had resigned his chair, Ehrenfest had succeeded him and in that capacity had guided the studies of Kramers. With all differences in human and scientific approach, Ehrenfest's uncompromising demand of rigour and honesty in scientific research and representation of arguments has, no doubt, exerted a deep influence on Kramers's development.

In Copenhagen, Kramers proved a most valuable asset to our small circle in which his rare gifts, combining a great power of penetrating to the root of problems with the mastery of mathematical methods, were from the first recognized. At the beginning, he assisted me in a most valuable way in completing work I had

*) This lecture was presented at the memorial meeting, organized by the Netherlands Physical Society, May 1952 in the Kamerlingh Onnes Laboratory, Leiden, Netherlands.

started in England, but, with remarkable independence, he soon performed, in rapid succession, a number of important investigations, one of which was to be accepted by the Leiden University as his doctor thesis in 1918. In the meantime the circle in Copenhagen was gradually enlarged by visits of other young physicists from abroad, and when the Institute for Theoretical Physics was completed in 1920, Kramers became the first assistant. Soon after, in 1924, he was appointed lecturer at the Institute, where his enthusiastic and inspired teaching was highly appreciated by the students and is still by many remembered with gratitude. The rapidly growing esteem which Kramers won in scientific circles in Denmark found also expression in his election, at an uncommonly young age, as member of the Danish Academy of Sciences and Letters, a year before he, in 1926, left the country to become professor of theoretical physics in Utrecht.

To give a right impression of Kramers's scientific achievements during his ten years' stay in Copenhagen, it may be proper to recall in a few words the situation of atomic physics in those days. The fundamental progress attained about the beginning of our century in this field exhibited two main aspects. On the one hand, the discovery of the electron, which revealed the atomistic nature of electricity, was proved to a wide extent capable of a harmonious incorporation into the classical scheme of electrodynamics, and the impressive power of Lorentz's electron theory had been emphasized by his successful explanation of Zeeman's discovery of the effect of magnetic fields on spectral lines. In fact, in revealing the close connection between the origin of atomic spectra and the electron binding in atoms, this explanation may be regarded as a landmark in the development of atomic theory. On the other hand, Planck's discovery of the universal quantum of action had revealed an atomistic feature of elementary physical processes which was quite foreign to the classical ideas of mechanics and electromagnetism. In particular, Einstein's ingenious application of quantum theory to explain the peculiar features of the photo-effect disclosed an unexpected dilemma regarding the propagation of radiation. Indeed, however useful the idea of the photon proved to be, it was at the same time clear that some recourse to the classical picture of electromagnetic waves was indispensable for the account of characteristic radiation phenomena.

The peculiar situation in atomic physics was further stressed by Rutherford's discovery, in 1911, of the atomic nucleus which was to be fundamental for the subsequent development in opening a whole new field of research. Notwithstanding the inherent simplicity of Rutherford's model of the atom, it was soon realized that the ideas of classical physics, when applied to the binding of electrons by the nucleus, were quite insufficient to account for the peculiar stability of atomic structures. A clue to the interpretation of the properties of atoms was offered, however, by the very existence of the quantum of action, which suggested that any change in the state of an atom demands a complete transition from one of its stationary quantum states to another and that in radiative reactions such a transition process is just accompanied by the emission or absorption of a photon.

These quantum postulates, which proved capable of accounting for the empirical laws of line spectra, were clearly irreconcilable with classical electron theory. Still, any attempt at a more comprehensive description of experience had evidently to make an essential use of classical concepts. In fact, the very definition of the mass and charge of the electron rests upon measurements interpreted in terms of ordinary electrodynamics. Progress, therefore, depended on the establishment of the closest possible correspondence between elements of the classical theories and a description based on the quantum postulates. An essential step in this direction was made by associating the general periodicity properties of electron motion in atoms with an ensemble of virtual oscillators corresponding to the radiative transition processes. In this way, it was found possible not only to ascertain quantitative connections between the typical constants appearing in the empirical spectral formulae and those characterizing the atomic model, but also to derive qualitative conclusions as to the intensity relations exhibited by the spectra. Especially after the successful application by Sommerfeld of the quantum theory to multiple periodic systems the correspondence principle proved helpful for the interpretation of the spectral selection rules.

It was about that time Kramers started to work in Copenhagen, and his first paper, which appeared in the Transactions of the Danish Academy and served as his Leiden thesis, deals just with the theoretical examination of the intensity relations of the components into which the hydrogen lines are split up under the influence of electric fields. Through an admirably elegant treatment of the complicated mechanical problem involved he was actually able to account with remarkable success for a number of the striking features of the intricate component pattern. With equal mastery he subsequently attacked the mechanical problem of the electron motion in the helium atom, although in this case his calculations rather indicated that a basis for a proper understanding of the duplexity of the helium spectrum was still wanting. Decisive progress, however, was achieved by Kramers in the application of the correspondence principle to the problem of the composition of the high frequency radiation emitted by the deflection of swift electrons. This well-known work proved essential for the interpretation of various experiments and has provided an important guidance in discussions of astrophysical problems.

Still, the greatest achievement of Kramers in the Copenhagen

[357]

years was his general dispersion theory. From the beginning, the dispersion problem had been in the centre of interest; thus I remember that the first time I had the great experience of meeting Lorentz personally our talks turned especially on these problems, the treatment of which had taken such an important position in classical electron theory. Up to his latest years Lorentz followed with deep interest and open mind the development on the new lines, but while he appreciated the general progress, he felt deep concern as to how it should be possible to combine in a consistent manner the dispersion phenomena with the quantum postulates. Surely, as regards emission and absorption processes Einstein had, in a famous paper from 1918, shown how a lucid derivation of Planck's radiation formula could be obtained on correspondence lines by formulating assumptions about the probabilities of spontaneous radiative transitions from higher to lower energy states and of transition processes induced by radiation between such states with absorption as well as emission of photons. It was first Kramers, however, who, in 1924, succeeded in clarifying the general connection between dispersion phenomena and the quantum conception of the origin of spectra.

Leaning on the empirical relations between spectral absorption and anomalous dispersion in the neighbourhood of spectral lines, which have found so suggestive an interpretation in Lorentz' electron theory, Kramers was able, with sharp intuition, to develop a general dispersion formula in which the classical concept of oscillator strength was replaced by the Einstein probabilities of radiative quantum processes. The progress consisted not only in the harmonious completion of the quantal treatment of radiation problems, but implied at the same time the prediction of new dispersion phenomena connected with the possibilities of induced transitions from excited atomic states in the direction of higher as well as lower energies corresponding to positive and negative oscillator strengths, respectively. On such lines the theory was still further developed by Kramers in collaboration with Heisenberg, who at that time had joined the Copenhagen group. This work actually proved a stepping stone for Heisenberg, who shortly afterwards in a most ingenious way accomplished a rational formulation of quantum mechanics in which all direct reference to ordinary pictures of mechanical motion was finally abandoned.

In the following period Kramers took an active part in the development and applications of quantum mechanics, to which another most fruitful approach had been opened independently by de Broglie's concept of matter waves and by Schrödinger's establishment of his famous wave-equation. I shall not dwell, however, on these important and many-sided investigations of Kramers, which mainly fall in his Utrecht and Leiden years, and will presently be described by Casimir, who as Kramers's assistant and colleague followed them so closely. I only wish to stress how, in spite of the energy and success with which Kramers in later years attacked and advanced problems in most different fields of atomic physics, the dispersion phenomena always remained an object of his deepest interest. Following up fundamental problems, with which Lorentz had already been occupied, regarding the proper separation between the reaction of an electron to an external electromagnetic field and the effect of the field surrounding it due to its own electric charge, Kramers not only elucidated essential points in the recent development of quantum electrodynamics, but was also able to introduce various refinements in the treatment of actual dispersion problems.

The universal recognition of Kramers's scientific work and authority naturally brought with it that in the course of time many responsible and exacting tasks were entrusted to him by the Dutch community. After the difficult years of the last world war when he proved himself so worthy of such trust, he went to U.S.A. with the Dutch delegation to the United Nations Atomic Energy Commission and was by general consent elected chairman of the Scientific Sub-Committee of this Commission. I remember most vividly meeting him in New York in the autumn of 1946 and learning about his untiring endeavours for promoting understanding about the necessity of common control of the formidable powers which through the development of atomic physics had come into the hand of man. In spite of disappointments we must all keep up the hope that such understanding in due time will be reached in the spirit in which Kramers to the admiration from all sides strove in those days. In the succeeding years Kramers exerted a great influence as President of the International Union of Pure and Applied Physics, and he also took a leading part in the Dutch Norwegian collaboration on the erection of an experimental uranium reactor. It was at the inauguration of this reactor in Norway last autumn that I saw Kramers for the last time and discussed with him the plans, in which we shared a warm interest, for the further extension of European co-operation in atomic physics.

Kramers always kept up a close relation with Denmark, where during his long stay he won so many friends, and where his frequent visits were most welcomed as a great pleasure and inspiration to us all. Not only did he on each occasion bring new knowledge and ideas relating to the development of physics, but his mind was constantly occupied by problems connected with every aspect of life. From his early youth Kramers had made profound studies in the most varied fields of knowledge and he was deeply susceptible to the richness art can offer us. Especially, he was entranced by music, and he played several instruments in a masterly way. I remember how, shortly after he came to Copenhagen, when we were working together and problems were hard, he sat down at a piano and played to me Beethoven's Eroica with its enchanting

[359]

and exalting tunes. It was also in Danish artistic circles that Kramers soon after his arrival met his wife, who was to bring him so much happiness and become so great a support to him. It is to her and to their children that our deepest sympathy goes in these days. To them as to every one of us the memories Hendrik Anthony Kramers has left us will, in spite of all pain which the feeling of what we have lost contains, remain a constant source of encouragement and strength.

XXXVI. FOREWORD

"Theoretical Physics in the Twentieth Century:
A Memorial Volume to Wolfgang Pauli"
(eds. M. Fierz and V. Weisskopf),
Interscience, New York 1960, pp. 1–4

See Introduction to Part II, p. [117].

[361]

Wolfgang Pauli, Copenhagen, 1923.

FOREWORD

Progress of physical science in this century is marked not only by the exploration of vast fields of experience, but equally by the development of new frameworks for the analysis and synthesis of experimental evidence. In this development, Wolfgang Pauli, to whose memory this volume is dedicated, played a prominent part by his own outstanding achievements as well as by the inspiration and stimulation which we all received from him.

Pauli's penetrating insight and critical judgment came early to the fore in his famous encyclopaedia article on relativity theory, which he published when he was only 20 years old and which is still one of the most valuable expositions of the basis and scope of Einstein's original conceptions. His early familiarity with this radical revision of fundamental physical concepts, and his mastery of the mathematical tools indispensable for the proper formulation of the new ideas, became the background for Pauli's important contributions to quantum physics.

While the theory of relativity, both as regards the principles and their application, had already in Einstein's hands reached a high degree of completion, the situation in quantum theory was very different indeed. Far from offering a general account of phenomena on the atomic scale, Planck's epoch-making discovery of the quantum of action rather represented a challenge to incorporate an entirely new elementary feature into a consistent description of physical phenomena. As is well known, the way to this goal was beset with many obstacles and was only gradually paved by the collaboration of a whole generation of physicists.

After his school days in Vienna, Pauli pursued his studies in Munich under the inspiration of Sommerfeld who with his unique mastery of the methods of mathematical physics exerted a deep influence on all his pupils. In later years, Pauli kept close contact with his old teacher of whom he often spoke with admiration and affection. When Pauli, after working with Born in Göttingen, came to Copenhagen in 1922, he at once became, with his acutely critical and untiringly searching mind, a great source of stimulation to our group. Especially he endeared himself to us all by his

1

intellectual honesty, expressed with candour and humour in scientific discussions as well as in all other human relations.

In those years, no comprehensive methods of quantum physics were yet developed and the interpretation of experimental evidence was guided mainly by the so-called correspondence argument which expressed the endeavour to uphold a description in classical terms to the utmost extent compatible with the individuality of atomic processes. By such a tentative procedure it proved possible in a more or less consistent manner to utilize spectral data to obtain a survey of the bindings of electrons in atoms, and in particular a first approach to an interpretation of the relationships between the physical and chemical properties of the elements.

I vividly remember discussions with Pauli, in which he expressed his dissatisfaction with the weak argumentation underlying the attempt to explain the peculiar stability of closed electron shells, so decisive for the periodicities exhibited by the properties of the elements when arranged according to their nuclear charge. The pertinence of his remarks would indeed be most strikingly vindicated by Pauli's continued work in the following years, resulting in the enunciation of the exclusion principle which expresses a fundamental property of systems of identical particles for which, as for the quantum of action itself, classical physics presents no analogy.

The ingenuity with which Pauli in those years mastered the application of correspondence arguments within their proper scope is illustrated by his beautiful analysis of the Compton scattering of radiation by free electrons. Inspired by Einstein's general statistical considerations of energy and momentum exchange in radiative processes, he showed that the scattering probability depends on the intensity of both radiation components involved in the process. The line followed in this work is indeed very closely related to the general dispersion theory, formulated by Kramers, which was to prove so important for the subsequent great developments.

To Pauli, with his abhorrence for any kind of ambiguity in physical theories, the advent of a rational quantum mechanics, excluding all irrelevant use of classical pictures, was a tremendous relief. In particular, it need hardly be recalled how this development permitted an harmonious incorporation of Pauli's exclusion principle in proper quantum statistics. The vigour with which Pauli threw himself into the exploration of the new methods, and the mastery of these which he soon acquired, are evidenced by his article on the foundations of quantum mechanics in *Handbuch der Physik* (1932), which retains a similar position in scientific literature as his old exposition of relativity theory.

Pauli's whole background made it inevitable that he would be deeply pre-occupied with the problem of adapting the fundaments of quantum physics to relativistic requirements. Not only did he from the beginning take a prominent part in the formulation of the quantum theory of electromagnetic fields, but also his contribution to the relativistic theory of the electron was most helpful in promoting the full understanding of its implications. In the elucidation of the apparent paradoxes to which the subsequent discussion of the measurability of field components and electric charges gave rise, Pauli's active interest was also a great stimulation.

In subsequent years, Pauli became ever more deeply occupied with the problems of the elementary particles and their associated quantized fields. At an early stage, he made a fundamental contribution to this development by the introduction of the concept of the neutrino, which ensured the upholding of the conservation laws in the β-decay of atomic nuclei. In this connection, it is also interesting to recall that Pauli, in 1926, was the first to draw attention to the source of information about nuclear spins and electromagnetic moments offered by the hyperfine structure of spectral lines.

In this memorial volume, experts in the various fields have surveyed Pauli's many-sided pioneering work and its background at the successive stages of development of physical science. When contemplating Pauli's great life-work, it is important to remember the inspiration he gave not least to the many pupils whom he gathered around him, first in Hamburg and later in Zürich where, apart from an interruption in the war years which he spent in Princeton, he worked for the last 30 years of his life. Still, through his participation in scientific symposia and his extensive correspondence with colleagues and friends, Pauli's influence reached much wider circles.

Indeed, everyone was eager to learn about Pauli's always forcefully and humorously expressed reactions to new discoveries and ideas, and his likes and dislikes of the prospects opened. We always benefited by Pauli's comments even if disagreement could temporarily prevail; if he felt he had to change his views, he admitted it most gracefully, and accordingly it was a great comfort when new developments met with his approval. At the same time as the anecdotes around his personality grew into a veritable legend, he more and more became the very conscience of the community of theoretical physicists.

Pauli's searching mind embraced all aspects of human endeavours. In Zürich he found colleagues sharing his many-sided interests, and his studies of historical, epistemological and psychological questions found expression in a number of highly suggestive essays. Also there he had the good fortune of meeting a life-companion who by her fine understanding of the power of his intellect and the integrity of his character gave him that peace of mind which he needed so much in his great research and teaching activity. With Wolfgang Pauli's death we have lost not only a brilliant and inspiring fellow worker, but a true friend who to many of us appeared as a solid rock in a turbulent sea.

NIELS BOHR

XXXVII. RECOLLECTIONS OF PROFESSOR TAKAMINE

"Toshio Takamine and Spectroscopy",
Research Institute for Applied Optics, Tokyo 1964, pp. 384–386

See Introduction to Part II, p. [118].

RECOLLECTIONS OF PROFESSOR TAKAMINE

It has been a great pleasure to learn that colleagues and friends of the late Professor Toshio Takamine have planned to edit a memorial pamphlet with recollections of the eminent physicist, and I welcome this opportunity to express my admiration for Toshio Takamine's scientific work and recount a few reminiscences about our personal contact.

After many fruitful spectroscopic researches in Japan and various laboratories abroad, Toshio Takamine came to work with us in Copenhagen in 1920, shortly after the creation of our institute designed for the promotion of intimate cooperation between experimentally and theoretically working physicists. Our group was at that time still small, and with his wide experience in spectroscopy, Takamine's participation in the work proved most valuable.

The study of the effect of electric and magnetic fields on spectral lines had in those years required great interest as a consequence of the development of the quantum theory of atomic constitution. In this connection, the investigation which Takamine, together with the Danish physicists, H. M. Hansen and W. S. Werner, performed on the Zeeman and Stark effect in the mercury spectrum brought much new information for the testing and elucidation of the theoretical view points.

Not only to his collaborators in this work, but to us all at the institute, Takamine endeared himself by a kindness and helpfulness from which we received a deep impression of the refinement of Japanese culture. Indeed, already in those years, the basis was laid for a close friendship which was strengthened through Takamine's most welcome visit to Copenhagen a few years later when he was accompanied by his charming young wife.

On our unforgettable visit to Japan in 1937 it was a special pleasure to my wife and me to meet again our Japanese friends who had in the course of the years come to work in Copenhagen and who were actively partaking in the great scientific developments in Japan. One of our treasured recollections is the beautiful day we spent in Takamine's hospitable home together with many old friends who all had kept such faithful relations with us.

In later years we heard regularly from Takamine and his wife, and the thought of how, in spite of gradually failing heatlh, he kept up his alert scientific interest and broad human attitude, will remain an inspiring and encouraging example to all.

Copenhagen, March 23, 1961

Niels Bohr
UNIVERSITETETS INSTITUT FOR TEORETISK FYSIK

[368]

XXXVIII. THE INTERNATIONALIST

UNESCO Courier **2** (No. 2, March 1949) 1, 7

For A. Einstein

See Introduction to Part II, p. [118].

THE INTERNATIONALIST (1949)

Versions published in English and Danish

English: The Internationalist
  A   UNESCO Courier **2** (No. 2, March 1949) 1, 7

Danish: Niels Bohr hylder Albert Einstein
  B   Politiken, 15 March 1949

  B is more extensive than *A*.

THE INTERNATIONALIST

By
Professor Niels BOHR

IT is most natural and appropriate that the United Nations Educational, Scientific and Cultural Organization should pay a tribute to Albert Einstein on the occasion of his seventieth birthday. Indeed, for the whole of humanity Einstein's name stands pre-eminently for that search to extend our knowledge and deepen our understanding which is not only the spirit and object of science, but which forms the very foundation of all human civilization.

Through Albert Einstein's work the horizon of mankind has been immeasurably widened, at the same time as our world picture has attained a unity and harmony never dreamed of before. The background for such achievement is created by preceding generations of the world-wide community of scientists and its full consequences will only be revealed to coming generations.

Man's endeavours to orient himself in his existence beyond the immediate necessities of life may be traced back to the widely spread birth-places of our civilization like Mesopotamia, Egypt, India and China and, above all, to the small free communities in Greece, where arts and science rose to a height unsurpassed for long ages. During the Renaissance, when all aspects of human culture again flourished, most intense and fruitful contacts between scientific endeavours all over Europe took place, as we are reminded by the names of Copernicus, Tycho Brahe, Kepler, Galileo, Descartes, Pascal and Huygens, men of many countries whose achievements created the basis of the edifice of which Newton's genius is the pinnacle.

New Insight

THE great advance of natural philosophy at that time, which came to exert a deep influence on all human thinking, consisted above all in the attainment of a rational description of mechanical phenomena based on well-defined principles. It must, however, not be forgotten that the idea of absolute space and time formed an inherent part of the basis of Newton's work and that also his well-known concept of universal gravitation constitutes an element so far not further explainable.

It was just at these points that Einstein initiated quite a new development which, in an unforeseen manner, has deepened and rounded our views and given us new insight and power of comprehension.

The way leading to this turning point was paved by the development during the nineteenth century, of our knowledge of the electromagnetic phenomena which has brought such a great increase in human facilities and created the modern means of world-wide communication.

This development was furthered by an ever more active international co-operation, the extent of which is recalled by such names of many nations as Volta, Cersted, Faraday, Maxwell, Hertz, Lorentz and Michelson. Gradually, however, the growth of knowledge in this new field disclosed more and more clearly the difficulties and paradoxes inherent in absolute space-time description.

A quite new outlook was here opened by the genius of Einstein, who changed the whole approach to the problems by exploring the very foundation for the description of our experience. Thus, Einstein taught us that the concept of simultaneity of events, occurring at different places was inherently relative, in the sense that two such events which to one observer appear simultaneous, may seem to follow each other in time from the standpoint of another observer.

This recognition of the extent to which the account of phenomena depends essentially on the motion of the observer proved, in the hands of Einstein, a most powerful means of tracing general physical laws valid for all observers.

In the following years, Einstein even succeeded in attaining a viewpoint wide enough to embrace the gravitational phenomena, by extending his considerations to the comparison of the effects experienced by observers with accelerated movement relatively to each other. Out of Einstein's novel approach to the use of space and time concepts grew gradually a wholly new attitude towards cosmological problems, which has given most fertile inspiration for the exploration of the structure of the universe.

Although simplicity and beauty are the principal marks of Einstein's fundamental ideas, the detailed treatment of complex problems often demands the use of abstract mathematical methods like non-Euclidean geometry. As often before, it has here been most fortunate that such tools were ready as the fruit of the work of older mathematicians.

The names of Gauss, Lobachevsky, Bolyai, Riemann, Ricci and Minkowski here again remind us of the fertility of international co-operation in all fields of science. And the same may be said of Einstein's other outstanding work.

For example, his explanation of the irregular motion of small bodies in liquids, based on the ideas of Maxwell, Boltzmann, Smoluchowski and Gibbs, made it possible for Jean Perrin accurately to count the atoms of which substances are built.

Dr. Bohr

We find ourselves today in a new epoch in physical science, in which experimental discoveries and theorical methods have led to a rapidity and fecundity of progress m a d e possible only by international co-operation of an unprecedented activity and extent.

It is not possible in this occasion to disentangle the contributions of individual workers, but mention must be made of the guidance, at almost every step, which Einstein has given us by his Relativity theory and by his analysis of elementary quantum phenomena.

Altogether, this short exposition of Einstein's scientific achievements aims at giving an impression of the extent to which his originality of outlook has made him an innovator in science. At the same time, I have attempted to remind you that all scientific endeavours are parts of a great common human enterprise.

The gifts of Einstein to humanity are in no way confined to the sphere of science. Indeed, his recognition of hitherto unheeded assumptions in even our most elementary and accustomed concepts means to all people a new encouragement in tracing and combating the deeprooted prejudices and complacencies inherent in every national culture.

With his human and noble personality, characterized equally by wisdom and humour, Einstein himself has through all his life, and not least in these latter years, worked for the promotion of international understanding. On his seventieth birthday evidence of the veneration and gratitude our whole generation owes to him will reach Einstein from many sides, and we all want to express the wish that the hopes for which he has lived and worked may be fulfilled to the benefit of all mankind.

XXXIX. OBITUARY

MINDEORD
Børsen, 19 April 1955

For A. Einstein

TEXT AND TRANSLATION

See Introduction to Part II, p. [119].

Mindeord

Ved Meddelelsen om Professor Albert Einsteins Død har Professor Niels Bohr fremsat følgende Mindeord:

Ved Albert Einsteins Død afsluttes et Liv i Videnskabens og Menneskehedens Tjeneste, saa rigt og frugtbart som faa i hele Kulturens Historie. Gennem Erkendelsen af hidtil upaaagtede Forudsætninger for Beskrivelsen af fysiske Fænomener i Rum og Tid skulde Relativitetsteorien udvide vor Horisont og skabe en Enhed i vort Verdensbillede langt ud over Fortidens dristigste Drømme.

For Einsteins Geni, kendetegnet ved et harmonisk Sammenspil mellem skabende Fantasi og logisk Klarhed, lykkedes det at afrunde den klassiske Fysiks Begrebsbygning og fremdrage almindelige Naturlove, der — som Sammenhængen mellem Masse og Energi — skulde faa afgørende Betydning for Belysningen af de nye Erfaringsomraader, som Udforskningen af Atomernes Verden i vor Tid har givet os Indblik i.

Paa denne Baggrund maa man med dybeste Beundring tænke paa det aabne Blik og den frie Indstilling, hvormed Einstein umiddelbart efter at det elementære Virkningskvantum var opdaget, bidrog til at klarlægge Følgerne af den overraskende Begrænsning af de tilvante fysiske Begrebers Gyldighedsomraade, som man her var kommet paa Spor efter, og som skulde danne Indledningen til en yderligere Omskabelse og Udvidelse af Naturbeskrivelsen.

Den Aand, der præger Einsteins videnskabelige Storværk, genspejlede sig paa smukkeste Maade i hans Stillingtagen til alle menneskelige Problemer. Ikke alene lagde han ved enhver Lejlighed — uanset den Ærbødighed, hvormed hans Person omfattedes fra alle Sider — en naturlig Beskedenhed for Dagen, men søgte, saavidt hans Indflydelse strakte, altid at komme vanskeligt stillede Medmennesker til Hjælp. Trods Smerten over, hvad vi har mistet, vil, for dem af os, der har haft den Lykke at lære Einstein nærmere at kende, Mindet om hans ædle Personlighed altid være en dyb og frisk Kilde til Styrkelse og Inspiration.

Niels Bohr.

[374]

TRANSLATION

Obituary

On the announcement of the death of Professor Albert Einstein, Professor Niels Bohr has presented the following obituary:

With the death of Albert Einstein, a life in the service of science and humanity which was as rich and fruitful as any in the whole history of our culture has come to an end. Through the recognition of hitherto unnoticed conditions for the description of physical phenomena in space and time, the theory of relativity should prove to extend our horizon and give us a world picture with a unity surpassing the boldest dreams of the past.

Einstein's genius, characterized by a harmonious interplay between creative imagination and logical clarity, succeeded in rounding off the conceptual edifice of classical physics and bringing forth general laws of nature which – like the relation between mass and energy – should prove to have decisive importance for the illumination of the new areas of experience into which the exploration of the world of atoms in our time has given insight.

On this background one must think with the deepest admiration of the open view and the free approach with which Einstein, immediately after the discovery of the elementary quantum of action, contributed to clarifying the consequences of the surprising limitation of the scope of the customary physical concepts, which one had got on the track of here and which should be the beginning of a further transformation and extension of the description of nature.

The spirit that marks Einstein's monumental scientific achievements was mirrored in the most beautiful way in his attitude in all human relations. Not only did he on every occasion – regardless of the reverence with which his person was held everywhere – behave with a natural modesty but sought always, as far as his influence reached, to help people in difficulties. Despite the grief for what we have lost, to those of us who had the good fortune to know Einstein well, the memory of his noble personality will always be a deep and fresh source of strength and inspiration.

Niels Bohr.

[375]

N. Bohr, A. Einstein, T. de Donder, O.W. Richardson, P. Langevin, P. Debye, A.F. Ioffe and B. Cabrera gathered in Brussels in 1932 to make plans for the 1933 Solvay Conference. Photograph by Queen Elisabeth of Belgium.

XL. ALBERT EINSTEIN 1879–1955

Sci. Am. **192** (1955) 31

See Introduction to Part II, p. [119].

SCIENTIFIC AMERICAN

JUNE, 1955

VOL. 192, NO. 6

ALBERT EINSTEIN: 1879-1955

Tributes by Niels Bohr and I. I. Rabi

With the death of Albert Einstein, a life in the service of science and humanity which was as rich and fruitful as any in the whole history of our culture has come to an end. Mankind will always be indebted to Einstein for the removal of the obstacles to our outlook which were involved in the primitive notions of absolute space and time. He gave us a world picture with a unity and harmony surpassing the boldest dreams of the past.

Einstein's genius, characterized equally by logical clarity and creative imagination, succeeded in remolding and widening the imposing edifice whose foundations had been laid by Newton's great work. Within the frame of the relativity theory, demanding a formulation of the laws of nature independent of the observer and emphasizing the singular role of the speed of light, gravitational effects lost their isolated position and appeared as an integral part of a general kinematic description, capable of verification by refined astronomical observations. Moreover, Einstein's recognition of the equivalence of mass and energy should prove an invaluable guide in the exploration of atomic phenomena.

Indeed, the breadth of Einstein's views and the openness of his mind found most remarkable expression in the fact that, in the very same years when he gave a widened outlook to classical physics, he thoroughly grasped the fact that Planck's discovery of the universal quantum of action revealed an inherent limitation in such an approach. With unfailing intuition Einstein was led to the introduction of the idea of the photon as the carrier of momentum and energy in individual radiative processes. He thereby provided the starting point for the establishment of consistent quantum theoretical methods which have made it possible to account for an immense amount of experimental evidence concerning the properties of matter and even demanded reconsideration of our most elementary concepts.

The same spirit that characterized Einstein's unique scientific achievements also marked his attitude in all human relations. Notwithstanding the increasing reverence which people everywhere felt for his attainments and character, he behaved with unchanging natural modesty and expressed himself with a subtle and charming humor. He was always prepared to help people in difficulties of any kind, and to him, who himself had experienced the evils of racial prejudice, the promotion of understanding among nations was a foremost endeavor. His earnest admonitions on the responsibility involved in our rapidly growing mastery of the forces of nature will surely help to meet the challenge to civilization in the proper spirit.

To the whole of mankind Albert Einstein's death is a great loss, and to those of us who had the good fortune to enjoy his warm friendship it is a grief that we shall never more be able to see his gentle smile and listen to him. But the memories he has left behind will remain an ever-living source of fortitude and encouragement.   —NIELS BOHR

XLI. EBBE KJELD RASMUSSEN: 12 APRIL 1901 – 9 OCTOBER 1959

EBBE KJELD RASMUSSEN:
12. APRIL 1901 – 9. OKTOBER 1959
Fys. Tidsskr. **58** (1960) 1–2

TEXT AND TRANSLATION

See Introduction to Part II, p. [119].

Ebbe Rasmussen.

Ebbe Kjeld Rasmussen.

12. april 1901 — 9. oktober 1959.

Ved professor Ebbe Rasmussens død har den fysiske videnskab mistet en fremragende forsker og det danske samfund en
personlighed, der ved sin saglige og retsindige indstilling og
hjælpsomme færd vandt venner i vide kredse, i hvilke han vil
blive dybt savnet.

Ebbe Rasmussens utrættelige og målbevidste forskningsarbejde, der påbegyndtes allerede i studietiden og fortsattes til
hans sidste leveår, omfattede især undersøgelser over grundstoffernes liniespektre og spektralliniernes finstruktur. Med me-
sterlig beherskelse af de spektrografiske metoder og en stedse
dybere indlevelse i de lovmæssigheder, som forsøgsresultaterne
bragte for dagen, lykkedes det ham ved disse undersøgelser at
give mange betydningsfulde bidrag til overblikket over elek-
tronbindingen i atomerne og kendskabet til de mekaniske og
magnetiske egenskaber hos atomkernerne, der har været så
væsentlig en ledetråd ved studiet af kernernes opbygning.

Den samme indstilling der kendetegnede Ebbe Rasmussens
videnskabelige forskning prægede, sammen med hans varme
interesse for sine medmennesker, hans så højt påskønnede
pædagogiske virksomhed og den aktive deltagelse i de mange
organisatoriske og administrative bestræbelser, hvortil han i
stadig stigende grad kaldtes. Fra den tid, Ebbe Rasmussen
først som assistent og dernæst som inspektør var knyttet til
Universitetets institut for teoretisk fysik, bevarer jeg i dyb taknemmelighed en levende erindring om hans råd og hjælp ved
udviklingen af den hastigt opvoksende institution. Allerede i de
år virkede han også som sekretær i Selskabet for naturlærens
udbredelse, hvor han trofast stræbte at videreføre de traditioner,
som Selskabets stifter, H. C. Ørsted, havde skabt.

[381]

Efter i en længere årrække at have virket som professor ved Landbohøjskolen, hvor han udførte et stort arbejde med tilrettelæggelsen af fysikundervisningen og tillige som efor på Højskolens kollegium med omsorg fulgte mange af de studerende, efterkom han for nogle år siden til sine gamle kollegers glæde indbydelsen til at vende tilbage til Universitetet. Med sine særlige forudsætninger måtte han her straks påtage sig et omfattende og ansvarsfuldt arbejde med planlæggelsen af Det fysiske laboratorium på det nye H. C. Ørsted institut, der oprettes såvel med henblik på udviklingen af de matematiske, fysiske og kemiske videnskaber som på den store tilgang af universitetsstuderende. Samtidig udnævntes han på grund af sit indgående kendskab til atomfysikken og dens anvendelser til medlem af Atomenergikommissionens forretningsudvalg, for hvilket hans menneskelige klogskab og praktiske indsigt blev en stor støtte.

Som årene gik, søgtes Ebbe Rasmussens bistand og råd af stedse større kredse, og trods de mange krævende hverv der efterhånden var lagt på hans skuldre, gik ingen fra ham uden at være styrket af hans sinds åbenhed for alt menneskeligt og hans trang til at hjælpe i enhver sag, der efter hans skøn indeholdt kim til fremgang. Den anseelse og tillid, der stod om hans personlighed som videnskabsmand og menneske, fandt et sidste udtryk ved hans valg til sekretær i Videnskabernes Selskab, men de forhåbninger, som Selskabets medlemmer stillede til hans virke i denne højt betroede stilling, skulle blive uopfyldt som følge af den snigende sygdom, der få måneder efter rev ham bort.

For hans nærmeste og de mange, for hvem han var så trofast en ven, betyder Ebbe Rasmussens død et smerteligt og uerstatteligt tab; men for os alle vil de rige minder han gav os forblive en levende kilde til mod og styrke.

NIELS BOHR

TRANSLATION

Ebbe Kjeld Rasmussen.
12 April 1901 – 9 October 1959.

On the death of Professor Ebbe Rasmussen, physical science has lost an eminent researcher and Danish society a personality who by his impartial and upright attitude and helpful conduct won friends in wide circles, in which he will be deeply missed.

Ebbe Rasmussen's untiring and dedicated research work, which began already in his student days and continued until his last years, comprised in particular investigations of the line spectra of the elements and the fine structure of spectral lines. With masterly command of spectrographic methods and an ever deeper understanding of the regularities brought to light by the experimental results, he succeeded with these investigations in making many important contributions to the overall picture of electron binding in atoms and to the knowledge of the mechanical and magnetic properties of atomic nuclei, which have been so essential a guiding line in the study of the constitution of nuclei.

The same attitude that characterized Ebbe Rasmussen's scientific research marked, together with his warm interest in his fellow beings, his so greatly appreciated educational efforts and the active participation in the many organizational and administrative endeavours to which he was called to an ever increasing degree. From the time when Ebbe Rasmussen, first as assistant and than as inspector,[1] was associated with the University Institute for Theoretical Physics, I hold in deep gratitude a vivid memory of his advice and help in the development of the rapidly growing institution. Already in those years he also acted as Secretary for the Society for the Dissemination of Natural Science where he faithfully endeavoured to continue the traditions that H.C. Ørsted, founder of the Society, had created.

After having worked for a long stretch of years as professor at the Royal Veterinary and Agricultural College, where he carried out a great effort in preparing the physics curriculum and, additionally, as warden of the College's hall of residence conscientiously looked after many of the students, he accepted some years ago, to the delight of his old colleagues, the invitation to return to

---

[1] Position with overall responsibility for the institute's buildings and equipment.

the University. With his special qualifications he should here immediately take on a comprehensive and responsible task in planning the Physical Laboratory at the new H.C. Ørsted Institute, which is being established with a view to the development of the mathematical, physical and chemical sciences as well as to the great increase in numbers of university students. At the same time, due to his thoroughgoing knowledge of atomic physics and its applications, he was elected member of the executive committee of the [Danish] Atomic Energy Commission, for which his human wisdom and practical insight were a great support.[2]

As the years passed, Ebbe Rasmussen's assistance and advice were sought by ever larger circles, and despite the many demanding obligations which, as time went on, were laid on his shoulders, no one left him without being strengthened by the openness of his mind for everything human and his urge to help in any matter which he deemed to contain a germ of progress. The respect and trust that surrounded his personality as scientist and human being found a last expression in his election as Secretary in the Academy of Sciences and Letters, but the hopes that the Academy's members laid in his activity in this position of great trust should remain unfulfilled as a result of the insidious disease which took him away a few months later.

For his closest family and the many for whom he was such a faithful friend Ebbe Rasmussen's death means a painful and irreplaceable loss; but for all of us the rich memories he gave us will remain a vital source of courage and strength.

NIELS BOHR

---

[2] On Bohr's involvement in the Danish Atomic Energy Commission, see Vol. 11, pp. [363] ff.

XLII. MAGISTER FRITZ KALCKAR

Politiken, 7 January 1938

TEXT AND TRANSLATION

See Introduction to Part II, p. [119].

Magister Fritz Kalckar.

Den unge begavede Fysiker Magister Fritz Kalckar, Professor Niels Bohrs nære Medarbejder, er Natten til i Gaar pludselig død ramt af en Hjerneblødning, kun 27 Aar gammel.

Professor Niels Bohr skriver her om den unge Videnskabsmand.

Magister Fritz Kalckars pludselige Død betyder et stort og smerteligt Tab for alle, der kendte ham, og ikke mindst for enhver af os, der daglig arbejdede sammen med ham paa Universitetets Institut for Teoretisk Fysik.

Straks i sin Studietid, hvor han hurtigt aabenbarede sine rige Evner, tiltrak han os alle ved sin friske og oprindelige Menneskelighed. Umiddelbart efter at han paa udmærket Maade havde bestaaet Magisterkonferensen, begyndte han sin Deltagelse i det videnskabelige Arbejde paa Institutet og fuldendte allerede i Løbet af det første Aar et teoretisk Arbejde om Spredning af Elektroner ved Sammenstød med Atomer, der paa smukkeste Maade forklarede nogle Forsøgsresultater, som Prof. Werner, der den Gang arbejdede paa Institutet, netop havde fundet.

I det følgende Aar udførte han sammen med en af de udenlandske Gæster paa Institutet Dr. Teller en overordentlig interessant Undersøgelse over paramagnetiske Luftarters katalytiske Virkning ved Omdannelsen mellem Brintmolekylernes saakaldte Orto- og Paramodifikation, en Undersøgelse hvis Resultater især tillod en nøjagtig Beregning af Forholdet mellem den tunge og den lette Brintkernes magnetiske Momenter. Efter dette Arbejdes Fuldendelse kastede Kalckar sig med stor Energi over de aktuelle Problemer vedrørende Atomkernernes Reaktioner, som imidlertid var blevet et Hovedemne for Arbejdet paa Institutet, og hans værdifulde Medvirken herved førte hurtigt til et intimt personligt Samarbejde med mig, hvoraf de første Resultater offentliggjordes for kun faa Maaneder siden.

At dette Samarbejde, til hvis Fortsættelse vi begge havde glædet os meget, saa brat skulde afbrydes, er en Tanke, som det er svært for mig at forsone mig med. Ligesom alle paa Institutet vil jeg daglig savne hans friske Arbejdsglæde og den fine Forstaaelse af alle Menneskelivets Sider, som vandt ham Venner overalt. Jeg vil ogsaa ofte tænke paa den store Hjælp og Støtte, han var for mig paa den Rejse til Amerika, som vi foretog sammen sidste Foraar, og hvor han ved sit samtidigt aabne og beskedne Væsen forstod straks at komme paa den fortroligste Fod med de unge Fysikere, med hvem vi diskuterede de aktuelle Problemer, der optog os begge. For Kalckars Vedkommende afsluttedes denne Rejse med et frugtbart Studieophold i Berkeley og Pasadena, hvorunder han fik Lejlighed til at fuldføre to smukke Afhandlinger om Atomkerneproblemer i Samarbejde med den fremragende amerikanske Teoretiker Professor Oppenheimer og en af dennes Elever Dr. Serber.

Trods sine knap 28 Aar havde Magister Kalckar saaledes allerede formaaet at skaffe sig et anset Navn i vide videnskabelige Krese, og han vil længe mindes og savnes som en af de mest sympatiske og lovende yngre danske Videnskabsmænd.

NIELS BOHR

TRANSLATION

Magister Fritz Kalckar.

The young talented physicist, Magister[1] Fritz Kalckar, Professor Niels Bohr's close collaborator, died suddenly the night before yesterday from a cerebral haemorrhage, only 27 years of age.

Professor Niels Bohr writes here about the young scientist.

The sudden death of Magister Fritz Kalckar is a great and painful loss for all who knew him and not least to each of us who worked with him daily at the University Institute for Theoretical Physics.

Already in his student days, when he rapidly showed his rich talents, he attracted all of us by his fresh and unspoiled human qualities. Immediately upon passing his master's examination with honours, he began his participation in the scientific work at the Institute, and already in the course of the first year he completed a theoretical paper on the scattering of electrons on collision with atoms,[2] which explained in the most beautiful manner some experimental results just obtained by Professor Werner,[3] who worked at the Institute at that time.

In the following year, together with Dr Teller,[4] one of the foreign guests at the Institute, he carried out an exceedingly interesting investigation of the catalytic effect of paramagnetic gases on the conversion between the so-called orthomodification and paramodification of hydrogen molecules, an investigation whose results allowed, in particular, an accurate calculation of the ratio between the magnetic moments of the heavy and the light hydrogen nucleus.[5] After the completion of this work, Kalckar devoted himself with great energy to the problems of current interest concerning the reactions of atomic nuclei, which in the meantime had become a main topic for the work at the Institute, and his

---

[1] Equivalent to "Master of Science".

[2] F. Kalckar, *Über die elastische Streuung von Elektronen in Argon und Neon*, Mat.–Fys. Medd. Dan. Vidensk. Selsk., **12**, No. 12 (1933).

[3] Sven Werner (1898–1984).

[4] Edward Teller (1908–2003), Hungarian-born American physicist.

[5] F. Kalckar and E. Teller, *On the Theory of the Catalysis of the Ortho–para Transformation by Paramagnetic Gases*, Proc. Roy. Soc. **A150** (1935) 520–533.

valuable participation therein led rapidly to close personal collaboration with me, the first results of which were published only a few months ago.[6]

That this collaboration, the continuation of which we had both looked forward to very much, should be interrupted so abruptly is a thought that it is difficult for me to reconcile myself with. Just as all others at the Institute, I shall miss daily his fresh joy in work and the fine understanding of all aspects of human life, which won him friends everywhere. I will also often think of the great help and support he was for me during the trip to America we made together last spring,[7] where, due to his at the same time open and modest personality, he was immediately able to be on closest terms with the young physicists with whom we discussed the problems of current interest that occupied us both. For Kalckar, this journey ended with a fruitful period of study in Berkeley and Pasadena, during which he had the opportunity to complete two beautiful papers on problems concerning atomic nuclei[8] in collaboration with Professor Oppenheimer,[9] an outstanding American theoretician, and Dr Serber,[10] one of his pupils.

In spite of his not quite 28 years, Magister Kalckar had thus already made a name for himself that is respected in wide scientific circles, and he will long be remembered and missed as one of the most sympathetic and promising young Danish scientists.

NIELS BOHR

---

[6] N. Bohr and F. Kalckar, *On the Transmutation of Atomic Nuclei by Impact of Material Particles*, Part I: *General Theoretical Remarks*, Mat.–Fys. Medd. Dan. Vidensk. Selsk. **14**, No. 10 (1937). Reproduced in Vol. 9, pp. [223]–[264].

[7] In 1937 Bohr visited the United States as part of a journey around the world together with his wife Margrethe and son Hans.

[8] F. Kalckar, J.R. Oppenheimer and R. Serber, *Note on Nuclear Photoeffect at High Energies*, Phys. Rev. **52** (1937) 273–278; *idem.*, *Note on Resonances in Transmutations of Light Nuclei*, Phys. Rev. **52** (1937) 279–282.

[9] J. Robert Oppenheimer (1904–1967).

[10] Robert Serber (1909–1979).

XLIII. [TRIBUTE TO RUSSELL]

"Into the 10th Decade: Tribute to Bertrand Russell",
Malvern Press, London [1962]

See Introduction to Part II, p. [120].

[389]

On the occasion of Lord Russell's ninetieth birthday I take pleasure in expressing my warm admiration for his great life-work, which has so deeply influenced intellectual culture in our age, and for the undaunted spirit with which he strives to promote human ideals.

NIELS BOHR

Niels and Margrethe Bohr's Golden Wedding, 1962.

4. FAMILY AND THE BROADER DANISH MILIEU

XLIV. [FOREWORD]

J. Lehmann, "Da Nærumgaard blev børnehjem i 1908",
Det Berlingske Bogtrykkeri, Copenhagen 1958, p. i

TEXT AND TRANSLATION

See Introduction to Part II, p. [120].

For D. B. Adler og hans Hustrus Slægt er det
en stor Glæde, at Johannes Lehmann's smukke
Skrift kan gives til alle, der med Interesse følger
det Børnehjem, der for 50 Aar siden indrettedes
paa Nærumgaard efter mine Bedsteforældres
Ønske og i Haabet om, at mange dér vilde kunne
vokse op under lykkelige Forhold svarende til de
lyse Minder, som deres egne Børn og Børnebørn bevarede om Hjemmet derude i de skønne Omgivelser.

TRANSLATION

For the descendants of D.B. Adler and his wife, it is a great pleasure that Johannes Lehmann's beautiful publication can be given to all who follow with interest the childrens' home which was established 50 years ago at Nærum-gaard, at the wish of my grandparents and in the hope that many there would be able to grow up under happy circumstances, corresponding to the bright memories that their own children and grandchildren kept of the home out there in the beautiful surroundings.

Niels Bohr

Niels Bohr and Hanna Adler, Tisvilde, 1946.

XLV. FOREWORD

FORORD
"Hanna Adler og hendes skole",
Gad, Copenhagen 1959, pp. 7–10

TEXT AND TRANSLATION

See Introduction to Part II, p. [120].

[397]

INDHOLD

FORORD

Denne bog, der på 100-års dagen for rektor Hanna Adlers fødsel udgives af en kreds af hendes elever, beretter om en rig livsgerning og vil for kommende slægter bevare mindet om en sjældent rank og ædel personlighed.

I bogens indledende, indholdsrige kapitel fremhæver Jens Vibæk så rigtigt, at en dybere forståelse af et liv og en gerning som Hanna Adlers kræver kendskab til de forhold, hvorunder hun voksede op og modtog sine første indtryk. Mange værdifulde oplysninger i denne henseende rummes i forfatterens kyndige redegørelse for D. B. Adlers virke i det danske samfund og fortællingen om det hjem, der skabtes af ham og hans engelskfødte hustru, der så inderligt delte hans mangesidige menneskelige bestræbelser og efter hans tidlige død levede for virkeliggørelsen af hans forhåbninger. Også Hanna Adlers egen livsindstilling fremtræder i klart lys gennem de af Jens Vibæk fremdragne breve til hendes moder fra den oplevelsesrige amerikarejse i 1892-93.

Med Hanna Adlers vågne og varme interesse for sine medmennesker og stræben efter at fremme en lykkelig udvikling af deres evner til et virksomt liv i samfundet måtte hun fra ganske ung øve stor påvirkning på alle, med hvem hun kom i berøring. Fra min tidligste barndom har jeg levende erindringer om hendes aktive og kærlige deltagelse i alt, hvad der angik hendes søskende og deres børn. Selvom min broder Harald og jeg ikke direkte hørte til eleverne på skolen, delte vi med dem „Moster Hanne"s opdragende indflydelse. Når hun kunne frigøre sig fra skolearbejdet, førte hun os om søndagene omkring på Københavns naturhistoriske og etno-

7

grafiske samlinger og kunstmuseer, og i sommerferierne på Næ-rumgård, hvor hun færdedes med os til fods eller på cykel i omegnens skove og enge, lærte vi både om naturen og om menneskelivet, medens hun i spøg og alvor talte med os om alt, der kunne fange vor fantasi.

I dette lille forord til mindebogen skal jeg dog ikke yderligere fordybe mig i mine egne erindringer. Den beundring og taknemmelighed, hvormed Hanna Adlers minde omfattes i familiens kreds, hvor hun stedse mere blev et samlende midtpunkt, har jo også fundet så smukt udtryk i den beskrivelse af hendes liv uden for skolen, som hendes nære slægtning, Paula Strelitz, har givet i et af bogens sidste afsnit. Netop som det deri træffende er fremhævet, gjaldt det om Hanna Adler, at hun på ganske enestående måde var den samme i alle livets forhold. Ja i den grad, at man tør sige, at enhver af eleverne på hendes skole, fra de yngste til de ældste, umiddelbart forstod hende og følte sig trygge og ansporedes, hvad enten hun på sin ligefremme og venlige måde fandt anledning til at rose eller dadle. Hvor frisk og betagende er ikke Ingrid Saxtorphs og Poul Billgrens skildringer af livet på skolen, hvor de tanker, som Hanna Adler enkelt og nøgternt fremsatte i sit foredrag i Pædagogisk Selskab i 1893, udfoldede sig så rigt; og med hvor sand humor har ikke også hendes mangeårige trofaste medarbejder J. Ostenfeld beskrevet det tillidsforhold, der bestod mellem Hanna Adler og skolens lærere.

Trods al medgang og forståelse for hendes reformatoriske bestræbelser måtte Hanna Adler dog også, som det med så dyb sympati er skildret af Mikael Sachs, gennem årene kæmpe for bevarelsen af de rammer for undervisningen, som hun fra første færd havde indført på skolen, og i denne forbindelse undgik hun heller ikke vanskeligheder og bekymringer, efter at hun i 70 års alderen i 1929 havde nedlagt sin rektorgerning. En stor glæde for hende var det imidlertid, at ånden på skolen bevaredes gennem de mange fremragende, hver på sin vis så personligt prægede lærere, der vedblev at være knyttet til denne, og blandt hvilke Thyra Eibe i det første års tid naturligt kom til at virke som rektor. Ligeledes følte

8

Rektor, mag. scient. frk. Hanna Adler, 1912. Foto: Julie Laurberg & Gad, København.

9

Hanna Adler i disse år stor taknemmelighed for den forståelse for skolens egenart, som den efterfølgende rektor Johannes Braae lagde for dagen.

De skygger, der i 30-erne formørkede livet i Europa, måtte gøre et dybt indtryk på Hanna Adler, hvis hele tænkemåde var så uforenelig med den mentalitet og de fordomme, der gav sig udtryk i de umenneskelige forfølgelser. Hun, der fra den første begyndelse i stilhed virksomt støttede bestræbelserne for at hjælpe dem, der søgte tilflugt her i landet, skulle jo selv, under tyskernes besættelse af Danmark, blive udsat for forfølgelse på grund af sin jødiske herkomst. På gribende måde fortæller Merete Bodelsen i bogens sidste afsnit om den helt eventyrlige aktion, der med så beslutsomt sammenhold iværksattes af skolens gamle elever, og hvorved det med støtte af vide kredse lykkedes at udvirke Hanna Adlers løsladelse fra den tyske arrestation. Dette vidnesbyrd om påskønnelse af hendes livsgerning og især om elevernes trofaste hengivenhed blev hende til største opmuntring i de mørke år, hvor hun med sædvanlig ro og værdighed bar alle tildragelser, uden tanke for sin egen sikkerhed og blot med ængstelse for den usikre skæbne som mange af hendes venner og slægtninge måtte friste.

I sine sidste år, hvor Hanna Adler til sin store glæde atter kunne samles med hele familien, levede hun med usvækket åndskraft og uformindsket hjælpsom deltagelse for sine venner og nærmeste og omfattede ikke mindst de yngste med en forståelse og kærlighed, der har givet dem minder, hvorpå de længe vil bygge.

NIELS BOHR

TRANSLATION

FOREWORD

This book, published by a circle of her pupils on the centenary of headmistress Hanna Adler's birth, tells of a rich lifework and will preserve the memory of an unusually upright and noble personality for coming generations.

In the comprehensive introductory chapter of the book, Jens Vibæk quite rightly emphasizes that a deeper understanding of a life and a work such as Hanna Adler's requires knowledge of the conditions under which she grew up and received her first impressions. Much valuable information in this regard is contained in the author's knowledgeable account of D.B. Adler's[1] activity in Danish society and the story of the home created by him and his English-born wife,[2] who so fully shared his many-faceted human endeavours and after his premature death lived for the realization of his hopes. Hanna Adler's own outlook on life also appears clearly through the letters to her mother from the eventful journey to America in 1892–93,[3] brought to light by Jens Vibæk.

With Hanna Adler's lively and warm interest in her fellow beings and the striving to encourage a successful development of their talents towards an active life in society, she was necessarily from an early age bound to exercise great influence on all with whom she came into contact. From my earliest childhood I have vivid recollections of her active and loving participation in everything to do with her brothers and sisters and their children. Although my brother Harald and I were not among the pupils in her school, we shared with them "Aunt Hanne"'s educational influence. When she was able to free herself from school work, she guided us on Sundays around Copenhagen's natural history and ethnographical collections and art museums, and on summer holidays at Nærumgård, where she went with us on foot or on bicycle in the neighbouring woods and meadows, we learnt both about nature and about human life, while she in jest and earnest talked with us about everything that might catch our imagination.

However, in this short preface to the memorial volume I shall not dwell further on my own memories. The admiration and gratitude with which the

---

[1] David Baruch Adler (1826–1878), Niels Bohr's maternal grandfather.
[2] Jenny, née Raphael (1830–1902).
[3] The letters are deposited in the NBA.

memory of Hanna Adler is held in the family circle, where more and more she became a uniting central figure, have of course also found such beautiful expression in the description of her life outside the school, which Paula Strelitz,[4] a close relative, has given in one of the last sections of the book. Just as expressed there so aptly, it was true for Hanna Adler that she in a quite unique fashion was the same in every aspect of life. Indeed, to such a degree that one dare say that every one of the pupils at her school, from the youngest to the oldest, understood her straight away and felt secure and was encouraged, regardless of whether she in her direct and friendly fashion found occasion to praise or to scold. Truly fresh and charming are the descriptions by Ingrid Saxtorph and Poul Billgren of life at the school, where the ideas that Hanna Adler stated simply and clearly in her lecture at the Pedagogical Society in 1893[5] were so richly developed; and J. Ostenfeld,[6] her faithful collaborator for many years, has described the relationship of trust existing between Hanna Adler and the teachers of the school with genuine humour.

Despite all the success and the appreciation of her endeavours for reform, Hanna Adler also had to fight throughout the years, as related with such deep sympathy by Mikael Sachs, to retain the framework for the teaching that she had introduced in the school from the very first, and in this connection nor was she free of difficulties and worries after she had retired from her position as headmistress in 1929 at the age of 70. It was, however, a great joy to her that the spirit at the school was kept through the many excellent teachers, all with their own personality, who remained associated with it, and among whom Thyra Eibe[7] in the first year as a matter of course would act as headmistress. Similarly, in these years Hanna Adler felt great gratitude for the appreciation of the school's special character, which Johannes Braae,[8] the subsequent headmaster, demonstrated.

The shadows that darkened life in Europe in the 1930s were necessarily bound to make a profound impression on Hanna Adler, whose entire way of thinking was so irreconcilable with the mentality and the prejudices expressed in the inhuman persecutions. She, who from the very beginning inconspicuously gave active support to the endeavours to help those who sought refuge in this country, should herself be subjected to persecution during the German

---

[4] (1892–1963).

[5] The speech, which deals with the situation in the United States with regard to co-education, is reproduced on pp. 82–93 in the book for which Bohr wrote this Foreword.

[6] Jens Ostenfeld.

[7] (1866–1955), the first woman in Denmark to obtain a university degree in mathematics (1895).

[8] (1887–1940).

occupation of Denmark because of her Jewish ancestry.[9] In a gripping fashion Merete Bodelsen[10] tells in the last chapter of the book about the quite incredible operation implemented with such resolute concord by the former pupils of the school, and whereby it was possible, with support from wide circles, to bring about Hanna Adler's release from German detention. This testimony of the appreciation for her lifework, and especially of the pupils' faithful devotion, was her greatest encouragement in the dark years when, with her usual calm and dignity, she bore all that happened without a thought for her own safety and only with anxiety about the uncertain fate which many of her friends and relations must suffer.

In her later years, when Hanna Adler to her great joy could again be together with all the family, she lived with unabated strength of mind and undiminished helpful sympathy for her friends and family and embraced not least the youngest of them with an understanding and love that have given them memories on which they will build for a long time.

NIELS BOHR

---

[9] Hanna Adler was interned by the Germans in October 1943, but was released one month later after a campaign by four hundred of her former pupils.
[10] (1907–1986), Danish art historian.

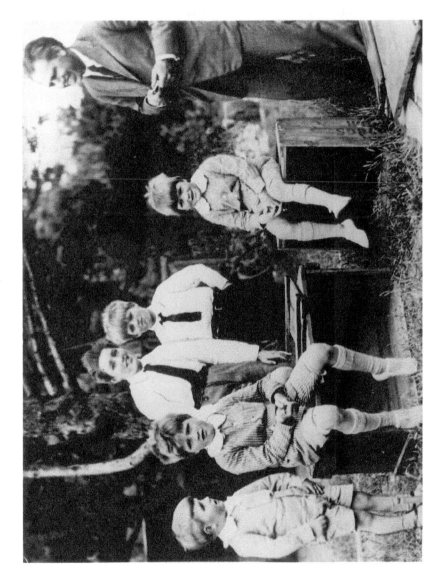

Niels Bohr at his summer house in Tisvilde with five sons, 1927. Left to right: Ernest, Erik, Christian, Hans and Aage.

XLVI. OBITUARY FOR CHRISTIAN ALFRED BOHR: BORN 25 NOVEMBER 1916 – DIED 2 JULY 1934

MINDEORD OVER CHRISTIAN ALFRED BOHR:
FØDT 25. NOVEMBER 1916 – DØD 2. JULI 1934
Private print, 1934

Includes Bohr's talk at the inauguration
of the Christian Alfred Bohr Memorial Fund, 25 November 1934

TEXT AND TRANSLATION

See Introduction to Part II, p. [122].

MINDEORD

OVER

CHRISTIAN ALFRED BOHR

FØDT 25. NOVEMBER 1916 — DØD 2. JULI 1934

Julius Paulsen pinx.   Sept. 1934

I.

Ved Afskeden fra Hjemmet

26. August 1934

Christians Mor og jeg er inderlig taknemmelige for, at vi, efter den lange tunge Ventetid, kan samles med hans Venner her, hvor han levede sine sidste saa lykkelige Aar, for at sige vor elskede Dreng et Farvel og en Tak fra det Hjem, han bragte saa megen Glæde, og i hvilket han efterlader saa stort et Savn.

Hvad Fryd og Haab det første Barn, der kommer til et ungt Hjem, bringer, behøver jeg ikke at sige; om det samles jo Forældrenes Drømme om en rigere Blomstring af alt, hvad de selv fra deres Barndomshjem og gennem deres eget Liv har lært at elske af smukt og godt. Dog først naar de af egen Erfaring har indset, hvor lidet de evner at fjerne hver Sten paa Barnets Vej, formaar Forældrene rigtigt at skatte, at en Dreng, med et kærligt og ridderligt Sindelag som Christians, allerede fra ganske lille forstod at være sin Mor og Far en trofast Kammerat i Medgang og Modgang og i fuldt Maal at blive sine yngre Brødre den Støtte, som kun en ældre Broder kan være.

Ikke alene for sine Forældre, men for hele Familien

3

skulde Christian fra første Færd blive en Kilde til den
største Lykke. Hvad Solskin har han ikke været for
sine saa højt elskede Bedsteforældre, som han skulde
miste saa tidligt, og for Moster Hanne, der gennem
sin trofaste, gengældte Kærlighed fra hans første Dag
til hans sidste har været en tredie Bedstemor for ham. —
Det er ogsaa med dyb Taknemmelighed, at Margrethe
og jeg tænker paa al den Glæde og Støtte, Christian har
haft af det varme Venskab, der vistes ham af saa mange
af vore gode Venner, og ikke mindst af den Kreds af
nære Venner, blandt hvem han befandt sig, da han
saa pludselig skulde blive revet bort fra os, og hvor han
lige til det allersidste Øjeblik følte sig saa lykkelig og
tryg.

At sige, at Christian var sine Forældre en kærlig
Søn, og at han var en ridderlig Fører for sine Brødres
Flok, vilde dog kun give et fattigt Begreb om, hvad
han betød for sit Hjem. Med sit Lune og sin Fantasi
og sin medfølende Forstaaelse var han ikke alene frem
for nogen den Sol, der lyste paa sine Brødres Vej, men
alt som Aarene gik, var han i stadig højere Grad Kil-
den til Fornyelse for os alle. Uden at hans Samfølelse
med Familien svækkedes, bragte Christian stedse stærkere og friskere Pust til vort Hjem fra den Ungdom,
der vokser op til eget Liv omkring os, og han forstod
at hente Impulser saavel fra Samværet med Kamme-
raterne fra Skolen, der ligesom hans Lærere har vist
ham saa megen Venlighed og Forstaaelse, som fra Om-
gangen med de unge Venner, især fra kunstnerisk ind-

4

stillede Kredse, som han vandt sig i sine sidste Aar, og som bragte ham i Forbindelse med nye Sider af Livet.

I disse Aar udvikledes han fra Barn til Yngling paa skønneste Maade, og mange Strenge, der kunde stemmes baade til den dybeste Alvor og til den mest smittende Munterhed, tonede i hans Sjæls Harmoni. Evner, hvortil han tidligt havde vist Spirer, modnedes, og for hans Forældre var det en dyb Glæde at se den maalbevidste Energi, som Christian, trods sine vidtspændende Interesser, lagde ind i Skolearbejdet og især i Forberedelsen til den Afslutningsprøve, han netop naaede at bestaa saa smukt, og som skulde have aabnet ham et nyt Livsafsnit, til hvilket han saa frem med det frejdigste Haab. Baade vi og hans Venner knyttede da ogsaa store Forventninger til hans Fremtid, selv om ingen endnu kunde vide, hvorledes hans Bane vilde have formet sig. Med den vaagne Sans for almene naturvidenskabelige Problemer, som Christian tog i Arv fra Slægten, forbandt han jo med sit stemningsbevægede Sind og sin sikre Smag Muligheder, der maaske efterhaanden fra de Studier, som han først havde tænkt at hellige sit Liv, vilde have draget ham ind paa Kunstens Vej.

Sværere for os end vi kan sige, er det at skulle forsone sig med Tanken om, at vi aldrig mere skal se Christians varme Smil og høre hans muntre Stemme, og at de rige Løfter om en frugtbar Manddomsgerning, til Glæde for ham selv og os andre, ikke skal blive

5

opfyldt. Men omend Christian aldrig skulde naa ud over sin fejre Ungdom, har hans Liv og Stræben ikke været forgæves. Enhver af os indenfor Hjemmet og dem, der er nært knyttet dertil, har han jo skænket dyrebare, af Smerten lutrede Minder, der vil binde os endnu fastere til hverandre, og i hvilke han uden at ældes altid vil leve med os. — De mangfoldige Vidnesbyrd om Trofasthed imod hans Minde, for hvilke vi baade fra Christian og os selv siger hans unge og ældre Venner vor inderligste Tak, har tillige givet os en sikker Fortrøstning om, at han ogsaa udenfor vor lille Kreds vil vedblive at virke i sin egen Aand, saa sandt som enhver af os vel lever sit stærkeste Liv i sine Medmenneskers Tanker.

II.

Ved Mindelegatets første Uddeling

25. November 1934

Da Christian for faa Maaneder siden saa fuld af Mod og Haab blev Student, drømte hverken hans Mor og jeg eller nogen af de mange Venner, der er samlet med os paa denne Dag, hvor han vilde være fyldt 18 Aar, om, at vi her skulde staa tilbage alene med Erindringerne om ham. Men netop fordi han nu kun lever og virker i vore Tanker, har den Trofasthed, som fra saa mange Sider er vist hans Minde, været den største Trøst i vor Lod at maatte forsone os med det bitre Savn.

6

Hvor meget har ikke den Forstaaelse af Christians alvorlige Stræben, som hans Lærere paa Skolen har givet Udtryk for, betydet for os, ligesom de mange Vidnesbyrd om hvor trofast hans Skolekammerater har bevaret Erindringen netop om de Træk hos ham, der bundede dybest i hans Sind, hans Trang til at forstaa og støtte andre og hans Iver for og Glæde ved at lære alt nyt og smukt at kende. Ikke mindst tænker vi paa, hvor meget det vilde have glædet Christian, om han kunde vide, at hans gamle Kammerater har besluttet at føre den lille Foredragsforening, ved hvis Stiftelse han var saa virksom, og i hvis Trivsel han tog saa stor Andel, videre og bevare dens Karakter ved Optagelse af yngre Elever fra Skolen.

Det er ogsaa med inderlig Taknemmelighed, at vi har erfaret, at alle, der er knyttet til det daglige Arbejde paa den videnskabelige Virksomhed, midt i hvilken Christian færdedes saa stor en Del af sit korte Liv, har forstaaet, hvor meget det laa ham paa Sinde ikke i nogen Henseende at gælde for mere, end hans egne Fortjenester maatte berettige ham til, og hvor beredt han var til at benytte enhver Mulighed til at dygtiggøre sig paa de forskelligartede, for en saadan Virksomhed lige nødvendige Arbejdsomraader. Hvilken Glæde var det ikke for Christian, efterhaanden som han voksede til, i stedse højere Grad at blive optaget som Kammerat blandt de unge Videnskabsmænd fra Ind- og Udland, som besøgte Instituttet, og som ikke alene har været en stadig Kilde til ny Kraft i

7

Arbejdet, men som tillige bragte saa mangt et Bud med sig om, hvad der i vor Tid bevæger Ungdommens Sind Verden over. Om den venlige Forstaaelse, Christian i sin unge idealistiske Stræben fandt hos disse højtbegavede unge Mennesker, som han saa op til med saa tillidsfuld Beundring, har vi modtaget mange Beviser, der har rørt os dybt. Især vil de skønne Broncevaser, som vore trofaste japanske Venner efter deres gamle Skikke har sendt til hans Grav, altid minde os om den Forbindelse med den fjerne Verden, som saa tidligt skulde bidrage til at bringe Eventyr ind i Christians Liv.

Fra første Færd havde Christian jo en aaben Sans for Livets mange Sider, og lige saa dyb hans Ærbødighed var for de i Videnskab og Kunst nedlagte evige Værdier, lige saa meget laa det ham paa Hjerte at se sit eget Ansigt i Tidens Spejl. De unge Venner, som Christian i sine sidste Aar saa nær sluttede sig til i fælles Trang til i den levende Kunst at genfinde Sindets inderste Stemninger, har paa den smukkeste Maade ladet os forstaa, hvor stærkt et Indtryk hans Varme og Alvor har efterladt hos dem. Den rene og dybe Glæde, som Christians Kærlighed til Kunsten beredte ham, vil vi ogsaa hver Dag genopleve ved de dejlige Billeder, hvori to af vore nære ældre Venner paa saa vidunderlig Maade har formaaet at fæstne deres Erindring om Christians tindrende Livsglæde og indtrængende Søgen, og som taler det stærkeste Sprog om Inderligheden af det Venskab og af den Forstaaelse, han mødte i vor Kreds.

8

I Dag tænker vi dog først og sidst paa det Legat, der bærer Christians Navn, og som hvert Aar paa hans Fødselsdag skal uddeles til Opmuntring og Støtte for et studerende eller kunstnerisk arbejdende ungt Menneske. Et smukkere Minde og et, der var mere i Christians egen Aand end det, der ved dette Legat er sat ham af hans kære gamle Tante, der hele Livet var ham saa god en Ven, kunde ikke tænkes. Med sin umiddelbare Forstaaelse af, at den største Tilfredsstillelse, Livet byder, maa søges i den Glæde og Hjælp, man kan give et Medmenneske, forenede Christian jo, saa ung han var, Overbevisningen om, at kun ærligt Arbejde paa at klare sig hvert stort eller lille Spørgsmaal, hvor fjernt dette end maatte synes at ligge Dagliglivets Krav, er vor eneste Vej til at ane en dybere Harmoni bag Tilværelsens Omskiftelser, og at det netop er heri, at den egentlige Tiltrækning af Studier og kunstneriske Bestræbelser bunder.

Omend Døden saa brat skulde tilintetgøre alle de Forventninger, vi nærede til Christians Fremtid, skulde Skæbnen heller ikke naa at skuffe noget af hans frejdige Haab, og for enhver af os, der stod ham nær, vil derfor Minderne om ham trods Smerten altid rumme ny Bestyrkelse til at stræbe efter det Ideal, der fyldte ham med saa sand Begejstring, og følge den Vej, ad hvilken han haabede selv at nærme sig det.

Den unge Maler, Mogens Andersen, der i Dag vil være den første Modtager af Legatet, har udtrykt dette saa smukt, da han i Sorgen over Christians Død skrev

9

til os, »vi maa arbejde med vort Arbejde. Christian arbejder da ogsaa, ellers ikke.« Paa det ægte Kunstnersind, som vi saavel gennem Christian som ved vore egne Erfaringer har lært at kende hos Mogens, bygger vi vort Haab om, at han i den saa krævende Gerning, han har valgt, vil kunne finde Tilfredsstillelse og skabe Glæde for andre, og vi vil alle her forene os i de bedste Ønsker for hans Fremtid.

Hvad Tanken om den smukke Maade, hvorpaa vor elskede Dreng ved Legatet skal mindes i de kommende Aar, betyder for Margrethe og mig, siger sig selv, og jeg vil derfor blot betone, at vi ved Legatuddelingens Tilknytning til Carlsberg altid vil tænke paa, hvor højt netop Christian skattede de Muligheder, som det nye Hjem her gav ham til at samles med unge Venner til gensidig Opmuntring og Belæring i de skønne traditionsrige Omgivelser. Endnu vil jeg gerne tilføje, at vi haaber, at enhver af dem, der efter os kommer til at bo herude, maa glæde sig over den Anledning, som Uddelingen af Mindelegatet vil give dem, til op gennem Aarene at bevare Forbindelsen med den Ungdom, der stadig bringer nyt Liv og nyt Haab med sig.

★

10

William Scharff pinx.   Okt. 1934

[419]

TRANSLATION

I.
On the Farewell from Home
26 August 1934

Christian's mother and I are deeply grateful that after the long heavy period of waiting we can now gather together with his friends here, where he lived his last so happy years, to bid our beloved boy a farewell and express our thanks from the home he brought so much joy and in which he is so greatly missed.

I need not say what delight and hope the first child brings to a young home; around it are gathered the parents' dreams of a richer blossoming of all that they themselves from the home of their childhood and through their own life have learned to love of what is beautiful and good. Still, only when they have learned from their own experience how little they are capable of removing every stone from the child's path are the parents truly able to appreciate that a boy, with a loving and chivalrous character such as Christian's, already from early childhood understood how to be a faithful comrade for his mother and father in happy and unhappy times and to the fullest extent to become the support for his younger brothers as only an elder brother can be.[1]

Not only for his parents, but for the whole family, Christian was from the very beginning to be a source of the greatest happiness. What sunshine has he not been to his greatly beloved grandparents whom he was to lose so early,[2] and for his Aunt Hanne[3] who, through her faithful and reciprocated love from his first to his last day has been a third grandmother for him. – It is also with deep gratitude that Margrethe and I think of all the joy and support Christian has had from the warm friendship shown to him by so many of our good friends, and not least from the circle of close friends among whom he was when he was so suddenly torn away from us and where right up to the very last moment he felt so happy and secure.

To say that Christian was a loving son for his parents and a chivalrous leader for his flock of brothers would, however, give only a poor idea of what he

---

[1] Christian (1916–1934) had five younger brothers: Hans (1918–), Erik (1920–1990), Aage (1922–), Ernest (1924–) and Harald (1928–1938).

[2] Christian's grandparents were, on the father's side, Christian (1855–1911) and Ellen Bohr, née Adler (1860–1930), and, on the mother's side, Alfred (1850–1925) and Sophie Nørlund, née Holm (1862–1926).

[3] Hanna Adler (1859–1947).

meant for his home. With his sense of humour and his imagination and his sympathetic understanding he was not only, more than anyone else, the sun that shone on his brothers' path, but as the years passed he was to an ever greater extent the source of new inspiration for us all. Without his fellow feeling with the family being weakened, Christian brought ever stronger breaths of fresh air to our home from the younger generation growing up to its own life around us, and he knew how to draw impulses from contacts with his schoolmates who, like his teachers, have shown him so much kindness and appreciation, as well as from the association with those young friends, especially from artistically oriented circles, whom he won during his last years and who put him in touch with new aspects of life.

During these years he developed from child to adolescent in the most beautiful way, and many strings, which could be tuned both to the deepest seriousness and to the most contagious cheerfulness, sounded in the harmony of his soul. Talents, of which he early had shown shoots, now ripened, and for his parents it was a profound joy to see the purposeful energy which, despite his wide interests, Christian put into his schoolwork, and especially into the preparation for the final examinations, which he just succeeded to pass so well and which should have opened a new chapter of his life, to which he looked forward with the most confident hope. Indeed, both we and his friends also had great expectations for his future, although nobody could know yet how his path in life would have formed itself. Together with the open sense for general scientific problems, which Christian inherited from his family, he combined – with his spirit, so easily stirred, and his unerring taste – possibilities which might eventually have drawn him into the sphere of art, away from the studies to which he had first intended to devote his life.

It is more difficult for us than we can say to reconcile ourselves to the thought that we shall never again see Christian's warm smile and hear his cheerful voice, and that the rich promises of fruitful accomplishments in his mature years to the benefit of himself and the rest of us will not be fulfilled. But although Christian should never reach beyond his fair youth, his life and striving have not been in vain. To each one of us within the home, and those closely attached to it, he has given precious memories, purified by the pain, which will bind us even closer to each other, and in which he will always live with us without aging. – The manifold testimonies of faithfulness to his memory, for which we thank his young and older friends most deeply on behalf of Christian and ourselves, have in addition given us the confident hope that also outside our small circle he will continue to be active in his own spirit, as truly as any of us may be said to live his strongest life in the thoughts of his fellow beings.

[421]

II.
On the First Award of the Memorial Grant
25 November 1934

When Christian matriculated only a few months ago so full of courage and hope, neither his mother and I nor any of the many friends gathered with us on this day, which would have been his 18th birthday, dreamt that we should be left here with only memories of him. But just because he is now only living and acting in our thoughts, the faithfulness to his memory that has been shown from so many sides has been the greatest support in our lot of having to reconcile ourselves with the bitter loss.

How much has the appreciation of Christian's serious striving, which his teachers at school have expressed, not meant for us, as well as the many testimonies of how faithfully his schoolmates have kept the memory of just those qualities in him which were most deeply rooted in his spirit – his need to understand and support others and his eagerness for and joy in learning about everything new and beautiful. Not least do we think how much it would have pleased Christian if he could know that his old comrades have decided to continue the small discussion club, in whose establishment he was so active and in whose vigorous development he took such a large share, and to maintain its character by the admission of younger pupils from school.

It is also with true gratitude that we have experienced that everyone connected with the daily work at the scientific institution in the midst of which Christian moved for so much of his short life, has understood how much it was on his mind that he should not in any way count for more than his own merits might justify, and how ready he was to use every chance to acquire new skills in the diverse spheres of work which are all equally necessary for such an institution. What joy was it not for Christian, as he grew older, to be accepted to an ever increasing degree as a comrade among the young scientists from at home and abroad who visited the Institute and who have not only been a constant source of new strength in the work, but have, in addition, brought so much news of what occupies the minds of young people around the world in our time. Of the friendly understanding Christian in his youthful idealistic striving found among these highly-gifted young people, whom he looked up to with such trusting admiration, we have received many proofs which have touched us deeply. Especially the beautiful bronze vases, which our faithful Japanese friends, in accordance with their old customs, have sent to his grave, will always remind us of the connection with the distant world, which so early should contribute to bringing adventure into Christian's life.

From the very beginning Christian had an open awareness of the many facets of life, and just as deep as his veneration was for the everlasting values found in science and art, just as important was it for him to see his own image in the mirror of the time. The young friends with whom Christian in his last years joined in a common urge to recover the innermost moods in living art, have in the most beautiful way let us understand how strong an impression his warmth and seriousness have left with them. The pure and deep joy which Christian's love of art gave him, we will also continue to experience every day through the beautiful portraits in which two of our close older friends have succeeded in such a wonderful way to capture their memory of Christian's sparkling joy of life and earnest search, and which speak the strongest language of the depth of the friendship and of the appreciation he met in our circle.[4]

Today, however, we think first and foremost of the grant which is named after Christian, and which every year on his birthday is to be awarded as encouragement and support for a student or for a young person working with art. A more beautiful memorial, and one that was more in Christian's own spirit, than this grant, established for him by his dear old aunt who was such a good friend to him throughout his life, cannot be imagined. With his spontaneous recognition that the greatest satisfaction life can offer must be sought in the joy and help one can give to a fellow human being, Christian combined, so young he was, the conviction that only honest work to clarify for oneself every large and small question, no matter how remote this might seem from the demands of daily life, is our one way to get a feeling for a deeper harmony behind the vicissitudes of existence, and that it is precisely herein that the true attraction of studies and artistic endeavours has its roots.

While death should so abruptly destroy all the expectations we held for Christian's future, fate should not have time to disappoint any of his confident hope, and for each of us who were close to him, the memories of him will therefore, despite the pain, always hold new strength to strive for the ideal that filled him with such true enthusiasm and to follow the path by which he himself hoped to approach it.

Mogens Andersen,[5] the young painter who today will be the first recipient of the grant, has expressed this so beautifully when, in grief at Christian's death, he wrote to us: "We must work with our work. Christian is then also working, otherwise not". On the genuine artistic spirit which we have learnt to know in

---

[4] Julius Paulsen (1860–1940) and William Scharff (1886–1959), who made the portraits reproduced immediately before and after the reprint of Bohr's memorial speech, respectively.
[5] (1916–2003).

Mogens, both through Christian and from our own experience, we build our hope that he will be able to find satisfaction and create joy for others in the demanding vocation he has chosen, and we will all here join in the best wishes for his future.

What the thought of the beautiful way in which our beloved son shall be remembered in the coming years by the grant means for Margrethe and me speaks for itself, and I will therefore only emphasize that the awarding of the grant being linked to Carlsberg will always remind us of how highly Christian, in particular, treasured the opportunities, given by the new home here, for him to gather with young friends for mutual encouragement and inspiration in the beautiful surroundings, rich in tradition. Finally, I would like to add that we hope that each of those who will come to live out here after us may take pleasure in the occasion that the awarding of the grant will give them through the years for maintaining contact with the younger generation which continually brings with it new life and new hope.

XLVII. KIRSTINE MEYER, N. BJERRUM: 12 OCTOBER 1861 – 28 SEPTEMBER 1941

KIRSTINE MEYER, F. BJERRUM:
12. OKTOBER 1861 – 28. SEPTEMBER 1941
Fys. Tidsskr. **39** (1941) 113–115

TEXT AND TRANSLATION

See Introduction to Part II, p. [123].

KIRSTINE MEYER, N. BJERRUM:
12 OCTOBER 1861 – 28 SEPTEMBER 1941 (1941)

Versions published in Danish

A   Fys. Tidsskr. **39** (1941) 113–115
B   *Dr. phil. fru Kirstine Meyer Død*, Berlingske Aftenavis, 29 September
1941

The two versions agree with each other

Kirstine Meyer, f. Bjerrum.

12. Oktober 1861 — 28. September 1941.

Paa Dr. phil. Fru Kirstine Meyers 80 Aars Fødselsdag vil alle danske Fysikere med Taknemmelighed tænke paa, hvad det for enhver af os har betydet, gennem hendes dybtgaaende historisk-fysiske Studier at blive belært om Udviklingen og Sammenhængen indenfor mange af de mest forskellige Omraader af Fysikken og om Baggrunden for de Bidrag, der her fra Landet gennem Tiderne er ydet til vor Videnskab.

Allerede hendes tidligste Undersøgelser, der blev samlet i hendes kendte Værk om Temperaturbegrebets Udvikling, giver et uforglemmeligt Indtryk af, hvilken Belæring historiske Undersøgelser kan give af de Vanskeligheder, der maa overvindes, før Begreber hentet fra Dagliglivet kan opnaa den Afklaring, som muliggør deres Anvendelse i den eksakte Videnskab. Det var ogsaa netop i Forbindelse med disse Undersøgelser, at Fru Meyer gjorde sit første Fund af Ole Rømers Optegnelser, der satte hende paa Spor efter dennes afgørende, indtil da ganske forglemte Bidrag til Fastlæggelse af Termometrenes Temperaturskala. Den Opfattelse af Rømers Indflydelse paa Fahrenheit, hvortil Fru Meyer med saa fin historisk Sans var naaet, fandt som bekendt den smukkeste Bekræftelse ved den senere Udgivelse af Fahrenheits Breve. Foruden at bidrage saa væsentligt til Klarlæggelsen af Rømers alsidige eksperimentelle Forskervirksomhed lykkedes det hende jo ogsaa at sætte selve hans store Opdagelse af Lysets Tøven i nyt Perspektiv gennem Fundet og Udredningen af de Originaloptegnelser vedrørende Jupitermaanernes Forsinkelser, hvorpaa Rømer havde baseret sine Slutninger.

Naar man tænker paa den Forskertrang og Indlevelsesevne, som Fru Meyer ved disse Undersøgelser havde lagt for Dagen, kan det ikke betragtes som nogen Tilfældighed, at det atter blev hende, der skulde fremdrage og værdsætte H. C. Ørsteds Optegnelser om de talrige og sindrige Forsøg, som han anstillede i umiddelbar Tilknytning til sin Opdagelse af Elektromagnetismen, og som førte ham til saa klar Erkendelse af mange af de elektromagnetiske Virkninger, hvis Opdagelse hidtil er blevet tilskrevet andre. Foruden at give Verden en helt ny Forstaaelse af det brede Grundlag, hvorpaa Ørsteds videnskabelige Virksomhed hvilede, har Fru Meyer som ingen anden bidraget til at belyse den gennemgribende Betydning, Ørsteds alsidige og vidt forgrenede Virksomhed har haft for Naturviden-skabens Udvikling og Udbredelse herhjemme. De varme Ord, hvor-med Fru Meyer omtaler H. C. Ørsteds Opfattelse af Lærergerningen som en Kunst, der ikke alene kræver utrætteligt Arbejde med Stoffet, men som ogsaa stiller de største Krav til Selvudvikling, kan ikke undlade at faa enhver til at tænke paa den uvurderlige Indsats, som Fru Meyer netop ud fra samme Indstilling har gjort for Højnel-sen af Ungdommens Undervisning i Naturfagene herhjemme.

Med disse faa Linier har det jo ikke været Hensigten at give nogen nærmere Redegørelse for de omfattende og frugtbare Studier, som Fru Meyer med utrættelig Begejstring gennem saa mange Aar har fortsat trods sin store pædagogiske Virksomhed, men blot at minde om nogle Hovedlinier i det Livsværk, med hvilket hun paa en Maade, der er uden Sidestykke, har gavnet dansk fysisk Videnskab og dens Anseelse.

Niels Bohr.

TRANSLATION

Kirstine Meyer, n. Bjerrum.

12 October 1861 – 28 September 1941.

On D.Phil. Kirstine Meyer's 80th birthday all Danish physicists will remember with gratitude what it has meant for each of us to be taught, through her thorough studies of the history of physics, about the development of and relationships between many of the most divergent realms of physics and about the background of the contributions which through the ages have been made from this country to our science.

Already her earliest investigations, collected in her well-known work on the development of the concept of temperature,[1] give an unforgettable impression of the lesson that historical studies can give regarding the difficulties that must be overcome before concepts from everyday life can achieve the clarification necessary for their application in exact science. It was indeed precisely in connection with these studies that Mrs Meyer made her first discovery of Ole Rømer's notes which put her on the track of his important, until then completely forgotten, contribution to determining the temperature scale of thermometers.[2] This understanding of Rømer's influence on Fahrenheit,[3] reached by Mrs Meyer with such a fine sense of history, found, as is well known, the most beautiful

---

[1] Meyer's doctoral dissertation was first published in Danish as K. Meyer, *Temperaturbegrebets Udvikling gennem Tiderne samt dets Sammenhæng med vexlende Forestillinger om Varmens Natur*, Gjellerup, Copenhagen 1909. It was subsequently published in German as *Die Wissenschaft: Einzeldarstellungen aus der Naturwissenschaft und der Technik*, Band 48: K. Meyer, *Die Entwicklung des Temperaturbegriffs im Laufe der Zeiten sowie dessen Zusammenhang mit den wechselnden Vorstellungen über die Natur der Wärme*, Vieweg, Braunschweig 1913.

[2] Meyer, *Temperaturbegrebets Udvikling*, ref. 1, pp. 66–95.

[3] Daniel Gabriel Fahrenheit (1686–1736), German instrument maker and glass blower who worked mostly in The Netherlands.

confirmation in the later publication of Fahrenheit's letters.[4] Besides making such an important contribution to the elucidation of Rømer's many-sided experimental research activity, she also succeeded in placing his great discovery of the hesitation of light itself into a new perspective through the finding and the explanation of the original notes concerning the delays of Jupiter's moons on which Rømer had based his conclusions.[5]

When we think of the desire for knowledge and ability for sympathetic insight that Mrs Meyer had shown through these studies, it cannot be considered as a coincidence that, again, it was she who should call attention to and appreciate H.C. Ørsted's notes concerning the numerous and ingenious experiments which he made in immediate connection with his discovery of electromagnetism, and which led him to such a clear recognition of many of the electromagnetic effects whose discovery has hitherto been ascribed to others. Besides giving the world an entirely new understanding of the broad basis on which Ørsted's scientific work rested, Mrs Meyer has contributed as no other to illuminating the decisive importance which Ørsted's many-sided and widely-ramified activities have had for the development and dissemination of natural science in our country. The warm words with which Mrs Meyer refers to H.C. Ørsted's understanding of the teaching profession as an art, which not only requires untiring work with the subject matter, but also places the greatest possible demands on self-development, cannot but make everyone remember the invaluable effort Mrs Meyer has made, precisely on the basis of the same outlook, for the improvement of science education for the young generation in our country.[6]

It has not been my intention with these few lines to give a detailed account of the extensive and fruitful studies that Mrs Meyer has continued with untiring enthusiasm for so many years despite her great teaching activity, but only to bring to mind some main themes in the lifework with which in a way unparalleled she has benefited Danish physical science and its standing.

Niels Bohr.

---

[4] Bohr is referring in particular to a letter from Fahrenheit to his Dutch colleague Hermann Boerhaave (1668–1738) dated 17 April 1729 in which Fahrenheit describes his experience at Rømer's laboratory in 1708. The letter was found in Leningrad in 1929. It was translated from Dutch to German and published in E. Cohen and W.A.T. Cohen, *Daniel Gabriel Fahrenheit*, Verhandl. der koninklijke Akademie van Wetenschappen te Amsterdam, Afdeeling Natuurkunde Deel XVI, No. 2, Amsterdam 1931, where Meyer first read it. See K. Meyer, *Ole Rømer and Fahrenheit's Thermometers*, Nature **139** (1937) 585.

[5] K. Meyer, *Om Ole Rømers Opdagelse af Lysets Tøven*, Kgl. Dan. Vid. Selsk. Skr., 7. Række XII.1 (1915), pp. 107–145.

[6] See Meyer, *Ørsted*, referred to on p. [122], ref. 122.

XLVIII. HE STEPPED IN WHERE WRONG HAD BEEN DONE: OBITUARY BY PROFESSOR NIELS BOHR

HAN TRAADTE HJÆLPENDE TIL HVOR URET BLEV BEGAAET:
MINDEORD AF PROFESSOR NIELS BOHR
Politiken, 7 October 1949

For A. Friis

TEXT AND TRANSLATION

See Introduction to Part II, p. [124].

Han traadte hjælpende til hvor uret blev begaaet

Mindeord af professor Niels Bohr

PROFESSOR *Aage Friis'* bortgang betyder ikke alene et stort tab for dansk historisk forskning, men vil føles smerteligt i langt videre krese. Med sit aabne og varme sind og ukuelige mod til med alle kræfter at kæmpe for sin overbevisning maatte han træde hjælpende til overalt, hvor han følte, at uret var begaaet eller mennesker uforskyldt var bragt i nød.

Som formand først for Den danske komité til støtte for landflygtige Aandsarbejdere og senere for De samvirkende danske Emigranthjælpekomiteer arbejdede han utrætteligt for at lette tilværelsen her i landet eller muliggøre vidererejse til sikre steder for er stort antal mennesker, der i aarene før krigen blev hjemløse paa grund af fordomme og politisk forfølgelse. Hos de danske myndigheder, der i disse spørgsmaal befandt sig i en vanskelig situation, vandt Aage Friis tillid og respekt, og ved det samarbejde, der kom i gang mellem myndighederne og komiteerne, udførte han en gerning, der ikke alene blev til redning for mange mennesker, men tillige verden over skabte for vort land en sympati, hvis oprigtighed man ofte har haft lejlighed til at erfare. Alle, der sammen med Aage Friis deltog i dette arbejde, følte for ham en beundring, der hos mange af os udviklede sig til et dybt rodfæstet venskab.

Ogsaa efter krigen laa omsorgen for de mennesker, der havde fundet et tilflugtssted i Danmark, Aage Friis dybt paa sinde, og han tog trods sin høje alder og tiltagende svagelighed til det sidste levende del i bestræbelserne for at skaffe dem rimelige levevilkaar og for deres optagelse i det danske samfund. Disse bestræbelser fandt vidtgaaende tilslutning, men meget staar endnu tilbage for løsningen af den humane opgave, landet er blevet stillet over for. Naar vi føler os sikre paa, at opgaven vil blive løst paa rette maade, skyldes dette først og fremmest den aand, for hvilken Aage Friis var saa varm en talsmand og forkæmper.

Niels Bohr.

TRANSLATION

He Stepped in
Where Wrong had been Done

Obituary by Professor Niels Bohr

The death of Professor *Aage Friis* means not only a great loss for Danish historical research but will also be felt with sorrow in much wider circles. With his open and warm personality and invincible courage to fight with all his powers for his convictions, he felt obliged to step in wherever he felt that wrong had been done or people were brought into distress by no fault of their own.

As chairman of, first, the Danish Committee for the Support of Refugee Intellectual Workers and, then, the Joint Danish Committees for Aiding Emigrants, he worked tirelessly to ease life here in this country or to enable transit to safe places for a large number of people, who in the years before the war lost their home due to prejudice or political persecution. Aage Friis won trust and respect with the Danish authorities, who in these questions were in a difficult position, and through the cooperation that was started between the authorities and the committees he carried on an activity which not only became the rescue for many people, but, in addition, created sympathy for our country the world over, the sincerity of which one has often had the opportunity to experience. All who together with Aage Friis took part in this work felt an admiration for him, which in many of us developed into a deeply rooted friendship.

After the war, too, solicitude for the people who had found refuge in Denmark was deeply felt by Aage Friis, and despite his advanced years and increasing infirmity he took a lively interest unto the last in the endeavours for acquiring reasonable living conditions for them and for their integration into Danish society. These endeavours found wide-ranging support but much remains to be done for the solution of the humanitarian task that the country has been faced with. That we feel certain that the task will be solved in the right way is due first and foremost to the spirit for which Aage Friis was so warm an advocate and champion.

Niels Bohr.

XLIX. PROFESSOR NIELS BJERRUM 50 YEARS

PROF. NIELS BJERRUM FYLDER 50 AAR
Berlingske Tidende, 9 March 1929

TEXT AND TRANSLATION

See Introduction to Part II, p. [126].

Prof. Niels Bjerrum fylder 50 Aar.

Af Professor, Dr. *Niels Bohr.*

Professor Niels Bjerrum.

NAAR Professor Niels Bjerrum paa Mandag fylder 50 Aar, vil der ikke alene fra Kemikernes, men ogsaa fra Fysikernes Kreds sendes ham mange hjertelige Lykønskninger og taknemmelige Tanker. Som vel kendt, besidder vi i Professor Bjerrum og Professor Brønsted — der forleden under saa megen Anerkendelse rundede samme Mærkepæl — to af de ypperste Repræsentanter for den Videnskabsgren, der betegnes som den fysiske Kemi, og hvis Tilblivelse er Udtryk for den nære Forbindelse, der i de senere Aar atter er indtraadt mellem de gamle Søstervidenskaber Fysik og Kemi. Medens det til Tider kunde se ud, som om Fysikere og Kemikere var henviste til hver sit Arbejdsfelt, hvor de uden at behøve nøje Føling med hverandre forfulgte forskellige Sider af Naturfænomenerne, er der i dette Forhold som Følge af de atomistiske Forestillingers store Udvikling sket en gennemgribende Forandring. Navnlig lader den Opfattelse sig ikke opretholde, at Fysikens Opgave er Studiet af de almindelige Naturlove, medens Kemiens er Udforskningen af Ejendommelighederne hos de i Naturen forefindende Stoffer. I Stoffernes Ejendommeligheder ser vi, i Stedet for Naturens Luner, i stedse højere Grad Udtryk for almindelige Naturlove, samtidig med at vi erkender, at alle saadanne Love kun faar Indhold gennem deres Anvendelse paa de forefaldende Fænomener.

Beherskelsen af det mægtige Erfaringsmateriale, som Fysikere og Kemikere i Tidernes Løb har ophobet, fordrer vel stadig en vidtgaaende Arbejdsdeling med en tilsvarende Specialisering i Uddannelsen, men ogsaa her flyder Grænserne mere og mere ud. Netop den fysiske Kemis Tilblivelse er jo Vidnesbyrd om den udstrakte Anvendelse, som de tidligere fysiske Discipliner, Varmelæren og Elektricitetslæren har fundet ved Studiet af kemiske Processer, og som samtidig har været i saa høj Grad befrugtende for selve disse Discipliners Udvikling. Medens Brønsted med sine termokemiske Studier paa fremragende Maade har fortsat vore store hjemlige Traditioner fra Jul. Thomsens Dage, falder Bjerrums Hovedarbejder indenfor Elektrokemiens Omraade, hvor han som en af de ledende Forskere med stort Held har deltaget i Udbygningen af den af Arrhenius grundede elektrolytiske Dissociationsteori, der skulde faa saa store Følger for Kemiens seneste Udviklingsstadium.

Ved Betragtningen af Professor Bjerrums videnskabelige Gerning vil en Fysiker dog maaske allerførst tænke paa den betydningsfulde Indsats, han har gjort paa det Omraade af Atomteorien, der vender ud imod Lyslæren, og som vel er den fysiske Disciplin, der senest har fundet Anvendelse overfor kemiske Problemer. Bjerrum paaviste her, hvorledes vi i Beskaffenheden af den Lysstraaling, som en Luftart kan bringes til at emittere eller absorbere, har en yderst værdifuld Kilde til Oplysninger om Molekylernes Bevægelsesforhold. Absorbtionsliniernes Udbredning ved stigende Temperatur tillader ikke alene i store Træk at følge, hvorledes Molekylernes Omdrejningsbevægelser ændrer sig under Luftartens Opvarmning, men den nøjere Undersøgelse af Liniernes Struktur viser, hvorledes disse Drejninger ikke kan foregaa paa jævn foranderlig Maade, saaledes som man efter vore sædvanlige mekaniske Forestillinger maatte vente, men kun med visse bestemte Omdrejningstal i nøje Overensstemmelse med de nu alt beherskende kvantemekaniske Forestillinger, der tager sit Udgangspunkt fra Plancks epokegørende Opdagelse af det saakaldte Virkningskvantum, og som den Gang befandt sig paa et primitivt Udviklingsstadium.

Samtidig med at afgive en overbevisende Støtte for Kvanteteoriens Realitet skulde disse Bjerrums Undersøgelser lægge den første Grund til den senere saa højt udviklede og for kemiske Studier saa lovende Teori for Molekyl-spektrene. Trods den ved Kvanteteoriens fortsatte Udvikling naaede Erkendelse af vore sædvanlige Anskuelsesformers principielle Begrænsning ved Beskrivelsen af de elementære Naturprocesser vil Bjerrums Betragtninger altid staa som et lærerigt Vidnesbyrd om anskuelige Forestillingers Frugtbarhed indenfor Atomteorien.

Naar Bjerrum har været i Stand til at gøre denne Indsats paa et Omraade af Videnskaben, hvor de væsentligste Arbejder baade før og siden har været udført af Fagfysikere, skyldes det foruden en lykkelig Intuition tillige en for en Kemiker usædvanlig Interesse for og Kendskab til almene fysiske Problemer. Dette Bjerrums sjældne Overblik er ikke alene Baggrunden for hans egen videnskabelige Virksomhed, men er ogsaa gennem personlige Tankeudvekslinger kommet store Kredse af Fysikere og Kemikere til Gode. Ikke mindst vil vi i denne Forbindelse tænke paa hans mange i lige Grad af saglig Indsigt og videnskabelig Begejstring prægede Indlæg i Diskussionerne i vore kemiske og fysiske Foreninger. I Erkendelse af, hvad hans Virksomhed og Personlighed betyder for vore Videnskabers Trivsel herhjemme, sender danske Fysikere og Kemikere ham i disse Dage de varmeste Ønsker for Fortsættelsen af hans betydningsfulde Gerning.

TRANSLATION

Professor Niels Bjerrum 50 Years.

By Professor, Dr. *Niels Bohr*.

When Professor Niels Bjerrum has his 50th birthday on Monday, many heartfelt greetings and grateful thoughts will be sent to him not only from the chemists' circle but also from the circle of physicists. As is well known we have in Professor Bjerrum and Professor Brønsted[1] – who the other day with so much appreciation passed the same milestone – two of the most excellent representatives of the branch of science called physical chemistry, whose creation is an expression of the close relationship which in recent years has once again set in between the old sister sciences of physics and chemistry. While at times it might seem that physicists and chemists were relegated to their respective fields of work, in which, without requiring close contact with each other, they pursued different aspects of natural phenomena, a thorough change in this relationship has taken place as a result of the great development of the ideas about the atom. In particular, it has been impossible to maintain the opinion that the task of physics is the study of the general laws of nature while that of chemistry is the investigation of the peculiarities of the substances found in nature. In the peculiarities of substances we see, instead of the whims of nature, to an ever increasing degree the expression of general laws of nature, at the same time recognizing that all such laws only obtain meaning through their application to the phenomena occurring.

A full command of the huge amount of experimental results that physicists and chemists have collected in the course of time will probably still require a far-reaching division of labour with a corresponding specialization in education, but also here the boundaries are becoming more and more blurred. Precisely the creation of physical chemistry testifies to the widespread use that the sciences of heat and electricity, disciplines previously belonging to physics, have found in the study of chemical processes, and which at the same time have to such a great degree been fertile for the development of these disciplines themselves. While with his thermochemical studies Brønsted has continued, in an excellent way, our great traditions here at home from the days of Julius Thomsen,[2] Bjerrum's major works fall within the field of electrochemistry, where as one of the leading researchers he has taken part with great success in the extension of

---

[1] Johannes Nicolaus Brønsted (1879–1947).
[2] (1826–1909).

the theory of electrolytic dissociation established by Arrhenius,[3] which was to have such great consequences for chemistry's most recent stage of development.

When considering Professor Bjerrum's scientific work, a physicist might nevertheless perhaps think first of all of the important effort he has made in the area of atomic theory touching on optics, which is probably the discipline of physics that most recently has found application as regards chemical problems. Here Bjerrum showed how in the kind of light radiation that a gas can be made to emit or absorb we have an extremely valuable source of information about how molecules move. The broadening of the absorption lines with increasing temperatures does not only allow us to follow in general terms how the rotational movements of the molecules change during the heating of gases, but closer investigation of the structure of the lines shows how these rotations cannot change continuously as would be expected on the basis of our usual mechanical ideas, but only occur at certain fixed rotational frequencies in close agreement with the now all-dominating quantum-mechanical ideas which have their origin in Planck's epoch-making discovery of the so-called quantum of action, and which at that time was at a primitive stage of development.

At the same time as providing convincing support for the reality of quantum theory, these investigations by Bjerrum would lay the first foundation for the later so highly developed and for chemistry studies so promising theory of molecular spectra. Despite the recognition, reached by the continued development of the quantum theory, of the fundamental limitations of our usual forms of visualization in the description of the elementary processes of nature, Bjerrum's considerations will always stand as an instructive testimony of the fruitfulness of visualizable ideas in atomic theory.

That Bjerrum has been able to make this effort in a field of science where the most important contributions both before and since have been carried out by specialized physicists is due, in addition to a happy intuition, to an interest in and knowledge of general physical problems unusual for a chemist. Bjerrum's remarkable breadth of view is not only the background for his own scientific activity but has also, through personal exchanges of ideas, been beneficial for large circles of physicists and chemists. In this connection, we should not least remember his many contributions, marked to an equal degree by professional insight and scientific enthusiasm, to discussions in our chemical and physical associations. In recognition of what his activity and personality mean for the flourishing of our sciences here at home, Danish physicists and chemists send him in these days the warmest wishes for the continuation of his important work.

---

[3] Svante Arrhenius (1859–1927), Swedish physical chemist.

L. [FOREWORD]

"Niels Bjerrum: Selected Papers,
edited by friends and coworkers
on the occasion of his 70th birthday
the 11th of March 1949",
Munksgaard, Copenhagen 1949, p. 3

See Introduction to Part II, p. [126].

[443]

The scientific publications of Niels Bjerrum have initiated great advances of our knowledge and understanding in many fields of chemistry and physics and bear throughout witness of that same openness of outlook and balance of judgment, which his friends and colleagues admire so highly and which together with his straightforwardness and loyalty has secured him the confidence of the whole Danish community. Many important tasks have been entrusted upon him, and his fertile activities and the encouragement he has given to wide circles will be remembered with deep gratitude from most different sides on his 70th birthday. Deliberating how his colleagues best could contribute on this occasion, the committee has thought that an edition of a selection of the papers of Niels Bjerrum by which he has erected himself a lasting monument in science, would be the most fitting way to express the indebtedness we owe to him, and we feel assured one which will be warmly welcomed by chemists and physicists all over the world.

Niels Bohr.

LI. HIS MEMORY A SOURCE OF
COURAGE AND STRENGTH

HANS MINDE EN KILDE TIL MOD OG STYRKE
Politiken, 1 October 1958

For N. Bjerrum

TEXT AND TRANSLATION

See Introduction to Part II, p. [126].

[445]

Hans minde en kilde til mod og styrke

Niels Bohr, der helt siden ungdommen var knyttet nær til Niels Bjerrum, skriver disse mindeord:

Ved professor Niels Bjerrums død har den fysisk-kemiske videnskab mistet en af vor tids mest fremragende forskere og det danske samfund en af sine største og mest højagtede personligheder.

Allerede i Niels Bjerrums tidligste videnskabelige arbejder aabnedes nye udsyn for forstaaelsen af kemiske processer og stoffernes fysiske egenskaber, som tiltrak sig stor opmærksomhed og hvis frugtbare videreførelse af ham selv og andre skabte ham en anerkendelse i den videnskabelige verden, der fandt udtryk i talrige æresbevisninger fra universiteter og videnskabelige selskaber.

Den samme indstilling, der kendetegnede Niels Bjerrums rige og mangesidige videnskabelige virke, prægede ogsaa hans deltagelse i ledelsen af danske forskningsinstitutioner og industrielle virksomheder. Hans gerning fulgtes stedse af trivsel og fremgang, og med sin retsindighed, indlevelsesevne og arbejdskraft omfattedes han overalt med en tillid og beundring, der medførte, at mange af vort samfunds mest ansvarsfulde hverv i aarenes løb blev ham betroet.

I vide krese vil Niels Bjerrum blive dybt savnet, og vi, der havde den lykke igennem mange aar at staa ham nær, har mistet en trofast ven, hos hvem man altid kunne hente raad og finde forstaaelse. Med sorg tænker vi paa aldrig mere at kunne se ham og tale med ham, men det minde, som han efterlader, vil altid forblive en kilde til mod og styrke.

Niels Bohr.

TRANSLATION

His Memory a Source for Courage and Strength

Niels Bohr, who since his youth has been closely attached to Niels Bjerrum, writes these commemorative words:

With the death of Niels Bjerrum, physical–chemical science has lost one of the most eminent researchers of our time and Danish society one of its greatest and most respected personalities.

Already in Niels Bjerrum's earliest scientific papers new perspectives were opened for the understanding of chemical processes and the physical properties of substances, which attracted great attention and whose fruitful continuation by him and others earned him a recognition in the world of science that found expression in numerous honours from universities and science societies.

The same attitude which characterized Niels Bjerrum's rich and many-faceted scientific work also marked his participation in the leadership of Danish research institutions and industrial enterprises. His work was always followed by growth and progress, and with his integrity, sympathetic insight and capacity for work he was regarded everywhere with a trust and admiration which led to many of the most responsible tasks in Danish society being entrusted to him throughout the years.

Niels Bjerrum will be deeply missed in wide circles, and we who had the good fortune of being close to him throughout many years have lost a faithful friend from whom it was always possible to seek advice and find understanding. With sorrow we bear in mind that we will never see or talk with him again, but the memory which he leaves behind will always remain a source of courage and strength.

Niels Bohr.

Niels Bohr, Niels Bjerrum and Ole Chievitz on Chita, the sailing boat they owned together (1930).

LII. SPEECH AT THE MEMORIAL CEREMONY FOR OLE CHIEVITZ 31 DECEMBER 1946

TALE VED MINDEHØJTIDELIGHEDEN FOR OLE CHIEVITZ
31. DECEMBER 1946
Ord och Bild **55** (1947) 49–53

TEXT AND TRANSLATION

See Introduction to Part II, p. [126].

[449]

TALE VED MINDEHØJTIDELIGHED FOR OLE CHIEVITZ
31. DECEMBER 1946 (1947)

Versions published in Danish

A   Ord och Bild **55** (1947) 49–53
B   *Ole Chievitz* in *Ole Chievitz*, Nordisk Boghandel, Copenhagen 1956,
    pp. 7–14

The two versions agree with each other

Ole Chievitz.

Tale ved Mindehøjtideligheden for Ole Chievitz 31. December 1946

Af Niels Bohr

SORGEN over Ole Chievitz's Bortgang deles af hele det danske Folk, i hvilket Mindet om hans Færd i gode og onde Tider altid vil bevares i Taknemmelighed og Stolthed. Store Kredse blev det forundt at føle den Styrke og Varme, der udstraalede fra hans Personlighed, og mange var de, der omfattedes af hans trofaste Venskab. Som en af dem, der har haft den Lykke at staa ham nær gennem hele Livet, har jeg faaet Lov til her ved hans Baare at forsøge at give Udtryk for, hvilket Tab vi har lidt, og hvilken Rigdom han efterlader os.

I Ole Chievitz's første Barndomsaar boede hans Forældre ligesom mine i det gamle kirurgiske Akademi i Bredgade, hvor hans Fader, den højt ansete Anatom, havde Embedsbolig helt oppe under Taget af den skønne Bygning, indtil Familien nogle Aar efter Oles Fødsel paa Grund af de lidet tidssvarende Forhold maatte flytte. Fra Ungdommen af havde vore Fædre været nære Venner, men mest afgørende for Oles og mit Forhold blev det vel, at vi, i de sidste seks Aar før vi sammen blev Studenter fra Gammelholms Latinskole, sad Side ved Side ved fælles Skolepult. Rent umiddelbart vandt han allerede i Skoletiden alles Respekt gennem sin Uafhængighed og Sandhedskærlighed, og sin Evne til at gribe enhver Sags Kerne. Ved sin Trang til at kaste sig ind i Kampen for, hvad han mente var Ret, kunde Ole jo ofte blandt jævnaldrende blive en Stridens Fane, men alle forenedes vi i Beundringen for hans Ligefremhed og Frygtløshed, der sammen med hans Begejstring og Handlekraft gjorde ham til saa god en Kammerat.

I sit Hjem modtog Ole Chievitz Paavirkninger, der kom til at præge hans Væsen dybt, især sluttede han sig paa mange Maader inderligt til Faderen, der ikke alene var betydelig som Forsker og Tænker, men tillige besad levende historiske och kunstneriske Interesser. Som Følge af den uhelbredelige Sygdom, der skulde kalde ham alt for tidligt bort, var dog Oles Fader allerede i vore Drengeaar henvist til at leve i stedse større Tilbagetrukkenhed inden for Studerekammerets Vægge, medens hans Moder med saa megen Tapperhed og saa friskt et Livsmod kæmpede med Hverdagens Vanskeligheder. I disse Omgivelser modnedes Ole Chievitz tidligt til Ansvarsbevidsthed og til Fortrolighed med lyse og mørke Sider af Tilværelsens Kaar.

Hvor stort et Slag end Faderens Død var for Ole, fjernede den ham imidlertid ingenlunde fra Familiens mange Venner, der alle forstod, hvor meget der boede i ham. Saavel i det Bjerrumske og det Ipsenske Hjem som i mine Forældres omfattedes han med de varmeste Følelser og fandt blandt den yngre Slægt Venner for Livet.

Overalt, hvor han kom, bragte han friske Pust med sig, og ved alle de mange Strenge, der klang i hans aabne og følsomme Sind, blev hvert Samvær med ham til Glæde og Berigelse. Fra den Tid mindes jeg især en Rejse, som min Broder og jeg som unge Studenter foretog sammen med Ole, og hvor vi under Vandringer igennem Tyrols skønne Bjergegne og gamle stemningsfulde Smaabyer fik de stærkeste Indtryk af hans vidtfavnende Følelser for Mennesker og Naturog af den sande Mandighed, som saa tidligt kendetegnede hans hele Væsen.

For sine fra Hjemmet medbragte videnskabelige Interesser og Trang til at virke til Hjælp for Medmennesker, havde jo Ole Chievitz fundet det rette Samlingspunkt i det Studium, hvormed han forberedte sig til en Lægegerning. Den Ildhu og Ansvarsfølelse, han lagde for Dagen, skabte ham ikke alene Venner og Beundrere blandt Studiefællerne, men tillige vandt han hurtigt Universitetslærernes Tillid og Paaskønnelse. Især er det jo velkendt, hvor højt Professor Rovsing, der i mange Aar for Ole var saa beundret et Forbillede, værdsatte ham, og hvilke ansvarsfulde Opgaver han allerede tidligt betroede ham. Blandt Lægekammeraterne paa Hospitalet skulde Ole Chievitz jo ogsaa finde den Hustru, der ved sin fine menneskelige Forstaaelse og Fællesskabet i faglige Interesser blev ham saa stor en Støtte, og for hvem hans Bortgang nu, efter at de i de mange Aar har delt saavel lyse Glæder som dybe Sorger, er saa haardt et Slag og uerstatteligt et Tab, men som sammen med os andre besidder saa stor en Rigdom i Minderne.

Hvor meget alle vi, der stod Ole Chievitz nær, skylder ham, lader sig slet ikke udtrykke. Jeg mindes Aftener i Hjemmet, hvor Ole og Ingeborg Aar efter Aar samlede Ungdomsvennerne og dem, der senere kom til, og hvor han kunde lade sin særegne Humor spille imellem den dybeste Alvor og den frieste Kaadhed paa en Maade, der gjorde enhver saadan Aften til en forfriskende og uforglemmelig Oplevelse. Fra de Aar, hvor jeg sammen med ham, Niels Bjerrum og Holger Hendriksen tilhørte et lille Baadelag, der foretog Langfarter i Ferierne, har jeg ogsaa dyrebare Minder om det fortrolige Samliv ombord, hvor det altid sprudlende Væld i Oles rige Sind bragte os alle saa megen Fornyelse og Opmuntring. Især tænker vi dog paa ham som den forstaaende og trofaste Ven, hos hvem man kunde søge Raad og finde Støtte i alle Livets Tilskikkelser og hvem man ofte tyede til som en Klippe i Tilværelsens Malstrøm.

Paa mange Maader var Ole Chievitz noget ganske for sig selv, men samtidig forbandtes i hans hele Væsen paa forunderlig Vis alle de Træk, der kendetegner det søgende og kæmpende Menneske. Hvor lidenskabelig hans Stillingtagen end kunde være og hans Trang til at give den Udslag i Handling, besad han med al sin Styrke en hjertevarmende Mildhed, som han, omend ofte forgæves, kunde gøre sig store Anstrengelser for at skjule. Baade Retfærdighedssans og Kærlighedsfølelse var hos ham udviklet som hos faa, men han forbandt dermed den store Personligheds Evne til uden Vaklen at skønne, hvad Forholdene i hvert Øjeblik mest krævede af ham. Det laa ham ganske fjernt paa Forhaand at stræbe efter Samtykke; tværtimod kunde han mere eller mindre paa Skrømt søge at bringe Modsætninger frem. Ved saadanne Lejligheder, der altid gav Anledning til Selvprøvelse, mærkede man maaske stærkest Dybden og Varmen af hans Sind og kom, om muligt, til at holde endnu mere af ham.

Hans hele menneskelige Indstilling havde faste Rødder, der stadig søgte dybere. Ikke alene bragte Livet ham Erfaringer i rigt Maal, men han besad tillige en sjælden Indlevelse i alle Sider af vor Kultur og i dens Historie og et rigt Kendskab til dansk

»Jens.«
Ole Chievitz forklædt under Besættelsestiden, da han var »gaaet under Jorden« og lød Navnet »Jens«.

Digtekunst, hvorfra han saa ofte forstod at hente de rette Ord til at udtrykke sine Følelser og sætte Hændelser, der mødte ham, i stemningsfuld Belysning. Ogsaa med de andre nordiske Folk kendte Ole Chievitz en inderlig Samhørighed. Tidligt fandt han Venner i Norge og Sverige, og især vandt han mange finske Hjerter, da han i Landets svære Stund mod Slutningen af den forrige Verdenskrig drog derop for, med sine særlige Forudsætninger som Krigslæge, at bidrage til at mildne de Saaredes Lod. De Frasagn om det Mod, han der udviste under de farligste Forhold, og den Tryghed han spredte om sig, baade dengang og senere da der atter kaldtes paa ham, vil sent glemmes.

At Ole Chievitz, der allerede i en tidlig Alder havde erhvervet sig saa stort et

Mesterskab i Kirurgiens Kunst, ogsaa blev inddraget i Arbejdet paa det Felt, som Straalebehandlingen har aabnet for Bekæmpelsen af nogle af de mest snigende og frygtede Sygdomme, og paa hvilket vi her i Danmark fra Finsens Foregangsvirke har saa store Traditioner, maatte komme som en naturlig Følge af hans vidtspændende Interesse for Naturvidenskaben i alle dens Grene. Hvor ofte er ikke hans Venner og Kolleger blevet slaaet af den Selvstændighed, hvormed han selv paa Omraader, der kunde synes at ligge hans egentlige Virkefelt fjernt, kunde finde Problemer og give Impulser til deres Løsning. Det maatte fylde enhver med Beundring, at han kunde finde Tid til at dyrke saa omfattende Studier og Kraft til at gennemføre et frugtbart og mangesidigt Forskningsarbejde til Trods for sin store og krævende Hospitalsgerning.

Som den sande Læge følte han jo forud for alt Ansvaret for de Syge, der var betroet til hans Varetægt, og han indskrænkede sig derved ikke til Sygdomsbehandlingen alene, men tog af sit fulde Hjerte Del i Patienternes Skæbne paa en Maade, der har vundet ham utallige Menneskers dybeste Taknemmelighed. Af alle sine Medarbejdere paa Hospitalet var han, der ved den ringeste Forsømmelighed kunde være saa streng, men som samtidig stillede de største Krav til sig selv og altid var rede til at anerkende hver oprigtig Stræben hos andre, beundret og elsket; for de mange unge Læger og medicinske Studerende, som han gennem Aarene har vejledet, var han saavel det inspirerende Forbillede som den faderlige Ven. Inden for den danske Lægeverden indtog han en enestaaende Stilling ved den Agtelse og Tillid, han nød i stadig voksende Maal, og med Aarene blev mange vigtige Samfundsopgaver lagt paa hans Skuldre.

Da den nye Storm over Verden trak op, var han, der saa nøje kendte Krigens

Rædsler, utrættelig i at advare sine Medborgere, og det var en Selvfølge, at det blev ham, hvem det i første Linie blev betroet at organisere de nødvendigste Foranstaltninger for at skaane Menneskeliv, hvis Uvejret ogsaa skulde naa vort Land. Da Danmarks mørke Tid kom og fremmede Voldsmænd trængte herind, maatte der i Bestræbelserne for at opretholde de Traditioner, der for os gør Livet værd at leve, og for at genvinde vor Frihed, blive Opgaver for enhver, alt efter de Muligheder der laa den enkelte nærmest. For Ole Chievitz var der efter hele hans Natur intet Valg, og uanfægtet af, hvad andre mente om det Ansvar deres Stilling indebar og den Besindighed der krævedes for at naa det fælles Maal, maatte han følge sin indre Stemme, og netop herved kom han da vel ogsaa til at yde det største, han i denne Nødens Tid kunde give os.

Hvilken Plads kom han ikke til at indtage i det danske Folks Hjerte ved den Uforfærdethed, han stedse lagde for Dagen, og det Mod, han indgød andre ved sin ildnende Tale. Det gjorde jo ogsaa det dybeste Indtryk, at en Mand, æret og agtet som han, skulde høre til de første, der blev fængslet af Fjenden og under Voldsmagtens Tryk maatte tages i Forvaring og dømmes af danske Myndigheder. Da den Tid kom, hvor den aktive Modstand saa helt kom til at præge saavel Forholdene herhjemme som det Billede Danmark frembød udadtil, var jeg selv langt borte fra Landet, men ude i det fremmede fulgte vi med Betagelse og Stolthed det Mod, der med saa tunge Ofre, men ogsaa med saa store Frugter, udvistes af stedse voksende Skarer under Frihedsraadets Ledelse. Ole Chievitz's Indsats i disse Aar er nu, efter Befrielsen, kendt af alle, men ingen af os, der ved Omstændighederne dengang var afskaaret fra at følge hans Færd, kunde tvivle om, at han vilde være at finde i de Kæmpendes første Række.

I de vanskelige Tider, som vi saa lykkeligt har gennemstaaet, maatte der jo efter Forholdenes Art opstaa Brydninger stammende fra Forskel i Opfattelsen af, med hvilke Midler Modstandsviljen bedst gaves Udtryk; men naar vi ser tilbage paa de store Hændelser, træder dog alt i Baggrunden for Tanken om det ubrydelige Sammenhold der voksede frem uanset Fordelingen af Byrder og Opgaver svarende til enhvers Evne og Syn. Netop for Styrkelsen af den Aand, der kædede hele det danske Folk sammen, blev Ole Chievitz's Betydning saa stor. Da de Maal, til hvilke han havde viet hele sin Kraft, og for hvilke han havde vovet sit Liv, var naaet, vendte han, trods Kald fra mange Sider, tilbage til sin gamle Gerning, hvor han til det sidste virkede til Hjælp for lidende Medmennesker med vanlig Pligttroskab og en Selvforglemmelse, der maaske blev skæbnesvanger for ham.

De Minder, som Ole Chievitz efterlader, og som med skarpe Træk tegner Billedet af en Mand af reneste Sind og højeste Værd, vil altid leve iblandt os. Maatte nu, hvor alle atter kan samles i endrægtigt Arbejde paa Fredens Værk, den fælles Beundring og Taknemmelighed bidrage til at forene hele Folket i vort Land, som han elskede saa højt og tjente saa vel.

TRANSLATION

Speech at the Memorial Ceremony for Ole Chievitz 31 December 1946

By N. Bohr[1]

The sorrow over the death of Ole Chievitz is shared by the whole Danish people, with whom the memory of his deeds in good and bad times will always be kept in gratitude and pride. Wide circles were privileged to feel the strength and warmth which shone forth from his personality, and many were those who were embraced by his faithful friendship. As one of those who has had the good fortune of being close to him throughout a lifetime, I have been allowed here at his bier to try to express how great a loss we have suffered and how much richness he leaves behind for us.

In the earliest years of Ole Chievitz's childhood, his parents, like mine, lived in the old Academy of Surgery in Bredgade, where his father, the highly-respected anatomist,[2] had his official residence right up under the roof of the beautiful building until a few years after Ole's birth, when the family had to move because of the rather outdated conditions. Our fathers had been close friends from their youth, but what proved most decisive for the relationship between Ole and me was surely that in the last six years before completing our gymnasium education together at Gammelholm Grammar School[3] we sat side by side at a shared school desk. Already in his school days he immediately won the respect of all through his independence and love of truth and his ability to get to the heart of any matter. With his urge to throw himself into battle for what he thought was right, Ole could often, of course, be a bone of contention among his contemporaries, but we all united in the admiration for his straightforwardness and fearlessness, which together with his enthusiasm and vigour made him such a good friend.

In his home Ole Chievitz received influences that would make a deep impact on his personality; in particular he attached himself fervently in many ways to his father, who was not only prominent as a researcher and a thinker, but in addition had lively historical and artistic interests. As a result of the

---

[1] Translation of the figure caption on p. [453]: "'Jens.' Ole Chievitz in disguise during the occupation period, when he had 'gone underground' and answered to the name 'Jens'."

[2] Johan Henrik Chievitz (1850–1901).

[3] Bohr attended this school from 1891 until 1903, when he entered the university.

incurable disease that should cause his much too early death, Ole's father was, however, already in our boyhood years obliged to live in increasingly greater isolation within the study walls, while his mother with so much bravery and so fresh spirits struggled with the everyday problems.[4] In these surroundings Ole Chievitz matured early to responsibility and to familiarity with the light and dark sides of the conditions of life.

However great a blow his father's death was for Ole, it did not, however, distance him in the least from the family's many friends, who all appreciated how much he had in him. In the Bjerrum home and the Ipsen home,[5] as well as that of my parents, he was embraced with the warmest feelings and found friends for life in the younger generation. Wherever he came, he brought with him a breath of fresh air, and because of all the many strings that sounded in his open and sensitive mind, any time spent with him gave joy and enrichment. From that time I remember in particular a journey my brother and I made together with Ole as young students and where on walking tours through the beautiful mountain areas and old picturesque villages of Tyrol we received the strongest impressions of his all-embracing feelings for humanity and nature and of the true manliness that so early characterized his whole being.

For his scientific interests and his urge to do something for the benefit of fellow beings that he brought from his home, Ole Chievitz had found the right focus in the studies whereby he prepared to become a medical doctor. The zeal and sense of responsibility he displayed won him not only friends and admirers among his fellow students, but he also quickly gained the trust and praise of the University teachers. In particular, it is well known how greatly Professor Rovsing,[6] who for many years was such an admired ideal for Ole, appreciated him, and which responsible tasks he entrusted to him already at an early time. Among the medical friends at the hospital Ole Chievitz was also to find the wife who with her splendid human understanding and comradeship in professional interests became such a great support for him, and for whom his death now, after they have for many years shared bright joys as well as deep sorrows, is so hard a blow and irreplaceable a loss, but who together with the rest of us has so great a treasure in the memories.[7]

It is quite impossible to express how much all of us who were close to Ole Chievitz owe him. I remember evenings at their home where Ole and Ingeborg

---

[4] Paula, née Johansen (1857–1928).
[5] The lawyer Johannes Ipsen (1843–1910) and his wife, Christel, née Schultz (1857–1945).
[6] The surgeon Thorkild Rovsing (1862–1927).
[7] Ingeborg, née Jacobsen (1882–1955).

year after year gathered the friends of their youth and those who had come later, and how he was able to let his special humour play between the deepest seriousness and the freest giddiness in a way that made every such evening a refreshing and unforgettable experience. From those years when, together with him, Niels Bjerrum[8] and Holger Hendriksen,[9] I was joint owner of a boat that was used for long voyages in the holidays, I also have precious memories of the intimate life on board, where the always sparkling torrent in Ole's rich mind brought us all so much renewal and encouragement. In particular, though, we think of him as the understanding and faithful friend from whom one could seek advice and find support in all vicissitudes of life and on whom one often relied as a rock in the whirlpool of existence.

In many ways Ole Chievitz was something quite unique, but at the same time all those features that characterize the seeking and struggling human being combined in a peculiar way in his whole personality. However passionate his attitude might be and his urge to put it into action, he possessed, with all his strength, a heart-warming gentleness which he, though often in vain, could make great efforts to conceal. In him, both the sense of justice and the feeling of affection were developed as in few, but he combined with this the ability of a great personality to judge without wavering what the circumstances of each moment required of him the most. It was quite beyond him to seek approval in advance; on the contrary, he could more or less make a show of trying to bring forth contradictions. On such occasions, which always gave rise to self-examination, one sensed perhaps most strongly the depth and warmth of his mind and came to be even more fond of him, if possible.

His entire human attitude had fixed roots which continually sought deeper. Not only did life bring him experience in rich measure, but in addition he possessed a rare sympathetic insight into all aspects of our culture and into its history, as well as a rich knowledge of Danish literature, from which he so often understood how to bring the right word to express his feelings and to put events that met him into an evocative light. Also with the other Nordic peoples Ole Chievitz had a strong sense of belonging. Early on he found friends in Norway and Sweden, and, in particular, he won many Finnish hearts when, in the difficult time for the country towards the close of the First World War, he went up there with his special qualifications as military doctor to contribute to tempering the lot of the wounded. The stories about the bravery he displayed there under the most dangerous conditions, and the peace of mind he spread

[8] See Bohr's tribute to Bjerrum, reproduced and translated on pp. [437] ff.
[9] (1875–1955), chemical engineer.

around him both then and later when he was again called upon, will long be remembered.

That Ole Chievitz, who already at an early age had acquired such great mastery of the art of surgery, was also drawn into the work in the field that radiation treatment has opened for combating some of the most insidious and feared diseases, and for which we here in Denmark, because of Finsen's[10] pioneering work, have such great traditions, was bound to come as a natural result of his wide-ranging interest in science in all its ramifications. How often have not his friends and colleagues been struck by the independence whereby he, even in fields that might seem to lie far from his real field of work, could find problems and provide impulses for their solution. It was bound to fill everyone with admiration that he could find time to cultivate such extensive studies and find energy to carry through a fruitful and many-sided research effort despite his great and demanding hospital activity.

As a true doctor he felt, of course, before everything else, responsibility for the sick who were entrusted to his care, and he did not thus confine himself to the treatment of illness alone but involved himself wholeheartedly in the fate of the patients in a way that has won him the deepest gratitude of countless people. He, who on the slightest sin of omission could be so severe, but at the same time made the greatest demands on himself and was always ready to acknowledge every honest striving in others, was admired and loved by all his colleagues at the hospital; for the many young doctors and medical students whom he has guided throughout the years he was the inspiring example as well as the fatherly friend. Within the world of Danish medicine he assumed a unique position with the respect and trust he enjoyed in a constantly increasing measure, and as the years passed many important tasks of society were laid upon his shoulders.

When the new tempest over the world gathered,[11] he, who knew the horrors of war so intimately, was tireless in warning his fellow citizens, and it was a matter of course that it was he who in the first instance was entrusted with the organization of the most necessary precautions for protecting human lives if the storm should also reach our country. When Denmark's dark hour arrived and foreign assailants forced their way into the country, the endeavours to preserve the traditions which for us make life worth living and to regain our freedom were bound to hold tasks for everybody, all according to the possibilities lying closest to each individual. For Ole Chievitz, according to his

---

[10] Niels Finsen (1860–1904), the first Danish Nobel Prize winner (1903).
[11] Bohr refers to World War II.

whole nature, there was no choice, and indifferent to what others thought about the responsibility inherent in their position and to the level-headedness required to reach the common goal, he had to follow his inner voice and precisely thus did he surely come to offer the most that he could give us in this hour of need.

What place did he not come to fill in the hearts of the Danish people with the bravery he always displayed and the courage he instilled in others with his inflaming words. It also made the deepest impression that a man, honoured and respected as he, should be among the first who were imprisoned by the enemy and under the pressure of the usurper should be taken into custody and sentenced by Danish authorities. When the time came that the active resistance so completely came to dominate the conditions here at home as well as the picture Denmark presented abroad, I was myself far away from the country, but in foreign parts we followed with fascination and pride the courage which, with such heavy sacrifices but also such great rewards, was demonstrated by ever growing numbers under the leadership of the Danish Liberation Council.[12] Ole Chievitz's effort in those years is now, after the liberation, known to all, but none of us who because of the circumstances were then cut off from following his activities, could doubt that he was to be found in the front line of the fighters.

In the difficult times which we have so fortunately survived, under the existing conditions conflicts were bound to arise, stemming from differences of opinion as to by which means the will to resist could best be expressed; but when we look back on the great events, everything is overshadowed by the thought of the unwavering concord that emerged regardless of the division of burdens and tasks according to everyone's ability and outlook. Ole Chievitz's significance was so great precisely for the strengthening of the spirit that united the whole Danish people. When the goals, to which he had dedicated all his strength and for which he had risked his life, were reached, he returned to his old profession despite calls from many quarters, where until the last he worked to the benefit of suffering fellow humans with his usual dedication and a self-effacement, which perhaps proved fatal for him.

The memories Ole Chievitz leaves behind, and which with sharp contours paint the picture of a man of the purest mind and highest value, will always live among us. May now, when all can again gather in harmonious work for the cause of peace, the shared admiration and gratitude contribute to uniting the whole people in our country which he loved so dearly and served so well.

---

[12] Established in mid-September 1943, the Liberation Council was the coordinator of Danish resistance.

LIII. WRITER AND SCIENTIST

DIGTER OG VIDENSKABSMAND
"Festskrift til Niels Møller paa Firsaarsdagen 11. December 1939",
Munksgaard, Copenhagen 1939, pp. 80–81

TEXT AND TRANSLATION

See Introduction to Part II, p. [127].

[461]

Digter og Videnskabsmand

af NIELS BOHR

For Mennesker, der søger at frigøre sig for den Begrænsning, som sættes os ikke alene af Dagliglivets snævre Skranker, men allerede af selve vor til de daglige Tildragelser afpassede Forestillingskreds, frembyder der sig to tilsyneladende vidt forskellige Udveje.

Nogle formaar at bøde paa Erfaringernes Fattigdom ved at give sig Fantasien i Vold og paa Baggrund af rige, indre Oplevelser at opnaa, at det Slør, hvormed Sædvaner og Fordomme fordunkler vort Blik, saa at sige af sig selv glider til Side og tillader dem at se ind i nye og skønnere Verdener. Andre er henvist til bevidst at stræbe efter at udvide Omfanget af vor Viden og derigennem efterhaanden klarlægge de almindeligvis upaaagtede Forudsætninger, hvorpaa Anvendelsen af selv vore mest elementære Begreber hviler.

Oftest vil man maaske sige, at den første Vej er den frie Digtekunsts og den sidste den metodiske Videnskabs; men, som enhver ved, er det meget svært at trække skarpe Grænser for Anvendelsen af Ord som »Kunst« og »Videnskab«, og netop fra nogle af de største Kunstneres og Videnskabsmænds Selvbekendelser erindrer vi vist alle Ytringer som, at det er en Videnskab at skabe den levende Kunst eller en Kunst at lede Videnskaben ind paa nye Baner.

Gang paa Gang har jo ogsaa videnskabelig Forskning, hvad enten det angik Studiet af Menneskeslægtens Fortidsminder eller Forsøgene paa at fravriste Naturen dens Hemmeligheder, ført os ind i Æventyrriger, saa skønne som noget Fantasiland, og samtidigt afsløret dybtliggende Træk af vor Erkendelses Væsen, som hidtil

— 80 —

var forblevet skjult for den skarpeste logiske Analyse, omend en kunstnerisk Intuition ofte havde bragt os Anelser derom.

Jeg tror, at Niels Møller selv vil finde, at vi har forstaaet ham ret, naar vi som Grundtonerne i alt hans Værk fornemmer en Skønhedslængsel og Sandhedsstræben forenet ved en usvigelig Evne til stedse at bevare Harmonien mellem Indholdets Rigdom og Formens Klarhed. Det allersmukkeste og mest rammende Udtryk for hans Opfattelse af Menneskets Stilling i Tilværelsens store Skuespil, hvor vi paa samme Tid er Tilskuere og Medspillere, har han maaske givet i den Kantate, som, hver Gang den ved Universitetsfesterne synges til Carl Nielsens Musik, paany griber os saa stærkt ved sin dybe og ægte Stemning.

Hvis det er lykkedes mig med disse faa Ord blot nogenlunde at udtrykke, hvad der ligger mig paa Hjerte, behøver jeg næppe at tilføje, at det ikke blot er vor Beundring for Alsidigheden i Niels Møllers Livsgerning, men først og fremmest vor Forstaaelse af Enheden i hans stærke Personlighed, der gør det saa svært for os at sige, om vi hos ham bedst genkender Digteren eller Videnskabsmanden.

Niels Møller

6

TRANSLATION

Writer and scientist

by NIELS BOHR

For people, who try to free themselves from the limitation set not only by the narrow bounds of daily life but already by our very world picture adapted to everyday incidents, there appear to be two widely different resorts.

Some are capable of making up for paucity of experience by surrendering to imagination, and on the basis of rich inner perceptions to reach the stage that the veil by which custom and prejudice darken our view draws aside of its own volition, so to speak, and allows them to see into new and more beautiful worlds. Others are obliged to consciously strive towards widening the scope of our knowledge and thus step by step clarify the commonly disregarded preconditions upon which the use of even our most elementary concepts rests.

Most often one would perhaps say that the first path is that of the free art of writing and the latter that of methodic science; but as everyone knows, it is very difficult to draw precise boundaries for the use of words such as "Art" and "Science", and precisely from the confessions of some of the greatest artists and scientists, we can all probably remember expressions such as: to create living Art is a Science, or, to lead Science down new paths is an Art.

Time and time again scientific research, whether it involved the study of mankind's ancient monuments or the attempts to wrest from nature its secrets, has, of course, led us into fairytale worlds, as beautiful as any realm of imagination, and has at the same time revealed fundamental traits of the character of our knowledge, which had hitherto remained hidden to the sharpest logical analysis, even though an artistic intuition had often given us some inkling thereof.

I think that Niels Møller himself would agree that we have understood him correctly when, as the keynotes in all his work, we sense a longing for beauty and a striving for truth combined with an unerring ability to always preserve the harmony between the richness of the content and the clarity of the form. The most beautiful and most telling expression of his view of the position of humans in the great drama of life, where we are at the same time spectators and actors, he has perhaps given in the cantata which, every time it is sung to

Carl Nielsen's music at University ceremonies, once again moves us so strongly with its deep and genuine mood.[1]

If I have been at least moderately successful with these few words in expressing what I want to convey, I need hardly add that it is not only our admiration for the versatility in Niels Møller's lifework, but first and foremost our understanding of the unity in his strong personality, that make it so difficult for us to say whether we in him best recognize the writer or the scientist.

Niels Bohr

[1] N. Møller and C. Nielsen, *Kantate ved Universitetets Aarsfest, 29. Oktober 1908.*

LIV. MY NEIGHBOUR

MIN GENBO
"Halfdan Hendriksen: En dansk Købmand og Politiker",
Aschehoug, Copenhagen 1956, pp. 171–172

TEXT AND TRANSLATION

See Introduction to Part II, p. [128].

[467]

MIN GENBO

af professor, dr. phil. & sc. & techn. *Niels Bohr*.

Som et lille bidrag til denne bog, hvor medarbejdere og venner bringer Halfdan Hendriksen en hyldest på hans 75 års dag, vil jeg gerne gengive nogle ord, jeg udtalte ved Carlsbergfondets fest, da han fyldte 70 og nærmede sig afslutningen af sin virksomhed som Bryggeriernes administrerende direktør.

Det er et sjældent stort og rigt livsværk i det danske samfunds tjeneste, hvorpå Halfdan Hendriksen ved sin 70 års fødselsdag kan se tilbage. Den styrke og det udsyn, som han fra sin første ungdom på mangfoldige måder har lagt for dagen, har fra alle sider vundet en anerkendelse og en beundring, der har givet sig udtryk i de talrige tillidshverv, der har været lagt på hans skuldre.

Halfdan Hendriksens kloge og uforfærdede bestridelse af gerningen som handelsminister i de for landet så vanskelige år vil længe erindres, og det føltes da også i vide kredse som en lykke, at netop en mand som han efter krigen var rede til at overtage stillingen som administrerende direktør for Carlsberg Bryggerierne, der i vort samfund indtager en så enestående plads ikke alene på grund af virksomhedens store omfang, men især ved de rige midler, den skænker til fremme for danske bestræbelser på kulturens område.

Om Halfdan Hendriksens krævende og ansvarsfulde gerning har formanden for Carlsbergfondets direktion talt så smukke og vægtige ord, og fra en af hans nærmeste medarbejdere i Bryggeriernes ledelse har vi fået et dybt indtryk af den beundring og tillid, som man føler for ham inden for hele den store virksomhed.

For at tale om Halfdan Hendriksens ledelse af Bryggerierne har jeg jo ingen sådanne forudsætninger; men som en af dem, der har deres hjem på Carlsberg, vil jeg gerne sige et par ord om den menneskelige ånd, der lever derude og som er baggrund for den følelse af tryghed og hygge, som vi alle så umiddelbart føler.

171

Jeg tror, at der på Carlsberg altid har hersket en særlig menneskelig indstilling, der stammer fra stifternes store personligheder og hvis
fortsatte bevarelse hviler på virksomhedens trivsel og bevidstheden
om de indtjente midlers betydningsfulde anvendelse i hele samfundets
tjeneste. Fra første færd følte min hustru og jeg en hjemlig atmosfære,
da vi for snart 20 år siden fik vor bolig inden for Bryggeriernes område. At besættelsestiden også på Carlsberg fik forstyrrende virkninger, kunne jo ikke undgås, og efter krigens ophør krævedes derfor en
kraftig og lykkelig hånd til at bringe alt i lave og genoprette den gamle
harmoni.

Af hvad der i denne henseende er nået under Halfdan Hendriksens
ledelse, har vi alle, der lever derude, fået et stærkt indtryk. Når man,
omend uden at tage del i virksomheden, færdes på Bryggerierne,
mærkes overalt en arbejdsglæde, der ikke alene skyldes tillid til den
faglige ledelse, men også bevidstheden om en omsorg for alles vel, der
bunder i en menneskelig forståelse hos det øverste styre.

Alle, der kender den smittende livsglæde og opmuntrende venlighed, som Halfdan Hendriksen forbinder med så store krav til sig selv,
vil forstå, hvad der udstråler fra ham og genspejles på så mange
måder i hele virksomheden. I denne forbindelse tænker vi også på
den varme menneskelige støtte, som Halfdan Hendriksen har fundet
i sit eget hjem, således som det så smukt er kommet til udtryk i de
ord, der i aften har været rettet til fru Hendriksen. Vi, der bor og
færdes derude, føler trang til at udtrykke vor taknemmelighed over
for Dem begge, og ønsket om, at den forståelsens og samarbejdets
ånd, der lever på Carlsberg, også i kommende tider må bevares og
trives.

På Halfdan Hendriksens 75 årsdag vil jeg gerne til disse ord føje
de hjerteligste ønsker om, at han endnu i mange år må bevare sin
fulde kraft, hvorpå der stadig er lagt så stort beslag, og den livsglæde,
der er ham egen og hvormed han bringer opmuntring til så mange.
Jeg føler også trang til at være med til at bringe Halfdan Hendriksen
en personlig tak for den hjælpsomhed, hvormed han inden for de vide
felter, som han omfatter med så varm interesse, altid er rede til at
støtte enhver sag, hvis fremme efter hans overbevisning er til samfundets gavn.

Niels Bohr.

172

TRANSLATION

MY NEIGHBOUR
by Professor *Niels Bohr*, D.Phil., D.Sc., D.Tech.

As a small contribution to this book, where colleagues and friends bring Halfdan Hendriksen greetings on his 75th birthday, I would like to reiterate some words I spoke at the Carlsberg Foundation celebration when he had his 70th birthday and approached the end of his activities as the Managing Director of the Breweries.[1]

On his 70th birthday Halfdan Hendriksen can look back on an unusually great and rich lifework in the service of Danish society. The strength and the foresight that he has displayed in a great many ways since his early youth have from all quarters won a recognition and an admiration which have been demonstrated in the numerous honorary offices that have been laid on his shoulders.

Halfdan Hendriksen's wise and undaunted handling of the task as Minister of Trade in the years so difficult for the country will long be remembered,[2] and it was also considered fortunate in wide circles that precisely a man such as he was prepared after the war to take over the position of Managing Director for the Carlsberg Breweries, which occupies such a unique place in our society, not only because of the great size of the enterprise, but, in particular, because of the rich funds it donates for the advancement of Danish endeavours in the cultural sphere.

The Chairman of the Carlsberg Foundation's Board of Directors[3] has spoken such beautiful and weighty words about Halfdan Hendriksen's demanding and responsible work, and from one of his closest colleagues in the management of the Breweries we have received a deep impression of the admiration and trust which is felt for him inside the whole great enterprise.

I have, of course, no such qualifications to speak about Halfdan Hendriksen's management of the Breweries; but as one of those who have their home at Carlsberg, I would like to say a few words about the human atmosphere that is

---

[1] Only a manuscript for the entire present contribution exists in the BMSS. N. Bohr, [Manuscript], *Inventory of Relevant Manuscripts in the Niels Bohr Archive*, below, p. [531], folder 48.

[2] Hendriksen was Minister from October 1940.

[3] In 1951, when Bohr gave his talk, Johannes Pedersen filled this position.

found out there and is the background for the feeling of security and comfort which we all so spontaneously feel.

I think that at Carlsberg a special human attitude has always reigned, which stems from the great personalities of the founders, and whose continued presence rests on the prosperity of the enterprise and the consciousness of the meaningful use of the income earned for the benefit of society as a whole. From the very beginning my wife and I felt a homely atmosphere when we nearly 20 years ago were given our house in the grounds of the Breweries. It could not be avoided that the Occupation[4] had disturbing effects also on Carlsberg, and after the end of the war a strong and fortunate hand was therefore necessary to put everything right and reestablish the old harmony.

All of us who live out there have received a strong impression of what has been achieved in this regard during Halfdan Hendriksen's leadership. When one, even though not being part of the enterprise, goes around at the Breweries, a pleasure in work is felt everywhere, which is not only due to trust in the professional leadership, but also to the consciousness of a caring for the well-being of all, which arises from a human understanding on the part of the top management.

All who know the infectious happiness and encouraging friendliness that Halfdan Hendriksen combines with such great demands on himself, will understand what radiates from him and is mirrored in so many ways in the entire enterprise. In this connection we also think of the warm human support that Halfdan Hendriksen has found in his own home, such as it has been expressed so beautifully in the words addressed to Mrs Hendriksen[5] this evening. We who live and go around out there feel an urge to express our gratitude to both of you, as well as the wish that the spirit of understanding and cooperation that lives at Carlsberg may be preserved and thrive also in times to come.

On Halfdan Hendriksen's 75th birthday I would like to add to these words the most heartfelt wishes that he for many more years may keep his full energy, on which so many demands are still made, and the happiness that is special to him and whereby he brings encouragement to so many. I also want to join in bringing Halfdan Hendriksen personal thanks for the helpfulness whereby, within the wide areas that he embraces with such warm interest, he is always prepared to support any cause whose advancement in his opinion is to the benefit of society.

Niels Bohr.

---

[4] The German occupation of Denmark from 9 April 1940 to 5 May 1945.
[5] Frida, née Riis (1893–1961).

[472]

LV. A FRUITFUL LIFEWORK

ET FRUGTBART LIVSVÆRK
"Noter til en mand: Til Jens Rosenkjærs 70-aars dag"
(eds. J. Bomholt and J. Jørgensen),
Det Danske Forlag, Copenhagen 1953, p. 79

TEXT AND TRANSLATION

See Introduction to Part II, p. [128].

[473]

ET

FRUGTBART

LIVSVÆRK

Jens Rosenkjærs 70-aars dag føles af mange som en kærkommen lejlighed til at give udtryk for taknemmelighed
for hans utrættelige virke for at bringe oplysning ud til
vide kredse og for at fremme forstaaelsen af dennes betydning for højnelsen af samfundets kultur.

Enhver af os, der har haft den glæde at samarbejde under
den ene eller anden form med Jens Rosenkjær, beundrer og
værdsætter hans fine personlighed, hans aabne sind og hans
varme begejstring for de værdier, der betinger harmoni og
rigdom i menneskelivet.

Alle vil vi derfor paa denne dag bringe Jens Rosenkjær
en hjertelig lykønskning til hans frugtbare livsværk til gavn
for vort samfund og ønske ham endnu mange lykkelige
arbejdsaar.   NIELS BOHR

TRANSLATION

A
FRUITFUL
LIFEWORK

Jens Rosenkjær's 70th birthday is felt by many as a welcome opportunity to give expression to gratitude for his tireless activity to bring enlightenment out to wide circles and to advance the understanding of its importance for raising the cultural level of society.

Any of us who have had the pleasure of cooperating with Jens Rosenkjær in one way or another admires and values his fine personality, his open mind and his fervent enthusiasm for the values that assure harmony and richness in human life.

All of us will therefore on this day convey to Jens Rosenkjær heartfelt congratulations for his fruitful lifework to the benefit of our society and wish him many more happy working years.   NIELS BOHR

LVI. OBITUARY

MINDEORD
"Bogen om Peter Freuchen"
(eds. P. Freuchen, I. Freuchen and H. Larsen),
Fremad, Copenhagen 1958, p. 180

TEXT AND TRANSLATION

See Introduction to Part II, p. [128].

NIELS BOHR:

Mindeord

Som en af de mange, hvem venskabet med Peter Freuchen og læsningen af hans bøger har glædet og beriget, vil jeg gerne med et par ord bidrage til dette mindeskrift. Allerede i vore studenterdage fik min bror og jeg gennem samværet i Akademisk Boldklub et levende indtryk af Freuchens strålende oplagthed og mangesidige begavelse. Hans trang til spændende oplevelse skulle jo snart få fuld udløsning ved deltagelsen i den store forskerfærd i arktiske egne, der lededes af Knud Rasmussen, hvis stærke og varme personlighed Freuchen sluttede sig til med største beundring og hengivenhed. Det nære samliv med polareskimoerne og indtrykket af deres under de strenge livsvilkår udviklede særprægede kultur kom ikke alene til at gribe dybt ind i Freuchens skæbne, men åbnede tillige de kilder i hans modtagelige sind, der gav ham inspirationen til hans frodige forfatterskab. Hvadenten han i spøg og alvor beskrev sit eget livseventyr i Grønland, eller han i friere digtning behandlede emner som eskimoernes samkvem med amerikanske hvalfangere og Alaska-indianernes kampe med russiske kolonister, bragte han altid bud om sider af menneskelivet, som vi trods al forskel i kår og livsvaner kunne genfinde i os selv. Peter Freuchens vidtspændende interesser og frie, hjælpsomme indstilling førte ham i årenes løb som få andre i virksom forbindelse med vide kredse både hjemme og ude. Hvert sammentræf med ham, hvor sjældent og tilfældigt det end kunne være, blev derfor en forfriskende oplevelse, der gav anledning til eftertanke og i taknemmelighed bevaredes i erindringen.

Niels Bohr.

180

TRANSLATION

NIELS BOHR:

Obituary

As one of the many for whom the friendship with Peter Freuchen and the reading of his books have brought joy and enrichment, I would like to contribute with a few words to this memorial volume. Already in our student days my brother and I received, through being together in Academic Ball Club, a vivid impression of Freuchen's radiant spirit and his multi-faceted talent. His urge for exciting adventure would, of course, soon be fully satisfied by his participation in the great expedition in the Arctic regions, led by Knud Rasmussen, whose strong and warm personality Freuchen embraced with the greatest admiration and devotion. The close life together with the polar Eskimoes and the impression of their unique culture, developed under severe living conditions, were not only going to affect Freuchen's own fate deeply, but also opened those sources in his impressionable mind that gave him the inspiration for his rich authorship. Whether describing his own adventures in Greenland in jest and in seriousness,[1] or dealing in a freer artistic form with topics such as the relations of the Eskimoes with the American whalers and the battles of the Alaska Indians against Russian colonists,[2] he always told of aspects of human life which we can recognize in ourselves, despite all differences in conditions and habits of life. Peter Freuchen's wide-ranging interests and his independent helpful attitude led him, as few others, in the course of the years to an active contact with wide circles both at home and abroad. Each meeting with him, however rare and unexpected, therefore became a refreshing experience, which gave rise to reflection and was remembered with gratitude.

Niels Bohr.

---

[1] See, for example, P. Freuchen, *Arctic Adventure: My life in the Frozen North*, Farrar & Rinehart, New York 1935.
[2] See the novel P. Freuchen, *Eskimo*, Grosset & Dunlap, New York 1931, subsequently made into a Hollywood film (1933).

LVII. FAREWELL TO SWEDEN'S AMBASSADOR
IN COPENHAGEN

AFSKED MED SVERIGES GESANDT I KØBENHAVN
Politiken, 15 November 1941

To C.F.H. Hamilton

TEXT AND TRANSLATION

See Introduction to Part II, p. [129].

[481]

Afsked med Sveriges Gesandt i København

Friherre Hamilton forlader i Dag Danmark

Friherre Hamilton fotograferet i Gaar ved sit Arbejdsbord under Udøvelsen af sine sidste Embedsfunktioner i Danmark.

Den svenske Gesandt, Friherre *Hamilton*, forlader i Dag Danmark. *Politiken* har anmodet Professor Niels Bohr om i en Hilsen at give Udtryk for de Følelser, hvormed Sveriges Gesandt har været omfattet her i Landet. Til Professor Bohrs Hilsen føjer *Politiken* sin egen.

ENHVER, som i disse Aar paa den ene eller anden Maade har deltaget i Bestræbelserne for Opretholdelsen af den kulturelle Forbindelse mellem Sverige og Danmark, hvis Betydning netop nu er større end maaske nogen Sinde, har følt den dybeste Taknemmelighed over den forstaaelsesfulde og hjælpsomme Støtte, vi altid har fundet hos Friherre *Carl F. H. Hamilton*, der har virket i syv Aar som svensk Minister i København, og som nu forlader os for at overtage en betydningsfuld Post i sit Hjemland.

For alle de mange, der har haft Lejlighed til at lære Friherre Hamilton at kende, staar hans kultiverede og ridderlige Personlighed som et Symbol paa de smukkeste og værdifuldeste svenske Traditioner. Hans fine Dannelse og varme Menneskelighed lagde sig for Dagen i al hans Færd og gjorde enhver Samtale med ham til en Berigelse, hvad enten det drejede sig om en Sag, hvor man søgte Ministerens Raad og Bistand, eller Talen var om kunstneriske og videnskabelige Emner, som han omfattede med saa stor og alsidig en Interesse.

Den Hjertelighed og Venlighed, hvormed saavel Ministeren som Friherreinde Hamilton kom alle hernede i Møde, skabte ikke alene en dyb Sympati om dem selv, men bidrog tillige til stadig at holde os den Broderhaand for Øje, som fra Sverige udstrækkes imod os. De mange personlige Venner, som de har vundet i Danmark, vil ved deres Bortrejse følge dem med taknemmelige Tanker og vore varmeste Ønsker.

NIELS BOHR.

TRANSLATION

Farewell to Sweden's Ambassador in Copenhagen

Baron Hamilton leaves Denmark today[1]

Baron Hamilton, the Swedish ambassador, leaves Denmark today. Politiken has asked Professor Bohr to express in a greeting the feelings with which Sweden's ambassador has been regarded in this country.

Anyone who in one way or another has participated during these years in the endeavours to maintain cultural links between Sweden and Denmark, whose importance is greater at present than perhaps ever before,[2] has felt the deepest gratitude for the understanding and helpful support we have always found in Baron Carl F.H. Hamilton, who has served for seven years as Swedish Minister in Copenhagen and who now leaves us to assume an important position in his native country.[3]

For all the many people who have had the opportunity of becoming acquainted with Baron Hamilton, his cultivated and chivalrous personality is a symbol of the most beautiful and valuable Swedish traditions. His fine manners and warm humanity were apparent in all his actions and made any conversation with him rewarding, whether it was on a matter where one sought the Minister's advice and support or the talk was about artistic and scientific topics, which he embraced with so great and comprehensive an interest.

The sincerity and friendliness with which both the Minister and Baroness Hamilton[4] met everybody have not only created a deep sympathy for them personally, but have also contributed to constantly keeping in view the brotherly hand which is stretched out to us from Sweden. The many personal friends they have won in Denmark will on their departure follow them with grateful thoughts and our warmest wishes.

NIELS BOHR.

---

[1] Translation of the figure caption on p. [482]: "Baron Hamilton photographed yesterday at his desk as he performs his last official duties in Denmark."

[2] Bohr wrote this during the German occupation of Denmark, when Sweden was neutral.

[3] Hamilton (1890–1977) was appointed Governor of the Swedish County of Östergötland south-west of Stockholm, a post he held until 1956.

[4] Countess Margareta (Märta) De la Gardie (1893–1955).

LVIII. [TRIBUTE TO WEISGAL]

"Meyer Weisgal at Seventy" (ed. E. Victor),
Weidenfeld and Nicolson, London 1966, pp. 173–174

See Introduction to Part II, p. [130].

Meyer Weisgal and Niels Bohr at the Weizmann Institute, Israel, 1953.

From The World of Science

Eminent scientists need little introduction in this science-oriented day and age; Nobel laureates still less. There are few people in the enlightened world today remaining unaware of the identity of those who are extending the frontiers of human knowledge and experience into areas of the unknown which none but the boldly imaginative would have dreamed about even two decades ago.

Among these giants of the intellect was the late Niels Bohr, of Copenhagen, who was awarded the Nobel Prize for Physics in 1922 and has been hailed as one of the founding fathers of the Atomic Age. Professor Bohr first visited the Weizmann Institute in 1953, when he was made an Honorary Fellow, and again five years later as guest of honour at the opening of the Institute of Nuclear Science. He attended the London Conference of the Institute's Board of Governors in November 1961.

His tribute to Meyer Weisgal was written some years before he passed away.

One can hardly imagine a more striking evidence of the courageous and purposeful effort of creating a consolidated and respected position for the new community of Israel than the wonderful development of the Weizmann Institute of Science in Rehovoth.

At a place which only twenty-five years ago was a sandy desert, one finds now flower-beds and orange groves surrounding magnificently equipped laboratories in which eminent scientists are working with great success in many branches of research.

Everybody who in one or another way has had the privilege of a closer connection with the Weizmann Institute has received a deep impression of how much the success of the whole enterprise owes to Meyer Weisgal. Never has Weisgal been afraid of the problems which new plans of progress entailed, and always he found ways to overcome any difficulty. His ability to

173

obtain the required support rests not merely on his convincing argumentation and his confidence in the future of Israel, but above all on his power to convey his enthusiasm to his collaborators and to the friends of the Weizmann Institute.

In all such respects, Meyer Weisgal's special resource has been the broad human understanding which so often finds expression in that balance between humour and earnestness, which is so characteristic of his personality and which makes every meeting with him an unforgettable and stimulating experience.

NIELS BOHR

PART III

SELECTED CORRESPONDENCE

INTRODUCTION

The letters to and from Bohr, quoted in the Introductions to Parts I and II, are reproduced here in the original language, arranged in alphabetical order according to correspondents. Letters in Danish are followed by a translation.

Discretion has been used tacitly to correct "trivial" mistakes, e.g., in spelling and punctuation. An effort has been made, however, to preserve "characteristic" mistakes.

In the reproduction of the letters we have attempted to make the layout of letterheads etc. correspond as closely as possible to that of the original letters.

The list preceding the letters includes references to the pages in the Introductions where the letters are referred to. A page number in parentheses indicates that the letter is cited, not quoted.

The letters are held in the NBA, primarily in the Niels Bohr Private Correspondence.

The explanatory footnotes have been added by the editor.

CORRESPONDENCE INCLUDED

	Reproduced p.	Translation p.	Quoted p.
T.C. HODSON			
Bohr to Hodson, 24 March 1926	508	–	10
R.J.L. KINGSFORD			
Kingsford to Bohr, 24 March 1960	509	–	15
Bohr to Kingsford, 29 March 1960	510	–	15
KIRSTINE MEYER			
Meyer to Bohr, 25 October 1931	511	511	122
Bohr to Meyer, 11 October 1936	512	512	123
NIELS MØLLER			
Bohr to Møller, 11 March 1934	513	514	(127)
Bohr to Møller, 11 December 1934	514	515	127
IVAN SUPEK			
Bohr to Supek, 9 May 1959	516	–	105
Bohr to Supek, 1 June 1959	517	–	105
SVEN WERNER			
Werner to Bohr, 9 September 1942	518	519	14

EDWARD NEVILLE DA COSTA ANDRADE

BOHR TO ANDRADE, 18 February 1928
[Carbon copy]

February 18, [192]8.

Dear Andrade,

I am sorry to write so late, and even afraid that this letter will disappoint you. Ever since my return to Copenhagen and up to the very latest days, however, all my time has been occupied with duties not to be postponed. In the last days I have tried to do my best as regards the revision of the article for the Encyclopaedia Britannica. Still I have found it difficult due to the rapid progress of the subject to re-write the article according to present views. As the old article still indicates the limits of representing the subject when use is only made of the elementary classical concepts familiar to the general reader, I have therefore found it most convenient to return to my old proposal of writing a separate addendum. I enclose such an addendum, but of course I do not know if you and the directors will be satisfied with this procedure.[1]

> With kindest regards,
> Yours sincerely,
> [Niels Bohr]

PIERRE AUGER

BOHR TO AUGER, 26 February 1949
[Carbon copy]

February 26, [19]49.

Dear Auger,

I am sending you a contribution to the Einstein tribute by UNESCO which I have written in English. As you will see, I have on the whole confined myself

---

[1] The "addendum" became the last section, *Recent Progress*, in N. Bohr, *Atom*, Encyclopædia Britannica, 14th edition, Vol. 2, London and New York 1929–32, pp. 642–648. Reproduced on pp. [41] ff.

to an account of Einstein's great scientific achievements and their background. Of course, I do not know your plans, but I thought that it might be most practical that my contribution in the broadcasting preceded those of Hadamard[2] and Compton[3] who will probably speak more of Einstein's social activities.

I have had very little time to work on the manuscript and I am not too well satisfied yet. I propose, however, to make a Danish translation myself which will be broadcasted here and perhaps also in Norway and Sweden. Likewise, I would prefer that the record of the English text be spoken in here by myself and forwarded to the places which you may direct us. This procedure would also give me the opportunity to correct the text at smaller points. Will you please send me answer by cable so that I can immediately contact the Danish radio which will then at once arrange all technical problems.

With kind regards,
Yours,
[Niels Bohr]

Professor Pierre Auger
UNESCO
19, Avenue Kléber
Paris 16ᵉ.

BOHR TO AUGER, 3 March 1949
[Carbon copy]

March 3rd [19]49.

Dear Auger,

I hope that in due time you received the letter with my manuscript which was posted by airmail last Sunday. Until now I am without any answer, but I am cabling you this afternoon about my consultations with the Danish radio officials who are quite prepared to make both an English and a Scandinavian record of my talk as soon as we have your instructions. In this letter I am enclosing a new copy of my contributions in which, as you will see, I have introduced a few smaller corrections to the language.

---

[2] Jacques Hadamard (1865–1963), French mathematician.
[3] Arthur Holly Compton (1892–1962), American physicist.

On receipt of your letter regarding the vacancy in the physics section of the French Academy I have at once written to Cotton[4] and I need not say that it was a pleasure to me most warmly to recommend your candidature.

With kind regards,
Yours
[Niels Bohr]

Professor Pierre Auger
12 rue Emile Faguet
Paris XIV.

P.S.   Just this moment I have received your cable of yesterday[5] and I am very happy that you found my contribution suitable. The small corrections in the manuscript aim at improving the language and making the text a little more clear, and I should be grateful if these improvements could still be introduced. My reason for suggesting to have an English record taken up here was only that I have been a friend of Einstein for many years and that he would recognize my voice, but this point is, of course, of very small importance and has to be abandoned if not convenient.

NIELS BJERRUM

BOHR TO BJERRUM, 29 December 1934
[Handwritten]

Möckelsnæs[6]
29–12–1934.

Kære Bjerrum,
    Margrethe og jeg og Børnene sender Dig og Dine vore inderligste Ønsker for det kommende Aar. Næppe havde nogen af os tænkt at det Venskab der

---

[4] Aimé Auguste Cotton (1895–1951), French physicist.
[5] Auger's cable is held in the same file in the NBA as the present letter.
[6] Möckelsnäs Herrgård at the lake Möckeln in southern Sweden, where Bohr was spending the Christmas holidays, has a history beginning in the fifteenth century.

i saa mange Aar har forbundet os og vore Familier skulde kunne blive mere varmt og dog føler baade Margrethe og jeg at Minderne om Din Ellen[7] og vor Christian[8] for hele Livet vil knytte os alle endnu fastere til hverandre.

Din

Niels Bohr

Translation

Möckelsnæs[6]

29 December 1934

Dear Bjerrum,

Margrethe and I and the children send you and yours our most sincere wishes for the coming year. None of us could hardly have thought that the friendship that for so many years has tied us and our families together should be able to be warmer, and all the same both Margrethe and I feel that the memories of your Ellen[7] and our Christian[8] will for life tie us all even closer to each other.

Yours

Niels Bohr

BJERRUM TO BOHR, 2 January 1935
[Handwritten]

2.1.1935

Kære Bohr.

Jeg har været meget rørt og glad over de Ord, hvormed Du ledsagede Dine og Margretes Ønsker for det nye Aar. Christians Død hviler tungt paa mig, og det er ikke let at skulle vænne sig til Livet uden Ellen. I denne Tid er den bedste Hjælp at mærke Børnenes Hengivenhed og Vennernes Trofasthed og ganske særlig Din som Christians Far.

Gid det nye Aar maa forløbe godt for Margrethe og Dig og Eders Børn.

Din trofast hengivne

Niels Bjerrum

---

[7] Ellen Emilie, née Dreyer (1885–1934), had died on 5 November 1934.
[8] Christian Bohr (1916–1934), Bohr's eldest son, had died in July 1934.

Translation

2 January 1935

Dear Bohr

I have been very touched and happy about the words with which you accompanied your and Margrethe's wishes for the New Year. Christian's death weighs heavily on me, and it is not easy to have to get used to life without Ellen. In this time the best help is to feel the devotion of the children and the faithfulness of friends and quite especially yours as Christian's father.

May the New Year go well for Margrethe and you and your children.

Your sincerely devoted
Niels Bjerrum

FELIX BLOCH

BOHR TO BLOCH, 15 June 1932
[Carbon copy, which does not contain the symbols and quotations added by hand in the original letter, which seems to have been lost. They have been tentatively inserted by the Editor in accordance with Bohr's manuscript subsequently submitted to Die Naturwissenschaften (see pp. [21] ff.).]

15. Juni [19]32.

Kære Dr. Bloch,

Tak for Deres venlige Brev. Vi savner Dem alle meget; ikke mindst i disse Dage vilde jeg forfærdelig gerne kunne snakke med Dem om nogle Spørgsmaal vedrørende Metalledningsevnen, som jeg er blevet meget interesseret i. Det drejer sig nemlig om en Tanke vedrørende Supraledningsevnen, som jeg har faaet og ikke kan slippe, selv om jeg langtfra forstaar Overgangen mellem Supraledningsevnen og den almindelige Elektricitetsledning. Lad os betragte et isoleret Metalstykke ved absolut Nulpunkt for Temperaturer, og lad Elektronernes Tilstand være givet ved en Bølgefunktion

$$\Psi(0) = \varphi(x_1, \ldots, z_N; \sigma_1, \ldots, \sigma_N)e^{i\frac{\pi}{\hbar}E_0 t} \ ,$$

hvor φ er antisymmetrisk i Elektronernes Koordinater. Medens denne Tilstand er Strømfører, vil der være Løsninger, der kun afviger lidt fra den, og som kan skrives paa Formlen

$$\Psi(\delta) = (\varphi + i\delta\psi_x)e^{i\frac{\delta}{a}(x_1 + \cdots + x_N)}e^{\frac{i}{\hbar}(E_0 + \delta^2 \frac{\hbar^2 N}{2ma^2}A_x)t} \ ,$$

hvor δ er en lille Størrelse, og hvor $\delta\psi_x$ er en antisymmetrisk Funktion, der tilfredsstiller Ligningen

[many-particle Schrödinger equation][9]

der tillige tjener til Bestemmelse af A_x. En saadan Tilstand, der naturligvis kræver Tilførsel og stadig Borttagning af elektrisk Ladning ved Metalstykkets Begrænsning, svarer til en Strømstyrke, der er lig med

$$S_x = \delta \frac{\varepsilon\hbar N}{maV}A_x \ .$$

Paa Grund af den store Værdi af N/V kan denne Strømstyrke blive meget stor, selv om δ er overordentlig lille. Det forekommer mig nu nærliggende at antage, at den beskrevne Tilstand svarer til Supraledningsevnen, og paa Grund af den store Bølgelængde a/δ vil Strømmen ikke væsentligt forstyrres af thermiske Svingninger i Metalgitre. Derimod maa man antage, at den omhandlede koordinerede Bevægelse ved en vis Temperatur ophører at bestaa paa Grund af Elektronernes thermiske Forstyrrelser. Overgangen til en Tilstand, hvor Elektronerne bevæger sig ukoordinerede mellem hverandre, skulde der være nøje analog til en Smeltning af et fast Legeme, men det har hidtil ikke været muligt for mig at gøre mig Forstaaelsen af denne Overgangsproces klar.

Den Antagelse, at det ved Supraledningsevnen drejer sig om en koordineret Bevægelse af hele Elektrongitteret, er jo gammmel,[10] men ligesom først Kvantemekanikken har gjort det muligt at bringe Forestillingen om de "frie" Elektroners Tilstedeværelse i Metaller i nærmere Forbindelse med Erfaringen, synes det ogsaa, at man først gennem Kvantemekanikken paa den ovenfor antydede Maade kan forstaa, hvorledes de to Gitre kan bevæge sig gennem hverandre uden Modstand og væsentlig Deformation. Jeg skal være meget glad for at høre et Par Ord fra Dem, om hvordan De ser paa det hele. Iøvrigt har Rosenfeld

[9] The explicit form of the equation given by Bohr has been lost.
[10] Bohr refers to the work of the British physicist Frederick Lindemann. See p. [26], ref. 5.

og jeg været meget flittige og først for faa Dage siden faaet Romberetningen afsendt i endelig Form.[11] Vi har slidt i det og haft mange livlige Diskussioner om Straalingsproblemerne, men jeg nøjedes dog til Slut med at indføre et Par Bemærkninger, der kort antyder det Synspunkt, som vi allerede talte om ved Juletid, og som jeg beskrev i mit Brev til Pauli.[12] Hvordan gaar det forøvrigt med Deres eget Arbejde? Det vilde glæde mig meget ved Lejlighed at høre derom. Jeg selv har tænkt en Del over Stødproblemerne og navnlig over Atomsønderdelingen ved Protoner, som frembyder et smukt Eksempel paa Afvigelsen mellem den kvantemekaniske Behandling og den klassiske.[13]

Med mange venlige Hilsener til Heisenberg og Dem selv fra os alle,

Deres hengivne
[Niels Bohr]

Translation

15 June [19]32.

Dear Dr. Bloch,

Thank you for your kind letter. We all miss you very much; not least at the present time, I would terribly much like to be able to speak to you about some questions regarding the conductivity of metals, in which I have become very interested. It concerns an idea regarding superconductivity, which I have got and cannot let go of, although I am far from understanding the transition between superconductivity and normal conduction of electricity. Let us consider an isolated piece of metal at absolute zero temperature, and let the state of the

[11] Bohr refers here to the manuscript for the edited published version of his discussion remarks at the Volta Conference on nuclear physics held in Rome in October 1931: N. Bohr, *Atomic Stability and Conservation Laws* in *Atti del Convegno di Fisica Nucleare della "Fondazione Alessandro Volta", Ottobre 1931*, Reale Accademia d'Italia, Rome 1932, pp. 119–130. Reproduced in Vol. 9, pp. [99]–[114].

[12] Bohr to Pauli, 8 December 1931, BSC (24.2). Reproduced in W. Pauli, *Scientific Correspondence with Bohr, Einstein, Heisenberg a.o., Volume II: 1930–1939* (ed. K. von Meyenn), Springer, Berlin 1985, pp. 97–99 (Danish original), 99–103 (annotated German translation). In the letter Bohr described his recent discussions with Bloch in Copenhagen.

[13] Bohr's life-time interest in collision problems constitutes the topic of Vol. 8, which quotes this particular sentence of the letter on p. [219].

electrons be given by a wave function

$$\Psi(0) = \varphi(x_1, \ldots, z_N; \sigma_1, \ldots, \sigma_N)e^{i\frac{\pi}{\hbar}E_0 t} \ ,$$

where φ is antisymmetric in the coordinates of the electrons. While this state is conducting, there will exist solutions which only deviates a little from it and which can be written by the formula

$$\Psi(\delta) = (\varphi + i\delta\psi_x)e^{i\frac{\delta}{a}(x_1 + \cdots + x_N)}e^{\frac{i}{\hbar}(E_0 + \delta^2\frac{\hbar^2 N}{2ma^2}A_x)t} \ ,$$

where δ is small and where $\delta\psi_x$ is an antisymmetric function satisfying the equation

[many-particle Schrödinger equation][9]

which in addition serves to determine A_x. Such a state, which of course requires supply and continuous removal of electric charge at the boundary of the metal, corresponds to an amperage equal to

$$S_x = \delta\frac{\varepsilon\hbar N}{maV}A_x \ .$$

Due to the large value of N/V this amperage can become very large even if δ is extremely small. It now seems to me a likely assumption that the state described corresponds to superconductivity, and due to the large wavelength a/δ the current will not be appreciably disturbed by thermal oscillations in metal lattices. On the other hand, one has to assume that the coordinated motion in question at a certain temperature will cease to exist due to the thermal disturbances of the electrons. The transition to a state in which the electrons move in an uncoordinated manner among each other should there be exactly analogous to the melting of a solid body, but it has hitherto not been possible for me to reach a clear understanding of this transition process.

The assumption that superconductivity concerns a coordinated motion of the entire electron lattice is of course old,[10] but just as it was first with quantum mechanics that it was possible to bring the conception of the presence of "free" electrons in metals in closer connection with experience, so it also appears that it is first through quantum mechanics that one can understand, in the manner suggested above, how the two lattices can move through each other without resistance and appreciable deformation. I should be very pleased to hear a couple of words from you about how you look at all this. Moreover, Rosenfeld and I have been very industrious, and only a few days ago could we send off the Rome report in its final form.[11] We have worked hard and have had many lively discussions about the radiation problems, but in the end, however, I was content with introducing a couple of comments briefly indicating the point of view we

[501]

already spoke about at Christmas and which I described in my letter to Pauli.[12] How does your own work progress, by the way? I would be very pleased to hear about it sometime. I have given a good deal of thought myself to the collision problems, and especially to the disintegration of atoms by protons, which presents a beautiful example of the divergence between the quantum-mechanical and the classical treatments.[13]

With many kind regards to Heisenberg and yourself from all of us,

Yours sincerely,

[Niels Bohr]

OLE CHIEVITZ

CHIEVITZ TO BOHR, 20 April 1924
[Handwritten]

CANADIAN PACIFIC

	S.S. Empress of Asia.
20/4 [19]24.	Stillehavet 415 miles
	fra Yokohama.

Kære Niels.

Jeg føler Trang til sende dig et Par Ord. Jeg har det godt.

Grunden til jeg skriver, er nærmest den, at jeg længes efter at komme hjem og snakke med dig og lære noget.

Som Lekture har jeg haft Kramers og Holst's[14] Bog[15] og Hevesy's om Radioaktivitet[16] med, tidligere har jeg læst dem 2 Gange og har nu læst dem omhyggeligt 3' Gang (vist nok for første Gang er de studeret paa det store eller det stille Hav) og jeg synes, jeg begynder at "forstaa" lidt af det, og ogsaa hvilken Stordaad du har øvet i disse Spørgsmaal.

– Tillige tror jeg meget af det maa kunne overføres paa Biologien, specielt hele "Straalebiologien" maa tage sit Udgangspunkt fra disse Synspunkter, og

[14] Helge Holst (1871–1944), Danish physicist and author.

[15] H. Holst and H.A. Kramers, *Bohrs Atomteori, almenfatteligt fremstillet*, Gyldendal, Copenhagen 1922, second edition 1929. The first edition was translated into English as H.A. Kramers and H. Holst, *The Atom and the Bohr Theory of Its Structure: An Elementary Presentation*, Knopf, New York 1926, with an introduction by E. Rutherford.

[16] G. von Hevesy and F. Paneth, *Lehrbuch der Radioaktivität*, Johann Ambrosius Barth, Leipzig 1923. This first edition appeared in English as G. Hevesy and F.A. Paneth, *A Manual of Radioactivity*, Oxford University Press, Oxford 1926.

her føler jeg uhyre Lyst til at tage fat. – Jeg har specielt tænkt meget over de Fænomener, der fremkommer ved Lysfluorescens (Amøber, der udmærket taaler Dagslyset, dør, naar der sættes lidt Eosin til Vandet, og de saa kommer i Dagslys, hvorimod de lever videre i Eosinvandet, hvis de holdes i Mørke; – dette er tydet som de "sensibiliseredes" for de gule Straaler og dræbtes af disse, ligesom ellers af de ultraviolette. – Jeg har tænkt mig, der ved Lyset's Gennemgang i Eosinen sker en "Transformering" af Straalerne og samtidig Elektronfrigørelse og disse dræber.) – Disse og lignende Fænomener er vel rimeligvis nogenlunde overskuelige og maa først løses, før det kan nytte at undersøge hvad der sker ved Röntgendybdeterapier. – Tror du det er helt galt mine Ideer her. – Ogsaa glæder jeg mig til at tage fat paa det "lille" Röntgenrør.

Skade man har saa lidt Tid, det der ødelægger det for mig er jo Arbejdet paa Rovsing's Klinik[17] fra Kl 4–7 (8) men dels er det jo et økonomisk Spørgsmaal, og dels tror jeg, han daarlig i Øjeblikket helt kan undvære mig, – men jeg vil se at faa lidt mindre at gøre derude. –

Rejsen er gaaet udmærket hidtil, det var en interessant Tur gennem Canada. – I Montreal havde vi Snestorm. – Rocky Mountains var pragtfuldt og Vancouver meget smuk. – Det er en By paa c 200,000 Mennesker ikke 35 Aar gammel.

Turen over Stillehavet har været god nogle Dage lidt haardt Vejr. – Gennem den traadløse Avis saa jeg, der er stor Spænding mellem Japan og Amerika og at Neergaard[18] har trukket sig, hvem har efterfulgt ham? – Det kan jo næppe blive værre end det var, noget andet er om det bliver bedre.

Yokohama 21/4 [19]24.

Vi er nu lige naaet hertil efter en pragtfuld Morgensejlads mellem grønne Øer og Sunde og i Baggrunden høje Bjærge og over alt Fusijama, der virkelig ser saa mærkelig ud som paa Billederne.

Byen der tidligere var paa $\frac{1}{2}$ Million er nu Grushobe og Bølgeblikskure.[19] Hils alle.

Din hengivne
Ole.

---

[17] The surgeon Thorkild Rovsing (1862–1927) set up a private clinic in Copenhagen in 1899.
[18] The historian Niels Neergaard (1854–1936) had become Danish Prime Minister in 1920 and now resigned as a result of Denmark's economic problems.
[19] The great Kanto earthquake on 1 September 1923 had devastated Yokohama, as well as Tokyo.

Translation

CANADIAN PACIFIC

S.S.   Empress of Asia.

The Pacific Ocean 415 miles

from Yokohama.

20 April [19]24.

Dear Niels.

I want to send you a few words. All is well with me.

The reason I write is rather that I am longing to come home and talk to you and learn something.

As reading matter I have had Kramers and Holst's[14] book[15] and Hevesy's on radioactivity[16] with me, I have read them twice before and have now read them carefully for the third time (though it is probably the first time they have been studied on the Great or rather the Pacific Ocean) and I think that I am beginning to "understand" a bit of it, and also what a great achievement you have accomplished in these questions.

– Besides, I think it should be possible to transfer much of it to biology, especially all of "radiation biology" must take its starting point from these ideas, and here I am enormously keen to get going. I have particularly thought a lot about the phenomena which appear as a result of light fluorescence (amoeba, which survive well in daylight, die when a little eosin is added to the water and they are then brought into daylight, whereas they live on in the eosin water if they are kept in the dark; – this is explained as their having been "sensibilized" for the yellow rays and were killed by them, just as otherwise by the ultra-violet. – I have an idea that with the passage of light in the eosin a "transformation" of the rays takes place with simultaneous release of electrons and these kill.) – These and similar phenomena are probably more or less uncomplicated and must first be solved, before it is of any use to investigate what happens in deep therapy with X-rays. – Do you think my ideas are quite crazy here. – I also look forward to starting on the "little" X-ray tube.

Pity that one has so little time, what spoils it for me is of course the work at Rovsing's clinic[17] from 4–7 (8) o'clock but partly it is a financial question and partly I think he can hardly manage without me completely at the moment – but I will see to getting a bit less to do out there.

The journey has been excellent until now, it was an interesting trip through

Canada – in Montreal we had a blizzard – the Rocky Mountains were marvellous and Vancouver very beautiful. – It is a city of about 200,000 people – not yet 35 years old.

The voyage across the Pacific Ocean has been good – some days a little rough weather. – Through the wireless news I saw that there is great tension between Japan and America and that Neergaard[18] has resigned, who is his successor? – It can hardly get worse than it was, it is another thing if it will get better.

Yokohama 21 April [19]24.

We have now just arrived here after a marvellous morning voyage between green islands and sounds and high mountains in the background and over everything Mount Fuji, which really looks just as peculiar as in the pictures.

The city which was previously half a million is now heaps of gravel and corrugated iron huts.[19]

Greetings to everybody.

Yours sincerely,
Ole.

PETER FREUCHEN

BOHR TO FREUCHEN, [Christmas 1939]
[Handwritten]

Telegramadresse:"Fefor"   Rikstelefon

Fefor Høifjellshotell
CHR. F. WALTER
pr. Vinstra St. – Norge

Kære Freuchen,

Det er ikke fordi jeg har fulgt den venlige Anvisning hvormed De forsynede Deres sibiriske Æventyr at jeg først skriver saa sent for at takke for denne baade

[505]

saa morsomme og interessante Bog.[20] Tværtimod har jeg for rigtig at kunne nyde den og lære af den, ventet med at begynde paa den til jeg var færdig med al Travlheden i København og er kommet herop med Familien for at tilbringe en haardt tiltrængt Juleferie i Solen og Sneen.[21] Al den Tid vi ikke har været paa Ski har jeg med største Spænding fra Time til Time fulgt Dem paa Deres Farten til Lands og til Søs og i Luften i det vældige næsten ukendte Land. Hvor nyt og tankevækkende alt end er hvad De oplever og saa livligt beskriver, er det jo samtidig saadan at man Punkt for Punkt genkender de samme rørende Træk i den russiske Psyke, som man ved et hvert Samvær med Russere indtages af. De Grundfejl i Organisationen og den Vildfarelse af Folkemasserne som De saa klart belyser baade i og mellem Linierne, giver een jo ogsaa meget at tænke paa netop i disse Dage hvor den altovervejende Følelse hos os alle er Taknemmeligheden og Begejstringen for hvad et lille frit og samlet Folk kan udrette selv mod en saa uhyre talmæssig Overmagt.[22] Det ser virkelig ud til at lysere Tider for Menneskeheden alligevel kan komme ud af de store Ulykker og at vi alle igen engang skal kunne aande frit. Er der nogen man ønsker bedre Tider saa er det sandelig ogsaa det russiske Folk der under Tilværelsens Yderligheder har maattet lide som næppe noget andet. Det kan jo maaske synes en mærkelig og ubegrundet Optimisme, men jeg føler mig overbevist om at Tyranniets Dage i Europa er talte og at vi allerede i det kommende Aar vil se Begyndelsen til Bestræbelserne paa at bygge en helt ny Tilværelse op her i den gamle Verden. Hvordan det nu end vil gaa maa alligevel dette Haab faa Lov til at danne Baggrunden for de varmeste Ønsker om alt godt i det nye Aar for Dem og Deres Familie fra baade Margrethe og

Deres hengivne

Niels Bohr

der begge glæder os til snart at
se Dem og Deres Kone[23] en Aften hos os og
tale nærmere sammen om alt.

---

[20] P. Freuchen, *Sibiriske Eventyr*, Frederik E. Pedersen, Copenhagen 1939.

[21] Bohr and his family were spending Christmas 1939 in the mountains of southern Norway.

[22] Bohr is referring to the "Winter War", which erupted on 30 November 1939, when Stalin's Red Army attacked Finland.

[23] Magdalena, née Vang Lauridsen (1881–1960).

Translation

Dear Freuchen,

It is not because I have followed the friendly advice with which you signed your Siberian adventures that I only write so tardily to thank you for this both so amusing and interesting book.[20] On the contrary, to be able to really enjoy it and learn from it I waited to start on it until I was finished with all the busyness in Copenhagen and had arrived up here with the family to spend a sorely needed Christmas holiday in the sun and the snow.[21] All the time we haven't spent on skis I have with the greatest excitement followed you hour by hour on your journey on land and at sea and in the air in the enormous more or less unknown country. However new and thought-provoking everything is that you experience and describe so vividly, at the same time it is such that point by point one recognizes the same touching features of the Russian psyche, by which one is enchanted on every contact with Russians. The fundamental faults in the organization and the mistaken beliefs of the masses that you so clearly illuminate both in and between the lines, give of course much food for thought precisely at this time, when the predominant feeling in all of us is gratefulness and enthusiasm for what a small and united people can achieve even against such an enormous superiority in numbers.[22] It really looks as though brighter times for humanity can emerge all the same from the great catastrophes and that we all again at some time will be able to breathe freely. If there is anyone for whom one wishes better times, then it is in truth the Russian people who under the extremes of existence have had to suffer as scarcely any other. This might perhaps appear to be a strange and unfounded optimism, but I feel convinced that the days of tyranny in Europe are numbered and that already in the coming year we will see the beginning of the endeavours to build up a completely new existence here in the Old World. No matter what happens, may this hope all the same be allowed to form the background for the warmest wishes for all the best in the new year for you and your family from both Margrethe and

Yours sincerely,

Niels Bohr

who both look forward to
seeing you and your wife[23] with us one evening soon and
talking about everything in more detail.

T.C. HODSON

BOHR TO T.C. HODSON, ENCYCLOPÆDIA BRITANNICA, 24 March 1926
[Carbon copy]

24th of March [192]6.

Dear Sir,

I enclose the manuscript of my article on the atom, which I hope may suit your purpose.[24] At the same time I want to apologize very much for the delay in its completion which I hope has not caused you too serious inconvenience. This delay is mainly due to overstrain from which I have been suffering the last months, but besides I have found the work more difficult than I anticipated due to the rapid progress within this field of science. In fact the development within the very latest months of the ideas of the quantum theory as well as of the constitution of the electron make it necessary to be very cautious, when wishing to give an account of the atomic theory which may still afford a reasonable impression of the subject after a few years. As regards references to literature I do not know what your wishes are. If it is in conformity with the other articles I would propose that some references to monographs on the subject were added at the end of the paper. I suppose this can easily be done in the proof which I shall take care to return without delay.

Yours sincerely,
[Niels Bohr]

P.S. On your request I enclose a photograph.

---

[24] The article was published as N. Bohr, *Atom*, Encyclopædia Britannica, 13th edition, Suppl., Vol. 1, London and New York 1926, pp. 262–267. Reproduced in Vol. 4, pp. [657]–[663].

R.J.L. KINGSFORD

KINGSFORD TO BOHR, 24 March 1960
[Typewritten]

UNIVERSITY PRESS CAMBRIDGE

Secretary:	*Assistant Secretaries:*
R.J.L. KINGSFORD, M.A.	M.H. BLACK, M.A.
Education Secretary:	R.A. BECHER, M.A.
BORIS FORD, M.A.	A.K. PARKER, M.A.

Assistant to the Secretaries: MARGARET HAMPTON

TELEPHONE: 4226

RJLK/MB.   24 March, 1960.

Dear Professor Bohr,

We should like to consider again the possibility of reprinting the volume of Essays on *Atomic Theory and the Description of Nature* which we published for you in 1934, and I write to ask whether you see any objection to the book being re-issued and whether you would be willing to contribute a new preface, or a brief introduction of a few pages to put the book into the context of the present day.

If you cannot spare the time to do this yourself, we have it in mind that we might ask two or three of your ex-pupils to contribute commemorative tributes to you.

I should be grateful for an answer as soon as possible.

Yours sincerely,
R.J.L. Kingsford

Professor N. Bohr,
Gl. Carlsberg,
Copenhagen,
Denmark.

BOHR TO KINGSFORD, 29 March 1960
[Carbon copy]

March 29, 1960.

R.J.L. Kingsford, M.A.
Secretary
University Press Cambridge.

Dear Mr. Kingsford,

Thank you for your letter of March 24 with the proposal that the Cambridge University Press re-issue my volume of Essays on "Atomic Theory and the Description of Nature". As you will know, the collection of some of my later essays, mentioned in the Preface to the first edition, has in the meantime under the title "Atomic Physics and Human Knowledge" been published by John Wiley and Sons, New York.[25] Due to the close connection between the essays in the two volumes, various requests concerning reprinting of the former volume have come to my notice, and I am therefore glad that the book can now be made available again. I agree that, besides the original preface, the book be provided with a short preface to the second edition, and within a few weeks I shall send you a manuscript for it.[26] I also suggest that the text is kept unchanged except for a few improvements of the English translation of the original Danish text.

Yours sincerely,

Niels Bohr.

---

[25] N. Bohr, *Atomic Physics and Human Knowledge*, John Wiley & Sons, New York 1958. Reissued as *The Philosophical Writings of Niels Bohr, Vol. II*, Ox Bow Press, Woodbridge, Connecticut 1987.
[26] The new preface is reproduced on pp. [89] ff.

KIRSTINE MEYER

MEYER TO BOHR, 25 October 1931
[Handwritten]

d. 25/10 1931

Kære Professor Bohr.

Mange Tak for Deres Brev, som jeg blev meget glad for. I Onsdags Aftes holdt Fysisk Forening en smuk Fest for mig, og Professor Hansen[27] og andre sagde mig mange venlige Ord. Da jeg svarede herpaa, havde jeg egentlig mest Lyst til at takke mange enkelte blandt Fysikerne, som jeg skylder meget, men da jeg havde nævnet de ældste, opgav jeg det og takkede mere officielt "Fysisk Forening". Den jeg særlig vilde have takket var Dem. De, der er ung, har baade hjulpet mig "den gamle" og har altid sammen med Deres Kone vist mig stor Venlighed, og De, den berømte, har altid vist Interesse for og givet Støtte til det lavt liggende, jeg har kunnet yde. Jeg ved nok, at Elskelighed og Trofasthed er givet Dem i Vuggegave, men derfor er man dog lige taknemmelig mod *Dem*, naar man kommer ind under dens Virkningssfære.

Med venligst Hilsen til Fru Margrethe

Deres hengivne
Kirstine Meyer

Translation

25 October 1931

Dear Professor Bohr,

Many thanks for your letter which pleased me very much. On Wednesday evening the Physical Society held a beautiful celebration for me, and Professor Hansen[27] and others said many kind words to me. When I responded, I was really most inclined to thank many of the individual physicists, to whom I

---

[27] Hans Marius Hansen (1886–1956).

owe much, but after I had mentioned the oldest ones I gave up and thanked the "Physical Society" more officially. The one I especially wanted to thank was you. You, who are young, have both helped me "the old one" and always, together with your wife, shown me great friendliness, and you, the famous one, have always shown interest in and given support to the humble effort that I have been able to offer. I am well aware that you were born with the gifts of lovableness and loyalty but all the same one is equally grateful to *you* when one comes within their sphere of action.

With the most friendly greetings to Mrs Margrethe

Yours sincerely
Kirstine Meyer

BOHR TO MEYER, 11 October 1936
[Handwritten]

11–10–1936.

GL. CARLSBERG
VALBY

Kære Fru Meyer,

Paa Deres 75 Aars Fødselsdag, hvor De kan se tilbage paa saa frugtbart og særpræget et Livsværk i Videnskabens og Undervisningens Tjeneste, vil Margrethe og jeg gerne sende Dem vore hjerteligste Lykønskninger og vor inderligste Tak for det varme og trofaste Venskab, som De altid har vist os.

Lige fra min tidligste Barndom, da jeg først af Moster Hanne[28] lærte at se op til Dem, har jeg gennem den stadig nærmere Forbindelse, som fælles Interesser og fælles Venner har bragt os i, stedse stærkere følt, hvor meget Deres sandhedssøgende og retlinede Personlighed har betydet for hele Fysiklivet herhjemme og for mig selv.

Med de venligste Hilsner og de hjerteligste Ønsker fra os begge.

Deres hengivne
Niels Bohr

[28] Hanna Adler (1859–1947).

[512]

Translation

11 October 1936.

GL. CARLSBERG
VALBY

Dear Mrs Meyer,

On your 75th birthday, when you can look back on a such a fruitful and exceptional lifework in the service of science and education, Margrethe and I would like to send you our most heartfelt congratulations and our most sincere thanks for the warm and faithful friendship you have always shown us.

Right from my earliest childhood, when I first learnt from Aunt Hanne[28] to look up to you, I have through the ever closer connection, into which mutual interests and mutual friends have brought us, felt more and more strongly how much your truth-seeking and upright personality has meant for the whole life of physics here at home and for me personally.

With the most friendly greetings and the most heartfelt wishes from both of us.

Yours sincerely
Niels Bohr

NIELS MØLLER

BOHR TO MØLLER, 11 March 1934
[Handwritten]

11–3–1934.

Kære Niels Møller,

Margrethe og jeg takker mange Gange for den store Elskværdighed at sende os Deres smukke Udgave af Longinos,[29] som det for os begge har været en stor Oplevelse at lære at kende. Sjældent har jeg læst noget, der fra først til sidst virkede mere belærende og betagende end dette gamle Skrift, der med al dets Nøgternhed er saa fuld af Aand og Begejstring.

Deres hengivne
Niels Bohr

Translation

11 March 1934.

Dear Niels Møller,

Margrethe and I thank you many times for your great kindness in sending us your beautiful edition of Longinus,[29] which it has been a great experience to get to know for both of us. Seldom have I read anything that from first to last has been more instructive and enchanting than this old text, which with all its soberness is so full of spirit and enthusiasm.

Yours sincerely
Niels Bohr

BOHR TO MØLLER, 11 December 1934
[Handwritten]

11/12 1934.

Kære Niels Møller,

Margrethe og jeg sender Dem vore hjerteligste Lykønskninger til Deres 75 Aars Fødselsdag. Vi tænker med den dybeste Beundring paa den Kraft og Ildhu, hvormed De gennem de mange Aar har budt alle ydre og indre Vanskeligheder Trods og derved ikke alene har formaaet at udføre saa stort et Livsværk i den

---

[29] Dionysius Longinus [Pseudo-Longinus], *Den store Stil* [*Peri hypsous*/*On the Sublime*] (ed. N. Møller), Levin & Munksgaard, Copenhagen 1934.

af Dem saa højt elskede Digtekunsts Tjeneste men ogsaa har sat os alle saa stort et Eksempel som Menneske.

Vi er bedrøvede over at vi ikke kan være med naar De i Aften fejres i Deres Venners Kreds, men ikke mindst med Tanken paa Deres Trofasthed imod Christians[30] Minde, takker vi Dem inderligt for al Deres Venlighed imod Børnene og os selv.

Med de venligste Hilsner og bedste Ønsker fra os begge.

Deres hengivne
Niels Bohr

Translation

11 December 1934.

Dear Niels Møller,

Margrethe and I send you the most heartfelt congratulations on your 75th birthday. We think with the deepest admiration of the strength and zeal with which you throughout the many years have faced all the outer and inner difficulties and thereby not only have managed to carry out so great a lifework in the service of the art of writing that you love so dearly but have also set us all such a great an example as a human being.

We are sad that we cannot be present when you will be celebrated this evening in the circle of your friends, but not least with the thought of your faithfulness towards Christian's[30] memory, we thank you sincerely for your kindness towards the children and ourselves.

With the most friendly greetings and best wishes from both of us.

Yours sincerely
Niels Bohr

---

[30] See ref. 8.

IVAN SUPEK

BOHR TO SUPEK, 9 May 1959
[Carbon copy]

May 9, 1959.

Professor Dr. Ivan Supek
Institut teorijske fizike
Marulićev trg 19
Zagreb, Jugoslavija.

Dear Professor Supek,

I thank you for your kind letter of April 25 concerning the address which I had the pleasure to deliver at the Roger Bošković celebration before your Academy in Zagreb. I am sorry for the trouble I have caused you in this matter by not earlier having sent you a manuscript of the address and I quite realize that it will not be possible to reconstruct it from the tape recordings of my speech which under the circumstances I had to give in an informal manner without thorough preparation.[31] I intended, therefore, to send you a more concentrated account of the address which dealt with the relationship between the viewpoints of natural philosophy at Bošković's time and modern physical developments. Due to many obligations in connection with the reorganization and extension of the Institute for Theoretical Physics in Copenhagen I have, to my regret, not yet found time to write such a report and I am afraid that also in the next months I will be very occupied with other pressing duties. I shall therefore be grateful to learn at what time the memorial edition is planned to appear and shall then let you know of how the possibility stands for including some adequate contribution from my side.

---

[31] The transcript of the tape recording is deposited in the BMSS. See *Inventory of Relevant Manuscripts in the Niels Bohr Archive*, below, p. [532], folder 55.

With kindest regards and best wishes to yourself and the other friends in Zagreb who made our visit such a beautiful and unforgetful experience,

as always yours,

Niels Bohr.

BOHR TO SUPEK, 1 June 1959
[Carbon copy]

June 1, 1959.

Professor I. Supek
Institut Ruđer Bošković
Zagreb.

Dear Supek,

Since your most welcome visit to Copenhagen I have with the kind help of Rosenfeld tried to write a brief foreword to the Bošković volume, as you suggested. I shall be very glad to know whether you find that the enclosed pages may suit your purpose.[32]

With kind regards from us all to yourself and all common friends in Zagreb,

Yours,

Niels Bohr.

---

[32] The foreword is reproduced on pp. [231] ff.

SVEN WERNER

WERNER TO BOHR, 9 September 1942
[Typewritten]

DET FYSISKE INSTITUT   AARHUS, Den 9–9–1942
AARHUS UNIVERSITET   Tlf. 8801

PROFESSOR, DR. PHIL. S. WERNER

Kære Professor Bohr,

Jeg skriver til Dem paa Opfordring fra Gyldendal.[33] Jeg ved ikke om De husker, at jeg for et Aars Tid siden spurgte Dem, om De kunde tænke Dem at skrive et Par Linier foran i Gamow's Bog 'Mr. Thompkins in Wonderland', som vi har oversat, og som nu udkommer.[34] Motiveringen er, at Bogen jo er tilegnet Professor Bohr.[35]

Jeg kommer – paa Grund af et Møde paa Universitetet – til København den 14'. ds. Maaske maa jeg saa ringe til Dem og faa Svar?

Vi tænkte saa meget paa Dem i Sommer, da vi hørte om det Uheld, De var ude for under Brændehugningen.[36] Jeg haaber rigtignok, at det forlængst er overstaaet.

Med de venligste Hilsner

Deres hengivne
Sven Werner

---

[33] Major Danish publishing house.
[34] Bohr's contribution is reproduced and translated on pp. [77] ff.
[35] The book was dedicated to Niels Bohr and Lewis Carroll.
[36] Bohr had hurt his eye while chopping wood at his summer house in northern Zealand. The incident was reported in Danish newspapers but had no long-term effect on Bohr's health.

Translation

9 September 1942

Dear Professor Bohr,

I am writing to you on behalf of Gyldendal.[33] I don't know if you remember that about a year ago, I asked you if you would consider writing a few lines at the front of Gamow's book, 'Mr. Tompkins in Wonderland', which we have translated and will now be published.[34] The motivation is that the book, of course, is dedicated to Professor Bohr.[35]

Because of a meeting at the University, I will be coming to Copenhagen on the 14th of this month. Perhaps I could phone you and get an answer?

We thought about you so much this summer, when we heard about the accident you had when chopping logs.[36] I certainly hope this is well in the past.

With the most friendly greetings

Yours sincerely
Sven Werner

INVENTORY OF RELEVANT MANUSCRIPTS IN THE NIELS BOHR ARCHIVE

INTRODUCTION

The following is a list of folders in the microfilmed BMSS containing manuscripts by Bohr of special relevance for the present volume. The list is arranged chronologically and covers the years 1915–1962. As in earlier volumes, the list does not contain documents from other, not microfilmed, collections.

In the first line of each listed folder, its title has been assigned by the cataloguers, as has any date in square brackets. Unbracketed dates are taken from the manuscripts. A number in the margin indicates the page number on which a particular document has either been reproduced (R) or cited (C) or on which the publication (P) resulting from the manuscript is printed.

The next line indicates whether the documents in the folder are typewritten, carbon copies or handwritten, as well as the language used. Finally, the number of the microfilm on which the folder can be found is provided (e.g., "mf. 11").

Other relevant information about the folders – such as the contents of individual manuscripts and the relation of manuscripts among themselves or to documents reproduced in the present volume – is given in explanatory notes in small print.

[283] P

1 *Moseley* [1915]

Typewritten, carbon copy, handwritten, 4 pp., English, mf. 5.

A survey of Moseley's work. Reproduced in Vol. 2, pp. [427]–[430].

[115] C

2 *Rede an Sommerfeld* 22 September 1919

Handwritten [Ellen Bohr], 2 pp., German, mf. 8.

Title added by N. Bohr: "Tale holdt til Sommerfeld efter Foredraget i Fysisk Forening 22–9–1919" (Speech to Sommerfeld after his lecture before the Danish Physical Society 22 September 1919).

3 *Tale til Prof. Carl Størmer* 2 September 1925

Handwritten [N. Bohr], 2 pp., Danish, mf. 11.

Response to Carl Størmer, Oslo, after Størmer's lecture on the problem of the Northern Lights given before the Danish Physical Society.

[66] C

4 *Mme. Curie's Opdagelse af Radium* [12–14 October 1926]

Typewritten, carbon copy, handwritten [N. Bohr and Margrethe Bohr], 7 pp., Danish, mf. 11.

Draft of speech or article on the discovery of radium and its properties, prepared on the occasion of Mme. Curie's visit to Copenhagen, where she delivered three lectures before the Science Society of Denmark from 12 to 14 October 1926.

[147] P

5 *Kort autobiografi* [1927]

Typewritten, carbon copy, 1 p., Danish, mf. 11.

Short autobiography. Partial manuscript for published version.

[106] C

6 *Knud Rasmussens 50 års dag* [June 1929]

Typewritten, carbon copy, handwritten [N. Bohr], 4 pp., Danish, mf. 12.

Draft of speech to Knud Rasmussen on the occasion of his 50th birthday 7 June 1929.

[108] C

7 *Rede bei einem Fest für M. Planck* 10 October 1930

Typewritten, carbon copy, 5 pp., German, mf. 12.

Title: "Tale ved Fysisk Forenings fest for professor Planck 10.10.1930" (Speech at the Danish Physical Society's celebration for Professor Planck 10 October 1930).

[21], [29] R

8 *Supraleitung* [June 1932]

Typewritten, carbon copy, handwritten, proofs [F. Bloch, N. Bohr and L. Rosenfeld], 23 pp., proofs, German and Danish, mf. 13.

Title: "On the question of superconductivity". Notes etc. for an unpublished paper, originally intended to appear in Die Naturwissenschaften.

[267], [253] P

9 *Rutherford's Visit in Copenhagen* September 1932

Typewritten, carbon copy, 11 pp., English and Danish, mf. 13.

(a) Words of welcome at the party for Rutherford at Carlsberg. (b) Words of thanks at the end of Rutherford's lectures, given in the ceremonial hall of the university and the biophysical lecture room 14 and 15 September 1932 respectively. (c) Speech at the banquet of the Science Society of Denmark in honour of Lord Rutherford 15 September 1932. (d) Title: "Professor, Sir Ernest Rutherford og hans betydning for fysikkens nyere udvikling" (Professor Sir Ernest Rutherford and his significance for the recent development of physics). Manuscript for article in Politiken, 18 September 1920.

10 *Thanks to I. Langmuir* [19 December 1932]

Typewritten, carbon copy, 2 pp., English, mf. 13.

Short speech of thanks to Langmuir for his lecture, delivered before the Danish Physical Society [19 December 1932].

[339] P

11 *F. Paschen zum 70sten Geburtstag* 22 January 1935

Typewritten, carbon copy, 2 pp., German, mf. 14.

Manuscript of note written on the occasion of F. Paschen's 70th birthday.

[121] C

12 *Mindeord om V. Henriques* 7 December 1936

Typewritten, carbon copy, 3 pp., Danish, mf. 14.

Speech in memory of V. Henriques at the funeral ceremony in his home 7 December 1936.

[271] P

13 *Tribute to Rutherford* 20 October 1937

Handwritten [N. Bohr] (photoprint), handwritten [L. Rosenfeld], 3 pp., English, mf. 14.

Copy of manuscript of tribute to Rutherford, prepared in connection with the short speech, given at the Galvani Congress, Bologna 20 October 1937 on the announcement of Rutherford's death.

[273] P   **14 *Relations to Rutherford*   [November 1937]**
Carbon copy, handwritten [N. Bohr], 6 pp., English and Danish, mf. 14.

(a) Notes and part of draft for a note about Bohr's relations to Rutherford. (b) Title: "Notes about my relations to Rutherford". Enclosed with letter to A.S. Eve dated 27 November 1937 in which Bohr provided help with the writing of an obituary for Rutherford by Eve and J. Chadwick published in Proc. Phys. Soc. **50** (1938) 441

[111] C   **15 *Foredrag om Rutherford*   22 November 1937**
Typewritten, carbon copy, 3 pp., Danish, mf. 14.

Three first pages of notes of lecture about Rutherford, delivered before the Danish Physical Society, Copenhagen 22 November 1937.

[119] C   **16 *Mindeord om F. Kalckar*   [January 1938]**
Typewritten, carbon copy, 2 pp., Danish, mf. 15.

Speech at the funeral of F. Kalckar.

[49] P   **17 *Structure of matter*   [14 January 1938]**
Typewritten, carbon copy, handwritten [L. Rosenfeld and Margrethe Bohr], 9 pp., English, mf. 15.

Title: "Structure of matter". Manuscript and part of draft. The loose sheet contains an outline of the article and a list of future work.

[101] C   **18 *Mach-Feier in Prag*   February 1938**
Typewritten, carbon copy, 2 pp., German, mf. 15.

Manuscript of speech celebrating 100th anniversary of the birth of Ernst Mach.

[73] P   **19 *Various notes*   [1938]**
Carbon copy, handwritten, 2 pp., English and Danish, mf. 15.

(a) First page of report of discussion on nuclear physics introduced by Bohr. (b) Outline of the preface to the book: "Atomer og andre småting" by C. Møller and E. Rasmussen.

[53] P   **20 *Structure of matter*   [19 January 1939]**
Typewritten, carbon copy, handwritten [Erik Bohr], 2 pp., English, mf. 16.

Manuscript of article in Encyclopædia Britannica.

[113] C

21 *Tale til Martin Knudsen*   18 October 1939

Typewritten, carbon copy, 3 pp., Danish, mf. 16.

Manuscript of speech to M. Knudsen, given at the meeting of the Society for the Dissemination of Natural Science, 18 October 1939, on Knudsen's retirement as President of the Society.

[123] C

22 *Valdemar Poulsens 70-årsdag*   23 November 1939

Typewritten, carbon copy, 5 pp., Danish, mf. 16.

Title: "Festen i anledning af Dr. Valdemar Poulsens 70-års fødselsdag 23.11.1939" (The celebration on the occasion of Dr Valdemar Poulsen's 70th birthday on 23 November 1939).

[289] R

23 *Martin Knudsens 70-årsdag*   15 February 1941

Newspaper cutting, typewritten, carbon copy, 3 pp., Danish, mf. 16.

(a) Article in "Berlingske Aftenavis" 14 February 1941. (b) Title: "Tale ved festen for Martin Knudsen i Danmarks Naturvidenskabelige Samfund 10.3.1941" (Speech at the celebration for Martin Knudsen in the Science Society of Denmark 10 March 1941).

[425] P

24 *Kirstine Meyers 80-årsdag*   12 October 1941

Typewritten, carbon copy, 2 pp., Danish, mf. 16.

Manuscript of article.

[237] P

25 *Mindefest for Vitus Bering*   19 December 1941

Typewritten, carbon copy, 3 pp., Danish, mf. 16.

Title: "Tale ved mindefesten for Vitus Bering i Horsens 19.12.1941" (Speech given at the commemoration for Vitus Bering in Horsens 19 December 1941).

26 *Mindeord om Vilhelm Thomsen*   [30 January 1942]

Typewritten, carbon copy with handwritten corrections [N. Bohr], 2 pp., Danish, mf. 16.

Manuscripts of short speeches, given at the beginning and end of the meeting of the Royal Danish Academy of Sciences and Letters 30 January 1942, at which the centenary of Vilhelm Thomsen's birth was commemorated. One of them is the manuscript for the speech reproduced in Vol. 11, pp. [393]–[396].

[101] C

27   *Mindeord om Georg Brandes*   [4 February 1942]

Typewritten, carbon copy with handwritten corrections, 1 p., Danish, mf. 16.

Commemorative words about Danish author and cultural critic Georg Brandes on the occasion of the centenary of his birth.

[113] C

28   *Martin Knudsens 25-års jubilæum*   11 May 1942

Typewritten, carbon copy, 3 pp., Danish, mf. 16.

(a) Manuscript of speech to M. Knudsen on the occasion of his 25 years jubilee as Secretary of the Royal Danish Academy of Sciences and Letters, given at supper following the meeting of the Academy on 8 May 1942. (b) Manuscript of remarks at the closing of the meeting.

[107] C

29   *Mindefest for Vitus Bering*   13 May 1942

Typewritten, carbon copy, 3 pp., Danish, mf. 16.

Title: "Tale ved Statsradiofoniens mindefest for Vitus Bering 13.5.1942" (Speech given at the commemoration by the Danish national radio for Vitus Bering 13 May 1942). Slightly revised version of the manuscript in folder 25.

30   *Tak til Rektor C.E. Bloch*   29 October 1942

Typewritten, carbon copy, 2 pp., Danish, mf. 16.

Title: "Tale ved aftenfesten efter den akademiske lærerforsamling 29.10.1942" (Speech at the evening party following the academic assembly 29 October 1942). Speech to C.E. Bloch on his retirement as Vice-Chancellor of the University of Copenhagen.

31   *Mindeord om O. Jespersen*   May 1943

Typewritten, carpon copy, 3 pp., Danish, mf. 16.

(a) Short speech at the funeral of O. Jespersen 4 May 1943. (b) Commemorative words at the meeting of the Royal Danish Academy of Sciences and Letters 14 May 1943. (c) Commemorative words possibly prepared for a Danish newspaper.

[219] P

32   *Newton's principles and modern atomic mechanics*   15–19 July 1946

Typewritten, carbon copy, handwritten [N. Bohr, L. Rosenfeld, S. Rozental and unidentified], 48 pp., English, mf. 17.

(a) Manuscript, "Newton's principles and modern atomic mechanics", practically identical to the published version. (b) Envelope, marked "Newton 1946", containing notes and drafts for the paper.

[101] C   33   *400th anniversary of the birth of Tycho Brahe*   [14 December 1946]

Handwritten [Aage Bohr and S. Rozental], 7 pp., English and Danish, mf. 17.

Notes for speech given 14 December 1946 to the astronomers upon whom the University of Copenhagen had conferred an honorary doctorate on the occasion of the 400th anniversary of the birth of Tycho Brahe.

[449] P   34   *Mindeord om Ole Chievitz*   31 December 1946

Typewritten, carbon copy, 7 pp., Danish, mf. 17.

Title: "Tale ved mindehøjtidligheden for Ole Chievitz 31.12.1946" (Speech given at the commemoration for Ole Chievitz 31 December 1946). Exact manuscript for published article.

[275] P   35   *10th anniversary of Rutherford's death*   [7 November 1947]

Typewritten, carbon copy, 6 pp., English, mf. 17.

Manuscript of address read by C. Møller at the commemoration arranged by the World Federation of Scientific Workers in Paris 7 to 8 November 1947 on the 10th anniversary of Rutherford's death. Exact manuscript for published article.

[369] P   36   *Tribute to Einstein*   14 March 1949

Typewritten, carbon copies with handwritten corrections [S. Rozental], 13 pp., Danish and English, mf. 18.

(a) Manuscript of speech, broadcast by the Scandinavian countries 14 March 1949 on the occasion of Einstein's 70th birthday. Published in Politiken 15 March 1949. (b) English version of the above, slightly modified.

[57] P   37   *Atomerne og vor erkendelse*   1 April 1949

Typewritten, carbon copy, handwritten [Margrethe Bohr, N. Bohr, S. Rozental, B. Schultz and unidentified], 157 pp., Danish, mf. 18.

Notes concerning the lecture, Atomerne og vor erkendelse (Atoms and human knowledge) given in the programme for schools on the Danish radio 1 April 1949 and on the Norwegian radio 7 April 1949.

[303] P   38   *Mindeaften for Martin Knudsen*   5 October 1949

Typewritten, carbon copy, 4 pp., Danish, mf. 19.

Title: "Mindeaften for Professor Martin Knudsen arrangeret af Fysisk Forening og Selskabet for Naturlærens Udbredelse den 5.10.1949. Indledningsord af professor Niels Bohr" (Commemoration for professor Martin Knudsen arranged by the

[529]

Danish Physical Society and the Society for the Dissemination of Natural Science, 5 October 1949. Words of introduction by Niels Bohr).

39 *Atomerne og vor erkendelse* 1950
Typewritten, carbon copy, 8 pp., Danish, mf. 19.

Manuscript of the article "Atomerne og vor erkendelse" (Atoms and human knowledge), Vor Viden **33** (1950) 123. See also folder 37.

40 *Supraledning*   30 March 1952
Handwritten [J. Lindhard], 3 pp., Danish, mf. 20.

Title: "Bemærkninger om metalledningsevne og supraledning" (Remarks on Conductivity of Metals and Superconduction). See also folder 43.

[347], [353] P

41 *At H.A. Kramers's death* [24 April 1952]
Typewritten, carbon copy with handwritten corrections [S. Rozental], 12 pp., Danish and English, mf. 20.

Manuscripts of: (a) "Ved Hendrik Anton Kramers død", Politiken 27 April 1952. (b) Hendrik Anthony Kramers, Ned. T. Natuurk. **18** (1952) 161.

[101] C

42 *Papers dedicated to Max Born* [1952–1953]
Handwritten [A. Petersen], 20 pp., English and Danish, mf. 20.

Notes for two unfinished papers, dedicated to Max Born on the occasion of (a) his 70th birthday 11 December 1952 (special issue of Z. Phys., December 1952) and (b) his retirement from Tait Chair of Natural Philosophy at the University of Edinburgh, intended for the book, "Scientific Papers Presented to Max Born", Oliver and Boyd, Edinburgh 1953.

[37] P

43 *Superconductivity* [22–26 June 1953]
Typewritten, carbon copy, 3 pp., English, mf. 20.

Manuscript for the report of the discussion following H. Fröhlich's lecture on superconductivity and lattice vibrations, given at the Lorentz Kamerlingh Onnes Conference, Leiden, 22 to 26 June 1953. See also folder 40

[473] P

44 *Jens Rosenkjærs 70-årsdag*   [1953]
Handwritten [A. Bohr], 1 p., Danish, mf. 20.

Manuscript of contribution to volume of tributes to Jens Rosenkjær on the occasion of his 70th birthday. Preliminary, somewhat more extensive, manuscript for published article.

[377] P   **45 *Albert Einstein 1879–1955* [17 April 1955]**

Typewritten, carbon copy, 2 pp., English, mf. 21.

Commemorative words at the death of Albert Einstein 17 April 1955. Dated 26 April 1955.

[331] P   **46 *Mindeord ved H.M. Hansens død*   13 June 1956**

Typewritten, carbon copy with handwritten corrections [S. Hellmann and A. Petersen], 10 pp., Danish, mf. 22.

(a) Commemorative words about H.M. Hansen in the Danish newspaper Politiken 14.6.1956. (b) Tale ved mindehøjtideligheden på universitetet 21.6.1956 (Speech at the commemoration at the University of Copenhagen 21 June 1956). (c) Title: "Noter taget af Aage Bohr ved N. Bohr's mindetale over H.M. Hansen i Videnskabernes Selskab [15.3.1957]" (Notes taken by Aage Bohr of N. Bohr's commemorative speech about H.M. Hansen at the Royal Danish Academy of Sciences and Letters [15 March 1957].) Referred to in *Mødet den 15. marts 1957* and *H.M. Hansen 7. September 1886 – 13. Juni 1956*, Overs. Dan. Vid. Selsk. Virks. Juni 1956 – Maj 1957, pp. 55, 69 respectively.

[235] P   **47 *Atoms and Human Knowledge, Nicola Tesla*   [10 July – 7 November 1956]**

Typewritten, carbon copy and handwritten [A. Petersen], 37 pp., English and Danish, mf. 22.

(a) Tribute at opening ceremony of the Nicola Tesla Conference, Belgrade, 10 July 1956. (b) Manuscripts. Title: "Atoms and Human Knowledge". Footnote: "Except for certain points, which have been elaborated, this article presents the content of the author's address delivered without manuscript at the Nicola Tesla congress, July 1956." One of the manuscripts is dated 7 November 1956. (c) Notes for the report of the lecture. Dated 27 October 1956.

[467] P   **48 *Halfdan Hendriksens 75-års dag*   [12 November 1956]**

Typewritten, carbon copy with handwritten corrections [Margrethe Bohr and N. Bohr], 3 pp., Danish, mf. 22.

Contribution to a book about Halfdan Hendriksen, edited on the occasion of his 75th birthday. Dated 13 June 1956.

[108] C   **49 *100th Anniversary of J.J. Thomson*   [15 December 1956]**

Typewritten, carbon copy, 1 p., English, mf. 22.

Message read at the dinner 15 December 1956 given on the occasion of the centennial celebration of the birth of J.J. Thomson.

[207] P

50 *Selvbiografi* [1956]

Typewritten, carbon copy, 5 pp., Danish, mf. 22.

Early draft.

[83] P

51 *10th anniversary of "Nucleonics"* 6 June 1957

Carbon copy, 1 p., English, mf. 22.

Congratulations to the journal "Nucleonics" on its tenth anniversary. Published in "Nucleonics", September 1957, p. 89.

[21] P

52 *Supraledning* [14 September 1955 – 15 August 1958]

Handwritten [N. Bohr, L. Rosenfeld and A. Petersen], 65 pp., English, German and Danish mf. 22.

Superconductivity. Notes and draft for unpublished paper. Includes 5 pp. of 1932 paper draft entitled "Zur Frage der Supraleitung" and 4 pp. notes for a talk at Rice University, "Science and Humanity". See also folder 8.

[227] R, P
[231] P
[445] P

53 *Diverse manuscripter* 1958

Typewritten and handwritten [N. Bohr, Margrethe Bohr, S. Hellmann], 35 pp., Danish and English, mf. 23.

Various manuscripts: (a) The Unity of Knowledge, talk in January 1958 at Iowa State University. (b) Atoms and Human Knowledge, talk at Roosevelt University, 4 February 1958. (c) Notes for talk at Houston, Texas. (d) Niels Bjerrum, talk at funeral, 2 October 1958 and exact manuscript for obituary in Politiken 1 October 1958. (e) Talk at "Rudjer Bošković Celebrations", Zagreb 1958, and notes for Slovenski Porocevalec, Ljubljana, for National Day of the Yugoslav Republic, 29 November 1958. See also folder 55.

54 *Superconductivity (efterladte noter)* [11 December 1958 – 19 March 1959]

Handwritten [L. Rosenfeld], 49 pp., Danish, mf. 23.

Rough notes for a paper plus reprints of 3 articles on superconductivity, dated between 11 December 1958 and 19 March 1959.

[231] P

55 *Rudjer Bošković* 1959

Typewritten, carbon copy, 7 pp., English, mf. 23.

Two different drafts for published version of a talk delivered at the inauguration of the Rudjer Bošković Institute in Zagreb in October 1958. See also folder 53.

[397] P

56 *Hanna Adler-bogen*   [March 1959]

Typewritten, carbon copy with handwritten additions [N. Bohr, Margrethe Bohr and Aage Bohr], 17 pp., Danish, mf. 23.

Three drafts, at different stages, of a foreword for a book issued to commemorate the 100th anniversary of the birth of Hanna Adler.

[361] P

57 *Pauli Memorial Volume*   [6–8 October 1959]

Handwritten [N. Bohr] and unidentified, typewritten, 5 pp., English and Danish, mf. 23.

Drafts of and final manuscript for foreword to the Pauli Memorial Volume.

58 *Nord -u. Westdeutsch. Rundfunkverb.*   [24 August 1960]

Typewritten, carbon copy with handwritten corrections [J. Kalckar], 16 pp., German, mf. 24.

Radio interview of N. Bohr by "Prof. Haber", 24 August 1960. (a) Typescript with handwritten corrections. (b) Corrected typescript with carbon copy.

[367] P

59 *1961 Diverse manuscripter*   [March and August 1961]

Typewritten, carbon copy, 16 pp., English and German, mf. 25.

Various manuscripts: (a) Recollections of Toshio Takamine prepared for a book, dated March 1961, (b) German translation, "Die Einheit menschlicher Erkenntnis", of "The Unity of Human Knowledge", published in Zeitschrift Europa, August 1961 (page 1 is missing).

60 *Quantum of Action (efterladte noter)*   [22–23 December 1961]

Handwritten [L. Rosenfeld and J. Kalckar], 17 pp., Danish and English, mf. 25.

Talks and lectures. Notes and drafts of a paper "The Quantum of Action and [the] Stability in Nature", December 1961. Also short notes on Superconductivity, Thermodynamics and Relativity theory.

[279] P

61 *Taler og forelæsninger, 1961*   [2 September, 5 September, 5 October, 11 October 1961]

Typewritten, carbon copy and handwritten [L. Rosenfeld], 46 pp., Danish and English, mf. 25.

Drafts of talks. (a) Talk on the occasion of the 150th anniversary of Carlsberg brewer J.C. Jacobsen's birth, 2 September 1961. (b) Manuscript for talk at Rutherford

Jubilee International Conference, September 1961. (c) Welcoming address to George de Hevesy held at the Anniversary of the Association of Engineers (Denmark), 5 October 1961. (d) Address to University of Brussels on award of doctor's degree, 11 October 1961.

62 *Supraledning (efterladte noter)* [16 January 1961 – 31 August 1962]

Handwritten [N. Bohr, L. Rosenfeld, J. Kalckar], 68 pp., Danish and English, mf. 26.

Superconductivity (notes left behind): (a) "Om supraledning" including note to observers of "quantized plugs". (b) "The Measuring Problem in Atomic Physics and the Complementary Description of Quantum Mechanics". (c) Notes on superconductivity. (d) "Quantization and Stability of Currents in Superconductors", by A. Bohr and B.R. Mottelson.

[121] C

63 *Taler og forelæsninger 1962* [January 1962, 3 and 10 February, 10 April, 1 June 1962]

Typewritten, handwritten [N. Bohr and E. Rüdinger], 11 pp., Danish and English, mf. 26.

Talks and lectures 1962: (a) Talk at Erik Warburg's 70th birthday, 3 February 1962. (b) Introductory talk to "Skole TV" (school television), programme on atoms, 10 February 1962. (c) Notes for an introduction to the series "Kvantebøgerne". (d) Statement of regret of inability to attend the inauguration of the Tata Institute of Fundamental Research, Bombay, dated January 1962. (e) Discussion on Thermodynamics and Microbiology, Rockefeller Institute, University of Copenhagen, 1 June 1962.

BIBLIOGRAPHY OF NIELS BOHR'S PUBLICATIONS REPRODUCED IN THE COLLECTED WORKS

INTRODUCTION

The contents of the Niels Bohr Collected Works have been organized in part by chronology, in part by topic, and it may not be obvious where a particular item can be found. In this last volume, it may therefore be helpful for the reader to have available a bibliography of material reproduced in the entire Collected Works. Because the series claims completeness only with regard to Bohr's publications, and because the several unpublished manuscripts and letters are included mainly to illustrate these, it has been deemed appropriate to limit the bibliography to publications proper, understood in a wide sense, so that Bohr's master's and doctor's theses at the University of Copenhagen as well as abstracts of lectures held at the Royal Academy of Sciences and Letters, published in the Academy's Proceedings, are included.

The list on the following pages is restricted to the versions of Bohr's publications reproduced in the Collected Works; for any other version the reader is referred to the information in the relevant volume itself. The number of the volume in which a publication is reproduced, as well as the page numbers it occupies there, are provided on the right-hand side of the page. The list is organized chronologically according to publication date, although there may be exceptions to this rule within each year.

(1) *Determination of the Surface-Tension of Water by the Method of Jet Vibration*, Phil. Trans. Roy. Soc. **209** (1909) 281–317 — **1**, 25–65

(2) *An Account of the Application of the Electron Theory to Explain the Physical Properties of Metals*; *En Fremstilling af Elektrontheoriens Anvendelse til Forklaring af Metallernes fysiske Egenskaber*, Master's thesis, 1909
Translation — **1**, 131–161

(3) *On the Determination of the Tension of a recently formed Water-Surface*, Proc. Roy. Soc. **A84** (1910) 395–403 — **1**, 79–89

(4) *Studies on the Electron Theory of Metals*; *Studier over Metallernes Elektronteori*, Dissertation for the Degree of Doctor of Philosophy, V. Thaning & Appel, Copenhagen 1911
Translation — **1**, 163–290
— **1**, 291–395

(5) [Autobiography], "Festskrift udgivet af Kjøbenhavns Universitet i Anledning af Universitetets Aarsfest, November 1911", Schultz, Copenhagen 1911, pp. 76–77 — **12**, 133–135
Translation — **12**, 136

(6) *Note on the Electron Theory of Thermoelectric Phenomena*, Phil. Mag. **23** (1912) 984–988 — **1**, 439–444

(7) *On the Theory of the Decrease of Velocity of Moving Electrified Particles on Passing through Matter*, Phil. Mag. **25** (1913) 10–31 — **2**, 15–39; **8**, 47–71

(8) *On the Constitution of Atoms and Molecules*, Part I, Phil. Mag. **26** (1913), 1–25 — **2**, 159–185

(9) *On the Constitution of Atoms and Molecules*, Part II: *Systems containing only a Single Nucleus*, Phil. Mag. **26** (1913) 476–502 — **2**, 187–214

[538]

(10) *On the Constitution of Atoms and Molecules*, Part III: *Systems Containing Several Nuclei*, Phil. Mag. **26** (1913) 857–875 — **2**, 215–233

(11) *The Spectra of Helium and Hydrogen*, Nature **92** (1913) 231–233 — **2**, 273–276

(12) *On the Spectrum of Hydrogen*; *Om Brintspektret*, Fys. Tidsskr. **12** (1914) 97–114

English version in "The Theory of Spectra and Atomic Constitution", Cambridge University Press, 1922, pp. 1–19 — **2**, 281–301

(13) *Atomic Models and X-Ray Spectra*, Nature **92** (1914) 553–554 — **2**, 304

(14) *On the Effect of Electric and Magnetic Fields on Spectral Lines*, Phil. Mag. **27** (1914) 506–524 — **2**, 347–368

(15) *On the Series Spectrum of Hydrogen and the Structure of the Atom*, Phil. Mag. **29** (1915) 332–335 — **2**, 375–380

(16) *The Spectra of Hydrogen and Helium*, Nature **95** (1915) 6–7 — **2**, 383–388

(17) *Modern Electrical Theory* [book review], Nature **95** (1915) 420–421 — **12**, 19–20

(18) *On the Quantum Theory of Radiation and the Structure of the Atom*, Phil. Mag. **30** (1915) 394–415 — **2**, 389–413

(19) *On the Decrease of Velocity of Swiftly Moving Electrified Particles in Passing Through Matter*, Phil. Mag. **30** (1915) 581–612 — **2**, 57–90; **8**, 127–160

(20) *Henry Gwyn Jeffreys Moseley*, Phil. Mag. **31** (1916) 173–176 — **12**, 283–288

(21) *On the Model of a Triatomic Hydrogen Molecule*, Medd. Kgl. Vet. Akad., Nobel Inst. **5**, No. 28 (1919) — **2**, 471–488

[539]

(22) *On the Quantum Theory of Line-Spectra*, Part I: *On the general theory*, Kgl. Dan. Vid. Selsk. Skr., 8. Række, IV.1 (1918–1922), pp. 5–36   **3**, 65–102

(23) *On the Quantum Theory of Line-Spectra*, Part II: *On the hydrogen spectrum*, Kgl. Dan. Vid. Selsk. Skr., 8. Række, IV.1 (1918–1922), pp. 37–100   **3**, 103–166

(24) *On the Quantum Theory of Line-Spectra*, Part III: *On the spectra of elements of higher atomic number*, Kgl. Dan. Vid. Selsk. Skr., 8. Række, IV.1 (1918–1922), pp. 101–111   **3**, 167–184

(25) *Professor Sir Ernest Rutherford and his Significance for the Recent Development of Physics*; *Professor Sir Ernest Rutherford og hans Betydning for Fysikens nyere Udvikling*, Politiken, 18 September 1920   **12**, 253–257

    Translation   **12**, 258–261

(26) *Our Present Knowledge of Atoms*; *Unsere heutige Kenntnis vom Atom*, Die Umschau **25** (1921) 229–232

    Translation   **4**, 84–89

(27) *On the Question of the Polarization of Radiation in the Quantum Theory*; *Zur Frage der Polarisation der Strahlung in der Quanten-theorie* Z. Phys. **6** (1921) 1–9   **3**, 339–349

    Translation   **3**, 350–356

(28) *Geleitwort*, in N. Bohr, "Abhandlungen über Atombau aus den Jahren 1913–1916", Vieweg & Sohn, Brauschweig 1921   **3**, 303–324

    Translation   **3**, 325–337

(29) *Atomic Structure*, Nature **107** (1921) 104–107   **4**, 71–82

(30) *Atomic Structure*, Nature **108** (1921) 208–209   **4**, 175–180

(31) *On the Result of Collisions between Atomic Systems and Free Electrical Particles*; *Om Virkningen af Sammenstød mellem Atomsystemer og fri elektriske Partikler*, Det nordiske H.C. Ørsted Møde i København 1920, H.C. Ørsted Komiteen, Copenhagen 1921, pp. 120–121   **8**, 195–198

    Translation   **8**, 199–200

(32) *On the Explanation of the Periodic System*; *Om Forklaringen af det Periodiske System*; "Autoreferat av föredrag vid Andra Nordiska Fysikermötet i Upsala, den 24.–26. augusti", Edv. Berlings Boktryckeri A.B., Uppsala 1922, pp. 3–4; also Fys. Tidsskr. **20** (1922) 112–115

    Translation   **4**, 421–424

(33) *On the Series Spectra of the Elements*, "The Theory of Spectra and Atomic Constitution", Cambridge University Press 1922, pp. 20–60   **3**, 241–282

(34) *The Structure of the Atom and the Physical and Chemical Properties of the Elements*; *Atomernes Bygning og Stoffernes fysiske og kemiske Egenskaber*, Jul. Gjellerup, Copenhagen 1922   **4**, 181–256

English version in "The Theory of Spectra and Atomic Constitution", Cambridge University Press, Cambridge 1924, iii–vii, x, 61–138.   **4**, 257–340
The English version includes an additional appendix. The pages in BCW also include the (untranslated) preface to the Danish book as well as *Preface* (May 1922) and *Preface to Second Edition* (May 1924) of the English book

(35) *The Difference between Series Spectra of Isotopes*, Nature **109** (1922) 745   **3**, 453–454

(36) *The Seventh Guthrie Lecture, Physical Society of London: The Effect of Electric and Magnetic Fields on Spectral Lines*, Fleetway Press, London 1922, pp. 275–302   **3**, 415–446

(37) *On the Selection Principle of the Quantum Theory*, Phil. Mag. **43** (1922) 1112–1116   **3**, 447–452

(38) *On the application of the quantum theory to atomic problems*; *L'application de la théorie des quanta aux problèmes atomiques*, "Atomes et électrons", Rapports et discussions du Conseil de physique tenu à Bruxelles du 1er au 6 avril 1921, Gauthier-Villars et Cie, Paris 1923, pp. 228–247

    Translation   **3**, 357–380

(39) [Speech at the banquet], "Les Prix Nobel en 1921–1922", P.A. Norstedt, Stockholm 1923, pp. 102–104

    Translation   **4**, 26–27

(40) *Niels Bohr, ibid.*, pp. 126–127   **12**, 143–145

(41) *The Structure of the Atom*; *Om Atomernes Bygning*, *ibid.*, 1–37 (paginated independently)   **4**, 425–465

    English version in Nature **112** (1923) 29–44   **4**, 467–482

(42) *X-Ray Spectra and the Periodic System of the Elements*; *Röntgenspektren und periodisches System der Elemente* (with D. Coster), Z. Phys. **12** (1923) 342–374   **4**, 483–518

    Translation   **4**, 519–548

(43) *Line Spectra and Atomic Structure*; *Linienspektren und Atombau*, Ann. d. Phys. **71** (1923) 228–288   **4**, 549–610

    Translation   **4**, 611–656

(44) *The Correspondence Principle*, Brit. Ass. Adv. Sci., Report of the Annual Meeting, Liverpool, September 1923, London 1924, pp. 428–429   **3**, 575–577

(45) *The Spectra of the Lighter Elements*, Nature **113** (1924) 223–224

    Contribution at the British Association Meeting at Liverpool, September 1923

    Transcription (from a slightly shorter manuscript)   **3**, 578–579

(46) *On the Application of the Quantum Theory to Atomic Structure*, Part I: *The Fundamental Postulates of the Quantum Theory*, Proc. Cambr. Phil. Soc. (Suppl.) 1924 — **3**, 455–499

(47) *The Quantum Theory of Radiation; Über die Quantentheorie der Strahlung* (with H.A. Kramers and J.C. Slater) Z. Phys. **24** (1924) 69 (abstract only) — **5**, 97–98

    Translation — **5**, 98

(48) *The Quantum Theory of Radiation* (with H.A. Kramers and J.C. Slater), Phil. Mag. **47** (1924) 785–802 — **5**, 99–118

(49) *The Foundations of Modern Atomic Research; Grundlaget for den Moderne Atomforskning*, Fys. Tidsskr. **23** (1925) 10–17 — **5**, 125–135

    Translation — **5**, 136–142

(50) *On the Polarization of Fluorescent Light; Zur Polarisation des Fluorescenzlichtes*, Naturwiss. **12** (1924) 1115–1117 — **5**, 143–147

    Translation — **5**, 148–154

(51) *On the Law of Conservation of Energy; Om Energisætningen*, Overs. Dan. Vidensk. Selsk. Forh. Juni 1924 – Maj 1925, p. 32 — **5**, 173–174

    English version in Nature **116** (1925) 262 — **5**, 174

(52) *On the Behaviour of Atoms in Collisions; Über die Wirkung von Atomen bei Stössen*, Z. Phys. **34** (1925) 142–157 — **5**, 175–193

    Translation — **5**, 194–206

(53) *Atomic Theory and Mechanics; Atomteori og mekanik*, Mat. Tidsskr. B (1925) 104–107 — **5**, 241–245

    Translation — **5**, 246–248

(54) *Atomic Theory and Mechanics*, Nature **116** (1925) 845–852 — **5**, 269–280

(55) *Some Aspects of the Later Development of Atomic Theory*; *Nogle Træk fra Atomteoriens senere Udvikling*, Fys. Tidsskr. **24** (1926) 20–21
Report by H.A. Kramers on Bohr's lecture at the Third Nordic Physicists' Meeting, Oslo, 24 August 1925 **5**, 281–283
 Translation **5**, 284–286

(56) *Sir Ernest Rutherford, O.M., P.R.S.*, Nature (Suppl.) **118** (18 Dec 1926) 51–52 **12**, 263–266

(57) *Atom*, Encyclopædia Britannica, 13th edition, Suppl., Vol. 1, London and New York 1926, pp. 262–267 **4**, 657–663

(58) *Spinning Electrons and the Structure of Spectra* (Letter to the Editor), Nature **117** (1926) 264–265 **5**, 287–289

(59) *Sir J.J. Thomson's Seventieth Birthday*, Nature **118** (1926) 879 **12**, 251–252

(60) *Atomic Theory and Wave Mechanics* (Abstract); *Atomteori og Bølgemekanik*, Overs. Dan. Vidensk. Selsk. Forh. Juni 1926 – Maj 1927, pp. 28–29 **6**, 55–56
 English version in Nature **119** (1927) 262 **6**, 56

(61) *The Quantum Postulate and the Recent Development of Atomic Theory*; *Kvantepostulatet og Atomteoriens seneste Udvikling*, Overs. Dan. Vidensk. Selsk. Forh. Juni 1927 – Maj 1928, p. 27 **6**, 107–108
 English version in Nature **121** (1928) 78 **6**, 108

(62) *The Quantum Postulate and the Recent Development of Atomic Theory*, Atti del Congresso Internazionale dei Fisici 11–20 Settembre 1927, Como–Pavia–Roma, Volume Secundo, Nicola Zanichelli, Bologna 1928, pp. 565–598 **6**, 109–146
Includes discussion remarks on Bohr's contribution (pp. 589–598)

(63) *The Quantum Postulate and the Recent Development of Atomic Theory*, Nature (Suppl.) **121** (1928) 580–590   **6**, 147–158

(64) *Discrepancies between Experiment and the Electromagnetic Theory of Radiation*; *Discordances entre l'expérience et la théorie électromagnetique du rayonnement*, "Électrons et photons", Rapports et discussions du cinquième Conseil de physique tenu à Bruxelles du 24 au 29 Octobre 1927, Gauthier-Villars, Paris 1928, pp. 91–92
Discussion of A.H. Compton's contribution   **5**, 207–210

    Translation   **5**, 211–212

(65) *General Discussion at the Fifth Solvay Conference*, *ibid.*, pp. 253–256, 261–263, 264–265 (and unpublished manuscript)

    Translation   **6**, 99–106

(66) *Sommerfeld and the Theory of the Atom*; *Sommerfeld und die Atomtheorie*, Naturwiss. **16** (1928) 1036   **12**, 343–344

    Translation   **12**, 345–346

(67) *At Harald Høffding's 85th Birthday*; *Ved Harald Høffdings 85 Aars-Dag*, Berlingske Tidende, 10 March 1928   **10**, 305–307

    Translation   **10**, 308–309

(68) [Autobiography], "Studenterne MCMIII: Personalhistoriske Oplysninger", Berlingske Bogtrykkeri, Copenhagen 1928, p. 275   **12**, 147–149

    Translation   **12**, 150

(69) *Quantum Theory and Relativity* (Abstract); *Kvanteteori og Relativitet*, Overs. Dan. Vidensk. Selsk. Forh. Juni 1928 – Maj 1929, p. 24   **6**, 199–200

    English version in Nature **123** (1929) 434   **6**, 200

(70) *Professor Niels Bjerrum 50 Years*; *Professor Niels Bjerrum fylder 50 Aar*, Berlingske Tidende, 9 March 1929   **12**, 437–439

    Translation   **12**, 440–441

(71) *The Quantum of Action and the Description of Nature*; *Wirkungsquantum und Naturbeschreibung*, Naturwiss. **17** (1929) 483–486   **6**, 201–206

    English version in "Atomic Theory and the Description of Nature", Cambridge University Press, Cambridge 1934, pp. 92–101   **6**, 208–217

(72) *The Atomic Theory and the Fundamental Principles underlying the Description of Nature*; *Atomteorien og Grundprincipperne for Naturbeskrivelsen*, Beretning om det 18. skandinaviske Naturforskermøde i København 26.–31. August 1929, Frederiksberg Bogtrykkeri, Copenhagen 1929, pp. 71–83   **6**, 219–235

    English version in "Atomic Theory and the Description of Nature", Cambridge University Press, Cambridge 1934, pp. 102–119   **6**, 236–253

(73) *Introductory Survey* (with *Addendum* of 1931); *Indledende oversigt* (med *Tillæg* fra 1931), "Atomteori og Naturbeskrivelse", Festskrift udgivet af Københavns Universitet i Anledning af Universitetets Aarsfest 1929, Bianco Lunos Bogtrykkeri, Copenhagen 1929, pp. 5–19, and J.H. Schultz Forlag, Copenhagen 1958, pp. 23–25   **6**, 255–276

    English version in "Atomic Theory and the Description of Nature", Cambridge University Press, 1934, pp. 1–24 (includes additional unpaginated preface)   **6**, 277–302

(74) *Atom*, Encyclopædia Britannica, 14th edition, Vol. 2, London and New York 1929, pp. 642–648   **12**, 41–48

(75) *Maxwell and Modern Theoretical Physics*, Nature **128** (1931) 691–692   **6**, 357–360

(76) *The Magnetic Electron*; *L'électron magnetique*, "Le magnétisme", Rapports et discussions du sixième Conseil de physique tenu à Bruxelles du 20 au 25 octobre 1930, Gauthier-Villars, Paris 1932, pp. 276–280
Discussion of Wolfgang Pauli's lecture

    Translation   **6**, 347–349

(77) *Philosophical Aspects of Atomic Theory* (Abstract), Nature **125** (1930) 958   **6**, 351–352

(78) *The Use of the Concepts of Space and Time in Atomic Theory* (Abstract); *Om Benyttelsen af Begreberne Rum og Tid i Atomteorien*, Overs. Dan. Vidensk. Selsk. Forh. Juni 1930 – Maj 1931, p. 26   **6**, 353–354

    English version in Nature **127** (1931) 43   **6**, 354

(79) *Tribute to the Memory of Harald Høffding*; *Mindeord over Harald Høffding*, Overs. Dan. Vidensk. Selsk. Virks. Juni 1931 – Maj 1932, pp. 131–136   **10**, 311–318

    Translation   **10**, 319–322

(80) *On Atomic Stability* (Abstract), Brit. Ass. Adv. Sci., Report of the Centenary Meeting, London, 23–30 September 1931, London 1932, p. 333   **6**, 355–356

(81) *Faraday Lecture: Chemistry and Quantum Theory of Atomic Constitution*, J. Chem. Soc. (1932) 349–384
Vol. 9 contains extract only   **6**, 371–408; **9**, 91–97

(82) *Atomic Stability and Conservation Laws*, Atti del Convegno di Fisica Nucleare della "Fondazione Alessandro Volta", Ottobre 1931, Reale Accademia d'Italia, Rome 1932, pp. 119–130   **9**, 99–114

(83) *On the Properties of the Neutron* (Abstract); *Om Neutronernes Egenskaber*, Overs. Dan. Vidensk. Selsk. Virks. Juni 1931 – Maj 1932, p. 52   **9**, 119–121

    Translation   **9**, 121

(84) *Light and Life*, Congress on Light Therapy in Copen-   **10**, 27–35
hagen, 15 August 1932, Nature **131** (1933) 421–423,
457–459

(85) *The Limited Measurability of Electromagnetic Fields*   **7**, 53–54
of Force (Abstract); *Om den begrænsede Maalelighed*
af elektromagtiske Kraftfelter (with L. Rosenfeld),
Overs. Dan. Vidensk. Selsk. Virks. Juni 1932 – Maj
1933, p. 35

English version in Nature **132** (1933) 75   **7**, 54

(86) *On the Question of the Measurability of Electromag-*   **7**, 55–121
netic Field Quantities; *Zur Frage der Messbarkeit*
der elektromagnetischen Feldgrössen (with L. Rosenfeld), Mat.–Fys. Medd. Dan. Vidensk. Selsk. **12**,
No. 8 (1933)

English version in "Selected Papers of Léon   **7**, 123–166
Rosenfeld" (eds. R.S. Cohen and J.J. Stachel),
D. Reidel Publishing Company, Dordrecht 1979,
pp. 357–400

(87) *On the Correspondence Method in Electron Theory*;   **7**, 167–182
Sur la méthode de correspondance dans la théorie de
l'électron, "Structure et propriétés des noyaux atomiques", Rapports et discussions du septième Conseil
de physique tenu à Bruxelles du 22 au 29 octobre
1933, Gauthier-Villars, Paris 1934, pp. 216–228

Translation   **7**, 183–191
Translation of pp. 226–228   **9**, 129–132

(88) *Discussion Remarks, ibid.*, pp. 72, 175, 180, 214–215   **7**, 192–193;
(in Vol. 7), 287–288, 327–328, 329–330, 331, 334   **9**, 133–141
Translation   **7**, 193;
    9, 135–141

(89) *Obituary for Christian Alfred Bohr: Born 25 Novem-*   **12**, 407–419
ber 1916 – Died 2 July 1934; *Mindeord over Chris-*
tian Alfred Bohr: født 25. November 1916 – død 2.
Juli 1934, Private print, 1934

Translation   **12**, 420–424

(90) *Zeeman Effect and Theory of Atomic Constitution*, "Zeeman Verhandelingen", Martinus Nijhoff, The Hague 1935, pp. 131–134   **10**, 335–340

(91) *Conversation with Niels Bohr; Samtale med Niels Bohr*, Berlingske Aftenavis, 2 October 1935   **12**, 151–162

    Translation   **12**, 163–175

(92) *Quantum Mechanics and Physical Reality*, Nature **136** (1935) 65   **7**, 289–290

(93) *Friedrich Paschen on his Seventieth Birthday; Friedrich Paschen zum siebzigsten Geburtstag*, Naturwiss. **23** (1935) 73   **12**, 339–340

    Translation   **12**, 341–342

(94) *Can Quantum-Mechanical Description of Physical Reality be Considered Complete?*, Phys. Rev. **48** (1935) 696–702   **7**, 291–298

(95) *Properties and Constitution of Atomic Nuclei* (Abstract); *Om Atomkernernes Egenskaber og Opbygning*, Overs. Dan. Vidensk. Selsk. Virks. Juni 1935 – Maj 1936, p. 39   **9**, 149–150

    English version in Nature **138** (1936) 695   **9**, 150

(96) *Neutron Capture and Nuclear Constitution (1)*, Nature **137** (1936) 344–348   **9**, 151–156

(97) *Neutron Capture and Nuclear Constitution (2)*, Nature **137** (1936) 351   **9**, 157–158

(98) *Conservation Laws in Quantum Theory*, Nature **138** (1936) 25–26   **5**, 215–216

(99) *Properties of Atomic Nuclei; Atomkernernes Egenskaber*, "Nordiska (19. skandinaviska) Naturforskarmötet i Helsingfors den 11.–15. augusti 1936", Helsinki 1936, pp. 73–81   **9**, 159–171

    Translation   **9**, 172–178

(100) [*Speech at the Meeting of Natural Scientists in Helsinki, 14 August 1936*]; [*Tale ved naturforskermødet i Helsingfors, 14. August 1936*], ibid., pp. 191–192   **11**,501–504

Translation   **11**, 505–507

(101) *Causality and Complementarity*, Phil. Sci. **4** (1937) 289–298   **10**, 37–48

(102) *Transmutations of Atomic Nuclei*, Science **86** (1937) 161–165   **9**, 205–211

(103) *On the Transmutation of Atomic Nuclei; Om Spaltning af Atomkerner*, 5. nordiske Elektrotekniker-møde, J.H. Schultz Bogtrykkeri, Copenhagen 1937, pp. 21–23   **9**, 213–217

Translation   **9**, 218–221

(104) *On the Transmutation of Atomic Nuclei by Impact of Material Particles*, Part I: *General Theoretical Remarks* (with F. Kalckar), Mat.–Fys. Medd. Dan. Vidensk. Selsk. **14**, No. 10 (1937)   **9**, 223–264

(105) [Obituary for Rutherford], Nature **140** (1937) 752–753   **12**, 271–272

(106) [Obituary for Rutherford], Nature (Suppl.) **140** (1937) 1048–1049   **12**, 273–274

(107) *Nuclear Mechanics; Mécanique nucléaire*, "Actualités scientifiques et industrielles: Réunion internationale de physique–chimie–biologie, Congrès du Palais de la découverte, Paris, Octobre 1937, Vol. II: Physique nucléaire", Hermann et Cie, Paris 1938, pp. 81–82   **9**, 265–268

Translation   **9**, 269

(108) *On Nuclear Reactions* (Abstract); *Om Atomkernereaktioner*, Overs. Dan. Vidensk. Selsk. Virks. Juni 1937 – Maj 1938, p. 32   **9**, 287–289

Translation   **9**, 289

(109) *Magister Fritz Kalckar*, Politiken, 7 January 1938   **12**, 385–386
       Translation   **12**, 387–388

(110) *Nuclear Photo-effects*, Nature **141** (1938) 326–327   **9**, 297–299

(111) *Quantum of Action and Atomic Nucleus*; *Wirkungsquantum und Atomkern*, Ann. d. Phys. **32** (1938) 5–19   **9**, 301–317
       Translation   **9**, 318–329

(112) *Biology and Atomic Physics*, "Celebrazione del secondo centenario della nascita di Luigi Galvani", Bologna – 18–21 ottobre 1937-XV: I. Rendiconto generale, Tipografia Luigi Parma 1938, pp. 68–78   **10**, 49–62

(113) *Resonance in Nuclear Photo-Effects*, Nature **141** (1938) 1096–1097   **9**, 331–332

(114) *Analysis and Synthesis in Science*, International Encyclopedia of Unified Science **1** (1938) 28   **10**, 63–64

(115) *Science and its International Significance*, Danish Foreign Office Journal, No. 208 (May 1938) 61–63   **11**, 527–531

(116) *Symposium on Nuclear Physics, Introduction*, Brit. Ass. Adv. Sci., Report of the Annual Meeting, 1938 (108th Year), Cambridge, August 17–24, London 1938, p. 381 (Abstract)   **9**, 333–335
       Nature (Suppl.) **142** (1938) 520–521 (Report)   **9**, 336–337

(117) *Matter, Structure of*, Encyclopædia Britannica Book of the Year 1938, pp. 403–404   **12**, 49–52

(118) *The Causality Problem in Atomic Physics*, "New Theories in Physics", Conference organized in collaboration with the International Union of Physics and the Polish Intellectual Co-operation Committee, Warsaw, May 30th – June 3rd 1938, International Institute of Intellectual Co-operation, Paris 1939, pp. 11–30   **7**, 299–322

(119) [Two speeches for Rutherford, 1932], A.S. Eve,   **12**, 267–269
"Rutherford: Being the Life and Letters of the Rt
Hon. Lord Rutherford, O.M.", Cambridge University
Press, Cambridge 1939, pp. 361–363

(120) *Natural Philosophy and Human Cultures,*   **10**, 237–249
Congrès internal des sciences anthropologiques
et ethnologiques, compte rendu de la deuxième
session, Copenhagen 1938, Ejnar Munksgaard,
Copenhagen 1939, pp. 86–95

(121) *Reactions of Atomic Nuclei* (Abstract); *Om Atom-*   **9**, 339–340
kernernes Reaktioner, Overs. Dan. Vidensk. Selsk.
Virks. Juni 1938 – Maj 1939, p. 25
English version in Nature **143** (1939) 215   **9**, 340

(122) *Disintegration of Heavy Nuclei,* Nature **143** (1939)   **9**, 341–342
330

(123) *Resonance in Uranium and Thorium Disintegration*   **9**, 343–345
and the Phenomenon of Nuclear Fission, Phys. Rev.
55 (1939) 418–419

(124) *Mechanism of Nuclear Fission* (with J.A. Wheeler),   **9**, 359–361
Phys. Rev. **55** (1939) 1124

(125) *The Mechanism of Nuclear Fission* (with J.A.   **9**, 363–389
Wheeler), Phys. Rev. **56** (1939) 426–450

(126) *Nuclear Reactions in the Continuous Energy Region*   **9**, 391–393
(with R. Peierls and G. Placzek), Nature **144** (1939)
200–201

(127) *Matter, Structure of,* Encyclopædia Britannica Book   **12**, 53–54
of the Year 1939, pp. 409–410

(128) *The Fission of Protactinium* (with J.A. Wheeler),   **9**, 403–404
Phys. Rev. **56** (1939) 1065–1066

(129) *Writer and Scientist*; *Digter og Videnskabsmand*, "Festskrift til Niels Møller paa Firsaarsdagen 11. December 1939", Munksgaard, Copenhagen 1939, pp. 80–81 — **12**, 461–463

    Translation — **12**, 464–465

(130) *On the Fragments Ejected in the Disintegration of the Uranium Nucleus*; *Om de ved Urankernernes Sønderdeling udslyngede Fragmenter*, Overs. Dan. Vidensk. Selsk. Virks. Juni 1939 – Maj 1940, pp. 49–50 — **8**, 317–318

    Translation — **8**, 318

(131) *Meeting on 20 October 1939*; *Mødet den 20. Oktober 1939*, Overs. Dan. Vidensk. Selsk. Virks. Juni 1939 – Maj 1940, pp. 25–26 — **11**, 379–381

    Translation — **11**, 382–383

(132) *The Theoretical Explanation of the Fission of Atomic Nuclei* (Abstract); *Den teoretiske Forklaring af Atomkernernes Fission*, Overs. Dan. Vidensk. Selsk. Virks. Juni 1939 – Maj 1940, p. 28 — **9**, 409–410

    Translation — **9**, 410

(133) *Foreword*, C. Møller and E. Rasmussen, "The World and the Atom," Allen & Unwin, London 1940, p. 9 — **12**, 73–75

(134) *Scattering and Stopping of Fission Fragments*, Phys. Rev. **58** (1940) 654–655 — **8**, 319–321

(135) *Successive Transformations in Nuclear Fission*, Phys. Rev. **58** (1940) 864–866 — **9**, 475–479

(136) *Velocity-Range Relation for Fission Fragments* (with J.K. Bøggild, K.J. Brostrøm and T. Lauritsen), Phys. Rev. **58** (1940) 839–840 — **8**, 323–326

(137) *Meeting on 15 March [1940]*; *Mødet den 15. Marts [1940]*, Overs. Dan. Vidensk. Selsk. Virks. Juni 1939 – Maj 1940, pp. 40–41 — **11**, 385–386

    Translation — **11**, 387

(138) *Meeting on 20 September 1940*; *Mødet den 20.*   **11**, 389–390
Septbr. 1940, Overs. Dan. Vidensk. Selsk. Virks. Juni
1940 – Maj 1941, pp. 25–26
   Translation   **11**, 391

(139) *Disintegration of Heavy Nuclei* (Abstract); *Tunge*   **9**, 481–482
Atomkerners Sønderdeling, Overs. Dan. Vidensk.
Selsk. Virks. Juni 1940 – Maj 1941, p. 38
   Translation   **9**, 482

(140) *Danish Culture. Some Introductory Reflections*;   **10**, 251–261
Dansk Kultur. Nogle indledende Betragtninger,
"Danmarks Kultur ved Aar 1940", Det Danske
Forlag, Copenhagen 1941–1943, Vol. 1, pp. 9–17
   Translation   **10**, 262–272

(141) *Recent Investigations of the Transmutations of Atomic*   **9**, 411–442
Nuclei; *Nyere Undersøgelser over Atomkernernes*
Omdannelser, Fys. Tidsskr. **39** (1941) 3–32
   Translation   **9**, 443–466

(142) *Kirstine Meyer, n. Bjerrum: 12 October 1861 – 28*   **12**, 425–429
September 1941; *Kirstine Meyer, f. Bjerrum: 12. Ok-*
tober 1861 – 28. September 1941, ibid., 113–115
   Translation   **12**, 430–431

(143) *Eighth Presentation of the H.C. Ørsted Medal*   **11**, 469–472
[K. Linderstrøm-Lang]; *Ottende Uddeling af H.C.*
Ørsted Medaillen, Fys. Tidsskr. **39** (1941) 175–177,
192–193
   Translation   **11**, 473–475

(144) *Velocity–Range Relation for Fission Fragments*, Phys.   **8**, 327–333
Rev. **59** (1941) 270–275

(145) *Professor Martin Knudsen*, Berlingske Aftenavis, 14   **12**, 289–291
February 1941
   Translation   **12**, 292–294

(146) *The University and Research*; *Universitetet og Forsk-*   **11**, 533–546
ningen, Politiken, 3 June 1941
     Translation   **11**, 547–552

(147) *Farewell to Sweden's Ambassador in Copenhagen*;   **12**, 481–482
Afsked med Sveriges Gesandt i København, Politiken,
15 November 1941
     Translation   **12**, 483

(148) *Mechanism of Deuteron-Induced Fission*, Phys. Rev.   **9**, 483–484
59 (1941) 1042

(149) *Analysis and Synthesis in Atomic Physics* (Abstract);   **7**, 323–324
Analyse og Syntese indenfor Atomfysikken, Overs.
Dan. Vidensk. Selsk. Virks. Juni 1941 – Maj 1942,
p. 30
     Translation   **7**, 324

(150) *Meeting on 30 January 1942*; *Mødet den 30. Januar*   **11**, 392–394
1942, Overs. Dan. Vidensk. Selsk. Virks. Juni 1941
– Maj 1942, pp. 32–34
     Translation   **11**, 395–396

(151) [Tribute to Bering], "Vitus Bering 1741–1941",   **12**, 237–243
H. Hagerup, Copenhagen 1942, pp. 49–53
     Translation   **12**, 244–246

(152) *Memorial Evening for Kirstine Meyer in the Society*   **11**, 493–496
for Dissemination of Natural Science; *Mindeaften for*
Kirstine Meyer i Selskabet for Naturlærens Udbre-
delse, Fys. Tidsskr. **40** (1942) 173–175
     Translation   **11**, 497–499

(153) *Foreword*; *Forord*, G. Gamow, "Mr. Tompkins   **12**, 77–80
i Drømmeland", Gyldendal, Copenhagen 1942,
pp. 7–8
     Translation   **12**, 81–82

(154) *Meeting on 13 November 1942 on the 200th Anniver-*   **11**, 397–405
sary of the Establishment of the Academy; *Mødet*
den 13. November 1942 paa 200-Aarsdagen for Sel-
skabets Stiftelse, Overs. Dan. Vidensk. Selsk. Virks.
Juni 1942 – Maj 1943, pp. 26–28, 31–32, 36, 40–41,
44–48

    Translation   **11**, 407–414

(155) *Harald Høffding's 100th Birthday*; *Harald Høffdings*   **10**, 323–324
100-Aars Fødselsdag, ibid., pp. 57–58

    Translation   **10**, 325

(156) *Science and Civilization*, The Times, 11 August 1945   **11**, 121–124

(157) *A Challenge to Civilization*, Science **102** (1945) 363–   **11**, 125–129
364

(158) *Meeting on 19 October 1945*; *Mødet den 19. Oktober*   **11**, 415–416
1945, Overs. Dan. Vidensk. Selsk. Virks. Juni 1945
– Maj 1946, pp. 29–31

    Translation   **11**, 417

(159) *Meeting on 19 October 1945*; *Mødet den 19. Oktober*   **12**, 295–296
1945, Overs. Dan. Vidensk. Selsk. Virks. Juni 1945
– Maj 1946, pp. 31–32 [M. Knudsen's Retirement as
Secretary of the Royal Danish Academy]

    Translation   **12**, 297

(160) *On the Transmutations of Atomic Nuclei* (Abstract);   **9**, 485–486
Om Atomkernernes Omdannelser, Overs. Dan. Vi-
densk. Selsk. Virks. Juni 1945 – Maj 1946, p. 31

    Translation   **9**, 486

(161) *On the Problem of Measurement in Atomic Physics*;   **11**, 655–661
Om Maalingsproblemet i Atomfysikken, "Festskrift til
N.E. Nørlund i Anledning af hans 60 Aars Fødsels-
dag den 26. Oktober 1945 fra danske Matematikere,
Astronomer og Geodæter, Anden Del", Ejnar Munks-
gaard, Copenhagen 1946, pp. 163–167

    Translation   **11**, 662–666

(162) *Humanity's Choice Between Catastrophe and Happier Circumstances*; *Menneskehedens Valg mellem Katastrofe og lykkeligere Kaar*, Politiken, 1 January 1946   **11**, 149–151

   Translation   **11**, 152–154

(163) *A Personality in Danish Physics*; *En Personlighed i dansk Fysik* [H.M. Hansen], Politiken, 7 September 1946   **12**, 325–327

   Translation   **12**, 328–330

(164) *Speech at the Memorial Ceremony for Ole Chievitz 31 December 1946*; *Tale ved Mindehøjtideligheden for Ole Chievitz 31. December 1946*, Ord och Bild **55** (1947) 49–53   **12**, 449–456

   Translation   **12**, 456–460

(165) *Meeting on 25 April 1947*; *Mødet den 25. April 1947*, Overs. Dan. Vidensk. Selsk. Virks. Juni 1946 – Maj 1947, pp. 53–54   **11**, 419–422

   Translation   **11**, 423–424

(166) *Newton's Principles and Modern Atomic Mechanics*, "The Royal Society Newton Tercentenary Celebrations, 15–19 July 1946", Cambridge University Press, Cambridge 1947, pp. 56–61   **12**, 219–225

(167) *Problems of Elementary-Particle Physics*, Report of an International Conference on Fundamental Particles and Low Temperatures held at the Cavendish Laboratory, Cambridge, on 22–27 July 1946, Volume 1, Fundamental Particles, The Physical Society, London 1947, pp. 1–4   **7**, 217–222

(168) *Atomic Physics and International Cooperation*, Address at Symposium of the National Academy of Sciences, Present Trends and International Implications of Science, Philadelphia, 21 October 1946, Proc. Am. Phil. Soc. **91** (1947), 137–138   **11**, 131–134

(169) *Meeting on 17 October 1947 in Commemoration of King Christian X*; *Mødet den 17. Oktober 1947 til Minde om Kong Christian X*, Overs. Dan. Vidensk. Selsk. Virks. Juni 1947 – Maj 1948, pp. 26–29   **11**, 425–429

Translation   **11**, 430–432

(170) [Tribute to Rutherford], *Hommage à Lord Rutherford 7–8 Novembre 1947*, Pamphlet, World Federation of Scientific Workers 1948, pp. 15–16   **12**, 275–278

(171) *The Penetration of Atomic Particles through Matter*, Mat.–Fys. Medd. Dan. Vidensk. Selsk. **18**, No. 8 (1948)   **8**, 423–568

(172) *On the Notions of Causality and Complementarity*, Dialectica **2** (1948) 312–319   **7**, 325–337

(173) [Foreword], "Niels Bjerrum: Selected Papers, edited by friends and coworkers on the occasion of his 70th birthday on the 11th March 1949", Munksgaard, Copenhagen 1949, p. 3   **12**, 443–443

(174) *Greeting from Niels Bohr*; *Niels Bohr's Hilsen*, "Akademisk Boldklub [Academic Ball Club] 1939–1949", Copenhagen 1949, pp. 7–8   **11**, 675–678

Translation   **11**, 679–680

(175) *Meeting on 11 March 1949*; *Mødet den 11. Marts 1949*, Overs. Dan. Vidensk. Selsk. Virks. Juni 1948 – Maj 1949, pp. 45–46   **11**, 433–434

Translation   **11**, 435

(176) *Atoms and Human Knowledge*; *Atomerne og vor erkendelse*, Berlingske Tidende, 2 April 1949   **12**, 57–62

Translation   **12**, 63–70

(177) *Niels Bohr on Jest and Earnestness in Science*; *Niels Bohr om spøg og alvor i videnskaben*, Politiken, 17 April 1949   **12**, 177–186

Translation   **12**, 187–196

[558]

(178) *Professor Martin Knudsen Died Yesterday*; *Prof. Martin Knudsen død i gaar*, Politiken, 28 May 1949   **12**, 299–300

     Translation   **12**, 301–302

(179) *Martin Knudsen 15.2.1871–27.5.1949*, Fys. Tidsskr. **47** (1949) 145–147   **12**, 303–307

     Translation   **12**, 308–311

(180) *He Stepped in Where Wrong had been Done: Obituary by Professor Niels Bohr*; *Han traadte hjælpende til hvor uret blev begaaet: Mindeord af professor Niels Bohr* [A. Friis], Politiken, 7 October 1949   **12**, 433–434

     Translation   **12**, 435

(181) *The Internationalist* [Albert Einstein], UNESCO Courier **2** (No. 2, 1949), 1, 7   **12**, 369–372

(182) *Discussion with Einstein on Epistemological Problems in Atomic Physics*, "Albert Einstein, Philosopher–Scientist" (ed. P.A. Schilpp), The Library of Living Philosophers, Vol. VII, Evanston, Illinois 1949, pp. 201–241   **7**, 339–381

(183) [Obituary for M. Knudsen], Overs. Dan. Vidensk. Selsk. Virks. Juni 1949 – Maj 1950, pp. 61–65   **12**, 313–319

     Translation   **12**, 320–324

(184) [*Discussion Remarks*], "Les particules élémentaires", Rapports et discussions du huitème Conseil de physique tenu à Bruxelles du 27 septembre au 2 octobre 1948, R. Stoops, Bruxelles 1950, pp. 107, 125–127   **8**, 569–572
Discussion of lectures by R. Serber ("Artificial Mesons") and C.F. Powell ("Observations on the Properties of Mesons of the Cosmic Radiation")

(185) *Some General Comments on the Present Situation in Atomic Physics*, "Les particules élémentaires", *ibid.*, pp. 376–380   **7**, 223–228
Contribution to the general discussion

(186) *Open Letter to the United Nations, June 9th, 1950,* Schultz, Copenhagen 1950 — **11**, 171–185

(187) *Field and Charge Measurements in Quantum Electro-dynamics* (with L. Rosenfeld), Phys. Rev. **78** (1950) 794–798 — **7**, 211–216

(188) *Niels Bohr's deeply serious appeal; Niels Bohrs dybt alvorlige appel*, Politiken, 19 January 1951 — **11**, 447–449

Translation — **11**, 450–452

(189) *H.C. Ørsted*, Fys. Tidsskr. **49** (1951) 6–20 — **10**, 341–356

Translation — **10**, 357–369

(190) *The Epistemological Problem of Natural Science* (Abstract); *Naturvidenskabens Erkendelsesproblem*, Overs. Dan. Vidensk. Selsk. Virks. Juni 1950 – Maj 1951, p. 39 — **7**, 383–384

Translation — **7**, 384

(191) *Meeting on 2 February 1951; Mødet den 2. Februar 1951, ibid.*, 453–457 — **11**, 453–457

Translation — **11**, 458–461

(192) *Meeting on 19 October 1951; Mødet den 19. Oktober 1951*, Overs. Dan. Vidensk. Selsk. Virks. Juni 1951 – Maj 1952, pp. 33–34 — **11**, 437–438

(193) *Meeting on 16 November 1951; Mødet den 16. November 1951, ibid.*, p. 39 — **11**, 439–440

Translation — **11**, 441

(194) *Statement by Professor Niels Bohr in the current affairs programme on Danish National Radio, Monday 4 February 1952; Udtalelse af Professor Niels Bohr i radioens aktuelle kvarter, mandag den 4. februar 1952*, Videnskabsmanden: Meddelelser fra Foreningen til Beskyttelse af Videnskabeligt Arbejde **6** (No. 1, 1952), p. 3 — **11**, 463–464

Translation — **11**, 465–467

(195) [Discussion remarks to papers by W. Kohn and B. Mottelson], "Report of the International Conference sponsored by the Council of Representatives of European States for Planning an International Laboratory and Organizing Other Forms of Co-Operation in Nuclear Research, Institute for Theoretical Physics, Copenhagen, June 3–17, 1952" (eds. O. Kofoed-Hansen, P. Kristensen, M. Scharff and A. Winther), pp. 16, 19

    Transcription   **9**, 527–529

(196) *Medical Research and Natural Philosophy*, Acta Medica Scandinavica (Suppl.) **142** (1952) 967–972   **10**, 65–72

(197) *On the Death of Hendrik Anthony Kramers*; *Ved Hendrik Anton Kramers død*, Politiken, 27 April 1952   **12**, 347–349

    Translation   **12**, 350–352

(198) *Hendrik Anthony Kramers †*, Ned. T. Natuurk. **18** (1952) 161–166   **12**, 353–360

(199) *Electron Capture by Swiftly Moving Ions of High Nuclear Charge* (Abstract); *Elektronindfangning af hurtigt bevægede Ioner med høj Kerneladning*, Overs. Dan. Vidensk. Selsk. Virks. Juni 1951 – Maj 1952, p. 49   **8**, 579–581

    Translation   **8**, 581

(200) [Discussion Contribution on Superconductivity], "Proceedings of the Lorentz Kamerlingh Onnes Memorial Conference, Leiden University 22–26 June 1953", Stichting Physica, Amsterdam 1953, pp. 761–762   **12**, 37–39

(201) *A Fruitful Lifework*; *Et frugtbart livsværk*, "Noter til en mand: Til Jens Rosenkjærs 70-aars dag" (eds. J. Bomholt and J. Jørgensen), Det Danske Forlag, Copenhagen 1953, p. 79   **12**, 473–474

    Translation   **12**, 475

(202) *Ninth Presentation of the H.C. Ørsted Medal*   **11**, 477–481
[A. Langseth]; *Niende Uddeling af H.C. Ørsted*
Medaillen, Fys. Tidsskr. **51** (1953) 65–67, 80
    Translation   **11**, 482–485

(203) *[Preface]*; *[Forord]*, ...fra Thrige **6**, No. 1 (1953) 2–4   **11**, 555–558
    Translation   **11**, 559–560

(204) *Speech Given at the 25th Anniversary Reunion of*   **10**, 223–232
the Student Graduation Class 21 September 1928;
Tale ved Studenterjubilæet, 1903–1928, Private print,
1953
    Translation   **10**, 233–236

(205) *Address Broadcast on Danish National Radio, 16*   **11**, 681–684
October 1953; *Tale ved Statsradiofoniens udsendelse*
den 16. oktober 1953, "Det kongelige Teater,
Forestillingen lørdag den 17. oktober 1953", pp. 4–6
    Translation   **11**, 685–688

(206) *Physical Science and the Study of Religions*, Studia   **10**, 275–280
Orientalia Ioanni Pedersen Septuagenario A.D. VII
id. Nov. Anno MCMLIII, Ejnar Munksgaard, Copenhagen 1953, pp. 385–390

(207) *The Rebuilding of Israel: A Remarkable Kind of Ad-*   **11**, 689–693
venture; *Israels Genopbygning: Et Æventyr af ejen-*
dommelig Art, Israel **7** (No. 2, 1954) 14–17
    Translation   **11** 694–699

(208) *Electron Capture and Loss by Heavy Ions Penetrat-*   **8**, 593–625
ing through Matter (with J. Lindhard), Mat.–Fys.
Medd. Dan. Vidensk. Selsk. **28**, No. 7 (1954)

(209) *Address at the Opening Ceremony*, Acta Radiologica   **10**, 73–78
(Suppl.) **116** (1954) 15–18

(210) *Foreword*; *Forord*, "Johan Nicolai Madvig: Et min-   **12**, 247–248
deskrift", Royal Danish Academy of Sciences and
Letters and the Carlsberg Foundation, Copenhagen
1955, p. vii

     Translation   **12**, 249

(211) *Unity of Knowledge*, "Unity of Knowledge" (ed. L.   **10**, 79–98
Leary), Doubleday & Co., New York 1955, pp. 47–
62

(212) *Rydberg's discovery of the spectral laws*, Proceed-   **10**, 371–379
ings of the Rydberg Centennial Conference on
Atomic Spectroscopy, Lunds Universitets Årsskrift.
N.F. Avd. 2, Bd. 50, Nr 21 (1955) 15–21

(213) *Greater International Cooperation is Needed for*   **11**, 561–570
Peace and Survival, "Atomic Energy in Industry:
Minutes of 3rd Conference October 13–15, 1954",
National Industrial Conference Board, Inc., New
York 1955, pp. 18–26

(214) *Obituary*; *Mindeord* [Einstein], Børsen, 19 April   **12**, 373–374
1955

     Translation   **12**, 375

(215) *Albert Einstein 1879–1955*, Sci. Am. **192** (1955) 31   **12**, 377–378

(216) *The Physical Basis for Industrial Use of the Energy*   **11**, 571–584
of the Atomic Nucleus; *Det fysiske grundlag for in-*
dustriel udnyttelse af atomkerne-energien, Tidsskrift
for Industri (No. 7–8, 1955) 168–179

     Translation   **11**, 585–607

(217) *Physical Science and Man's Position*, Ingeniøren **64**   **10**, 99–106
(1955) 810–814

(218) *The Goal of the Fight: That We in Freedom May*   **11**, 701–702
Look Forward to a Brighter Future; *Kampens mål:*
At vi i frihed kan se hen til en lysere fremtid,
"Ti år efter", Kammeraternes Hjælpefond, Copen-
hagen 1955
    Translation   **11**, 703–704

(219) *My Neighbour*; *Min Genbo*, "Halfdan Hendriksen: En   **12**, 467–470
dansk Købmand og Politiker", Aschehoug, Copen-
hagen 1956, pp. 171–172
    Translation   **12**, 471–472

(220) *Atoms and Society*; *Atomerne og Samfundet*, "Den   **11**, 609–617
liberale venstrealmanak", ASAs Forlag, Copenhagen
1956, pp. 25–32
    Translation   **11**, 618–631

(221) *Mathematics and Natural Philosophy*, The Scientific   **11**, 667–672
Monthly **82** (1956) 85–88

(222) *A Shining Example for Us All*; *Et lysende forbillede*   **12**, 331–332
for os alle [H.M. Hansen], Politiken, 14 June 1956
    Translation   **12**, 333–334

(223) [Obituary for H.M. Hansen], Fys. Tidsskr. **54** (1956)   **12**, 335–337
97
    Translation   **12**, 338

(224) *On Atoms and Human Knowledge*; *Atomerne og den*   7, 395–410
menneskelige erkendelse, Overs. Dan. Vidensk. Sel-
sks. Virks. Juni 1955 – Maj 1956, pp. 112–124
    English version in Dædalus **87** (1958) 164–175   7, 411–423

(225) *Open Letter to the Secretary General of the United*   **11**, 191–192
Nations, November 9th, 1956, Private print, Copen-
hagen 1956

(226) *Autobiography of the Honorary Doctor*; *Selvbiografi*   **12**, 207–212
af æresdoktoren, Acta Jutlandica **28** (1956) 135–138
    Translation   **12**, 213–217

(227) *Greeting to the Exhibition from Professor Niels Bohr*;   **11**, 633–635
Hilsen til udstillingen fra professor Niels Bohr, Elektroteknikeren **53** (1957) 363
     Translation   **11**, 636

(228) [For the tenth anniversary of the journal "Nucleon-   **12**, 83–84
ics"], Nucleonics **15** (September 1957) 89

(229) *The Presentation of the first Atoms for Peace Award*   **11**, 637–644
to Niels Henrik David Bohr, October 24, 1957, Na-
tional Academy of Sciences, Washington, D.C. 1957,
pp. 1, 5, 18–22

(230) *Obituary*; *Mindeord*, "Bogen om Peter Freuchen"   **12**, 477–478
(eds. P. Freuchen, I. Freuchen and H. Larsen), Frem-
ad, Copenhagen 1958, p. 180
     Translation   **12**, 479

(231) *His Memory a Source of Courage and Strength*; *Hans*   **12**, 445–446
minde en kilde til mod og styrke [N. Bjerrum], Poli-
tiken, 1 October 1958
     Translation   **12**, 447

(232) *Quantum Physics and Philosophy – Causality and*   **7**, 385–394
Complementarity, "Philosophy in the Mid-Century,
A Survey" (ed. R. Klibansky), La nuova Italia ed-
itrice, Firenze 1958, pp. 308–314

(233) *Professor Niels Bohr on Risø*; *Professor Niels Bohr*   **11**, 645–647
om Risø, Elektroteknikeren **54** (1958) 238–239
     Translation   **11**, 648–652

(234) *Preface* and *Introduction*, "Atomic Physics and Hu-   **10**, 107–112
man Knowledge", John Wiley & Sons, New York
1958, pp. v–vi, 1–2

(235) *Physical Science and the Problem of Life*, "Atomic   **10**, 113–123
Physics and Human Knowledge", John Wiley &
Sons, New York 1958, pp. 94–101

(236) [Foreword], J. Lehmann, "Da Nærumgaard blev børnehjem i 1908", Det Berlingske Bogtrykkeri, Copenhagen 1958, p. i   **12**, 393–394

Translation   **12**, 395–395

(237) [Greeting on the National Day of the Republic of Yugoslavia], Slovenski Poročevalec, Ljubljana, 29 November 1958   **12**, 227–228

Manuscript   **12**, 229–230

(238) *R.J. Bošković*, "Actes du symposium international R.J. Bošković, 1958", Académie Serbe des sciences, Académie Yougoslave des sciences et des arts, Académie Slovène des sciences et des arts, Belgrade, Zagreb, Ljubljana 1959, pp. 27–28   **12**, 231–234

(239) *Tenth and Eleventh Presentation of the H.C. Ørsted Medal* [J.A. Christiansen and P. Bergsøe]; *Tiende og ellevte Uddeling af H.C. Ørsted Medaillen*, Fys. Tidsskr. **57** (1959) 145, 158   **11**, 487–489

Translation   **11**, 490–491

(240) [Tribute to Tesla], "Centenary of the Birth of Nikola Tesla 1856–1956", Nikola Tesla Museum, Belgrade 1959, pp. 46–47   **12**, 235–236

(241) *Foreword*; *Forord*, "Hanna Adler og hendes Skole", Gad, Copenhagen 1959, pp. 7–10   **12**, 397–402

Translation   **12**, 403–405

(242) *Foreword*; *Forord*, J.R. Oppenheimer, "Naturvidenskab og Livsforståelse", Gyldendal, Copenhagen 1960, pp. i–ii   **12**, 85–86

Translation   **12**, 87

(243) *Foreword*, "Theoretical Physics in the Twentieth Century: A Memorial Volume to Wolfgang Pauli" (eds. M. Fierz and V. Weisskopf), Interscience, New York 1960, pp. 1–4   **12**, 361–366

(244) *Quantum Physics and Biology*, Symposia of the Society for Experimental Biology, Number XIV: "Models and Analogues in Biology", Cambridge 1960, pp. 1–5   **10**, 125–131

(245) *Ebbe Kjeld Rasmussen: 12 April 1901 – 9 October 1959*; *Ebbe Kjeld Rasmussen: 12. april 1901 – 9. oktober 1959*, Fys. Tidsskr. **58** (1960) 1–2   **12**, 379–382
    Translation   **12**, 383–384

(246) *The Connection Between the Sciences*, Journal Mondial de Pharmacie **3** (1960) 262–267   **10**, 145–153

(247) *Physical Models and Living Organisms*, "Light and Life" (eds. W.D. McElroy and B. Glass), The Johns Hopkins Press, Baltimore 1961, pp. 1–3   **10**, 133–137

(248) *The Unity of Human Knowledge*, Revue de la Fondation Européenne de la Culture, July 1961, pp. 63–66   **10**, 155–160

(249) *Atomic Science and the Crisis of Humanity*; *Atomvidenskab og menneskehedens krise*, Politiken, 20 April 1961   **10**, 281–288
    Translation   **10**, 289–293

(250) *The Rutherford Memorial Lecture 1958: Reminiscences of the Founder of Nuclear Science and of Some Developments Based on his Work*, Proc. Phys. Soc. **78** (1961) 1083–1115   **10**, 381–420

(251) *Preface to the 1961 reissue*, "Atomic Theory and the Description of Nature", Cambridge University Press, Cambridge 1961, p. vi   **12**, 89–90

(252) *Meeting on 14 October 1960*; *Mødet den 14. oktober 1960*, Overs. Dan. Vidensk. Selsk. Virks. Juni 1960 – Maj 1961, pp. 39–41   **11**, 443–445

(253) *The Solvay Meetings and the Development of Quantum Physics*, "La théorie quantique des champs", Douzième Conseil de physique tenu à l'Université Libre de Bruxelles du 9 au 14 octobre 1961, Interscience Publishers, New York 1962, pp. 13–36   **10**, 429–454

(254) *Foreword by the Danish editorial committee*; *Den danske redaktionskomités forord*, I.B. Cohen, "Fysikkens Gennembrud", Gyldendals Kvantebøger, Copenhagen 1962, pp. 5–6   **12**, 91–94

    Translation   **12**, 95–96

(255) *The General Significance of the Discovery of the Atomic Nucleus*, "Rutherford at Manchester" (ed. J.B. Birks), Heywood, London 1962, pp. 43–44   **12**, 279–282

(256) *Address at the Second International Germanist Congress* [Copenhagen, August 22, 1960], "Spätzeiten und Spätzeitlichkeit", Francke Verlag, Bern 1962, pp. 9–11   **10**, 139–143

(257) [Tribute to Russell], "Into the 10th Decade: Tribute to Bertrand Russell", Malvern Press, London [1962]   **12**, 389–390

(258) *Light and Life Revisited*, ICSU Review **5** (1963) 194–199   **10**, 161–169

(259) *The Genesis of Quantum Mechanics*, "Essays 1958–1962 on Atomic Physics and Human Knowledge", Interscience Publishers, New York 1963, pp. 74–78   **10**, 421–428

(260) *Recollections of Professor Takamine*, "Toshio Takamine and Spectroscopy", Research Institute for Applied Optics, Tokyo 1964, pp. 384–386   **12**, 367–368

(261) [Tribute to Weisgal], "Meyer Weisgal at Seventy" (ed. E. Victor), Weidenfeld and Nicolson, London 1966, pp. 173–174   **12**, 485–488

INDEX

Words which describe main themes of the volume, such as "popularization", "tribute" or "physics", are not listed. Some particularly relevant items, such as "atomic physics", "Copenhagen, University of", "Institute for Theoretical Physics (Copenhagen)" and "Royal Danish Academy of Sciences and Letters", have been included, even though they appear often. There are index terms for some of Bohr's characteristic expressions, such as "complementarity", "harmony", "international cooperation", "natural phenomena" and "unity/concord".

All persons (other than Niels Bohr) and most institutions in the text are listed, whereas places are only indexed selectively.

When a term on a page is found only in a footnote, the page number in the index is followed by the letter n. References to sources (books, articles and archival material) in footnotes are not indexed.

Picture captions are fully indexed, with page numbers followed by the letter p.

It is hoped that the cross references will help the reader identify subjects that are expressed in two or more ways in the text.